Farm-made aquafeeds

Edited by
Michael B. New
Marlow, Buckinghamshire
United Kingdon

Albert G.J. Tacon
FAO Fisheries Department

Imre Csavas
FAO Regional Office for Asia and the Pacific
Bangkok, Thailand

Reprint by arrangement with the
Food and Agriculture
Organization of
the United Nations
by
Daya Publishing House®
A Division of
Astral International Pvt. Ltd.
New Delhi - 110 002

"This book was originally published by the Food and Agriculture Organization of the United Nations (FAO) as Farm-made Aquafeeds, FAO Fisheries Technical Paper, 343".

"The designations employed and the presentation of material in this publication do not imply the expression of any opinion, whatsoever on the part of the Food and Agriculture Organization of the United Nations concerning the legal status of any country, territory, city or area or of its authorities, or concerning the delimitations of its frontiers or boundaries."

'The designations 'developed' and 'developing' economies are intended for statistical convenience and do not necessarily express a judgment about the stage reached by a particular country, territory or area in the development process.'

First Indian Edition, 1997
ISBN: 978-81-7035-165-8

PREPARATION OF THIS DOCUMENT

In December 1993, an expert consultation on farm-made feeds was jointly organized by the FAO Regional Office for the Pacific and Asia (FAO-RAPA) and the ASEAN-EEC Aquaculture Development and Coordination Programme (AADCP). The proceedings of this meeting, the first to specifically address this topic, were edited by staff of AADCP and FAO. First published as FAO/RAPA Publication 1993/18 and AADCP Publication AADCP/PROC/5, the costs of production were shared by FAO-RAPA and AADCP. Demand was high and the original limited edition was only available in English. The book has been reprinted as an FAO document because of the importance of farm-made feeds in aquaculture and in order to broaden its distribution. Translation into other official FAO languages will also make the information available to more member states. Though the book concerns experience in the use of farm-made aquafeeds in Asia, it is also of particular relevance to Africa and Latin America.

Distribution:

FAO Fisheries Department
FAO Regional Fisheries Officers
FAO Aquaculture Projects
Authors

New, M.B.; Tacon, A.G.J.; Csavas, I.(eds.)
Farm-made aquafeeds.
FAO Fisheries Technical Paper. No. 343. Rome, FAO. 1994.
434 p.

ABSTRACT

This book is the proceedings of a meeting held in Bangkok in December 1992 on the use of farm-made feeds in Asia. It contains eleven country reviews of the topic, for Bangladesh, Cambodia, China, India, Indonesia, Malaysia, Nepal, the Philippines, Singapore, Thailand and Vietnam. Nine technical papers are also included. Three are on on-farm feed preparation and feeding strategies - for carps and tilapias, for catfish and snakehead, and for marine shrimp and prawns. Five other working papers are on economics, the selection of equipment, feed ingredients, formulation and on-farm management, and supplementary feeding in semi-intensive aquaculture, all directed at farm-made, rather than commercial feeds. The ninth working paper is a regional overview of aquafeeds in Asia. An analysis of the material in the eleven country papers is also presented. Emphasis is placed on the important role, hitherto not completely recognized, of farm-made aquafeeds to aquaculture production in Asia. While it was estimated that 50% of shrimp production from aquaculture comes from commercial feeds, only 10% of Asian finfish is currently produced in this way. Although statistical data is non-existent it was speculated that over 1 million t of farm-made feeds are produced annually in Asia and that about one third of Asian finfish and crustacean production is achieved partially through their use. The proceedings also include the official report of the meeting, with recommendations aimed primarily at those concerned with research and development on aquafeeds and those organizations that fund such work. A strong plea is made for more attention to be paid to the needs of the small-scale farmers already using, or with the potential to use, farm-made feeds. Commercial feedstuff manufacturers are generally able to fund their own research and development on complete feeds for use in intensive aquaculture.

CONTENTS

REPORT OF THE FAO/AADCP REGIONAL EXPERT CONSULTATION ON FARM-MADE AQUAFEEDS

(14-18 December 1992, Bangkok, Thailand)

Opening of the session

1. The FAO/AADCP Regional Expert Consultation on Farm-Made Aquafeeds was held at the premises of the FAO Regional Office for Asia and the Pacific (RAPA) in Bangkok, Thailand, from 14 to 18 December 1992.

2. The meeting was attended by 15 participants from ten Asian countries, three staff members of AADCP, and four staff members and three consultants of FAO. Six national/international organizations (ASA, AIT, BOBP, BP Nutrition, SEAFDEC AQD and SEAPRODEX) delegated eight participants to the Expert Consultation at their own expense.

3. Welcoming addresses were delivered by Mr. Imre Csavas, Regional Aquaculture Officer, on behalf of FAO and Mr. Michael B. New, Coordinator, on behalf of AADCP.

Organization of the meeting

4. Participants elected Dr. Felicitas P. Pascual as Chairperson of the Expert Consultation and Dr. M. C. Nandeesha as Rapporteur.

5. The draft agenda of the meeting was adopted without changes.

Papers presented

6. A total of 10 country papers on aquafeeds and feeding strategies were presented at the Expert Consultation, covering Bangladesh, Cambodia, China, India, Indonesia, Malaysia, Nepal, the Philippines, Thailand and Vietnam.

7. Another 9 technical papers covering crucial aspects of aquafeed manufacturing and use as well as farm-made feed preparation and feeding strategies for the most important species groups were also presented by acknowledged experts in these special fields.

Summary of findings

8. The Consultation recognized the fact that the bulk of Asian finfish and crustacean aquaculture production is currently realized in semi-intensive pond farming systems. The great majority of these farming systems, and in particular freshwater non-carnivorous finfish production (which accounts for over 80% of the total finfish production in Asia) depend upon the use of farm-made feeds. Some intensive systems, notably the cage culture of marine fish and some carnivorous freshwater finfish (*e.g.* snakehead and catfish species), also use farm-made feeds. Only about 10% of Asian finfish production is based on the use of commercial feeds, whereas almost 50% of shrimp production depends on commercial feeds.

9. It was realized that there is a lack of an exact definition for farm-made aquafeeds. For the purpose of this Expert Consultation farm-made feeds were defined as feeds in pellet or other forms, consisting of one or more artificial and/or natural feedstuffs, produced for the exclusive use of a particular farming activity, not for commercial sale or profit.

10. The Consultation recognized that there is an increasing tendency for farmers to utilize commercial feeds formulated as nutritionally complete diets in semi-intensive pond-farming systems. However, the Consultation reiterated that the nutrition and feeding of finfish and crustaceans in semi-intensive pond farming systems are complex and poorly understood; little or no information being available on dietary nutrient requirements of the cultured species in such systems. To a large extent this is due to the difficulties of quantifying the contribution of naturally available food organisms in the overall nutritional budget of pond-raised finfish or crustaceans.

11. The Consultation recognized that farm-made feeds facilitate the use of locally available agricultural products and wastes of agro-processing industries that would otherwise have limited use within the community. In this respect, their use in farm-made feeds has significant environmental advantages. However, the Consultation also recognized the current dependence of commercial and, to a lesser extent, farm-made feeds upon animal protein sources and the need to identify and utilize alternative protein sources which are both inexpensive and sustainable.

12. The Consultation recognized that farm-made feeds are potentially cheaper than commercial aquafeeds and that there is scope for a reduction in feed and farm production costs using appropriate on-farm feed management techniques for semi-intensive and some intensive aquafarming systems.

13. The Consultation recognized that farm-made feeds fill an important niche for small-scale farmers which is not covered by commercial feedstuff manufacturers. However, it was also noted that those farmers whose initial success was based on the use of farm-made aquafeeds often shift to the use of commercial feeds. Thus improvements in the production and use of farm-made feeds would ultimately benefit the feedstuff industry as well.

Recommendations

14. This Consultation being the first which has dealt with farm-made aquafeeds, participants unanimously recommend that the proceedings of the Consultation should be published, with an analysis of the information in the country papers, and circulated widely to governments, international agencies, and potential donors to:

- alert them to the importance of farm-made feeds to small-scale aquaculture and to the need for technical support; and to
- persuade donors that future aid concerning feed development should concentrate on farm-made feeds; public sector funding should primarily assist small-scale farmers, not feed manufacturers.

15. The approach towards this topic should be "bottom-up" rather than "top-down"; preconceived ideas should not be forced on farmers but

- improvements on farmers' existing practices should be developed, and
- the fact that improvements can lead to greater profitability without compromising environmental quality, should be demonstrated.

16. Recognizing that farm-made feeds can utilize locally available ingredients, simple and cheap methods of increasing their nutritional value for fish and crustacea should be developed, with particular reference to digestibility, removal of toxic substances, and palatability.

17. Bearing in mind the needs of small-scale farmers,

- the development or improvement of simple and cheap machinery for farm-made feed production should be encouraged; and
- improved techniques for on-farm processing and storage developed.

18. Considering that farm-made feeds, whether for intensive aquaculture (*e.g*. marine fish in cages or snakehead in ponds) or for semi-intensive systems (such as pond culture of freshwater prawns), feed advisers should formulate feeds based on:

- locally available ingredients;
- the nutritional requirements of the specific farming system rather than mimicking commercial feed specifications;
- the minimal use of vitamin premixes, binders and other expensive ingredients to maximize cost effectiveness;
- improved knowledge of the role of natural food organisms in semi-intensive farming based on systems which optimize pond fertilization; and
- an emphasis on the quality of the ingredients and of the final product.

19. Bearing in mind that the potential for improved profitability in small-scale aquaculture is greater through improvements in feeding strategy rather than through perfect dietary composition, developmental work should focus on:

- feeding frequency;
- methods of feed presentation, including the separate feeding of energy-providing ingredients from high protein mixed feed;
- the investigation of other two-component systems (*e.g*. alternation of feeds of different composition, alternation of different feeding rates);
- biomass assessment of the cultured species;
- reduction of feed wastage, in the farming enclosure and during manufacture and storage; and
- farmer-friendly sensory methods of assessing ingredient quality.

20. In view of the need for training of farmers in simple formulation and in ingredient choice, aquafeed processing, storage and on-farm feed management,

- village level training should be organized in local languages;
- instructional videos developed; and
- simple booklets prepared (*e.g*. in the FAO "Better Farming" series) for farmers and training manuals for trainers which can easily be translated into local languages.

AQUAFEEDS IN ASIA - A REGIONAL OVERVIEW

Michael B. New[1] and Imre Csavas[2]

[1] ASEAN-EEC Aquaculture Development Programme (AADCP)
P.O. Box 1006, Kasetsart Post Office, Bangkok 10903, Thailand

[2] FAO Regional Office for Asia and the Pacific
Phra Atit Road, Bangkok 10200, Thailand

NEW, M.B. and I. CSAVAS. 1993. Aquafeeds in Asia - a regional overview, p.1-23. *In* M.B. New, A.G.J. Tacon and I. Csavas (eds.) Farm-made aquafeeds. Proceedings of the FAO/AADCP Regional Expert Consultation on Farm-Made Aquafeeds, 14-18 December 1992, Bangkok, Thailand. FAO-RAPA/AADCP, Bangkok, Thailand, 434 p.

INTRODUCTION

A major purpose of the meeting at which this paper is being presented is to exchange views and experiences on economical feeding strategies and feed production. The emphasis is on farm-made feeds but the country review papers will summarize national feed ingredient availability, industrial and on-farm aquafeed manufacturing, on-farm feed formulation and feeding strategies and will provide details of information on the major institutions involved in aquafeed research and development, national feed regulations, current aquafeed problems and constraints and trends foreseen. We hope that this consultation will provide unique information on these topics, which will be published by FAO and AADCP.

We do not, in our introductory paper, intend to pre-empt the information which we expect to gain, either from the country papers or from the documents which we have invited other speakers to prepare. Rather, we aim to present a broad overview of the current scale of aquafeed production in Asia and some thoughts on future trends.

SCALE OF AQUACULTURE

Total global aquaculture production in 1990 was 15.3 million tons, with a value of US$ 26.5 billion. 84 % of the global production and 78 % of the value was attributable to Asia (Figure 1). The average annual growth rate in aquaculture production (1980-90) is running at 7.3 % in Asia and the Pacific, compared to

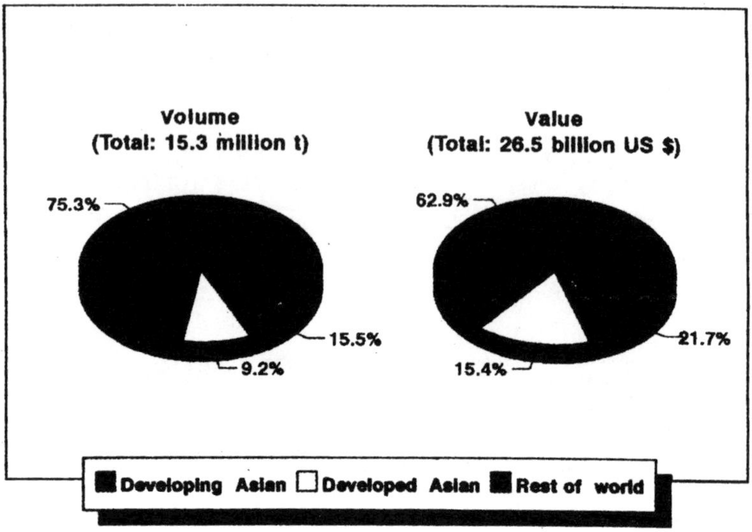

Figure 1. Aquaculture production in 1990 by volume and by value (FAO 1992)

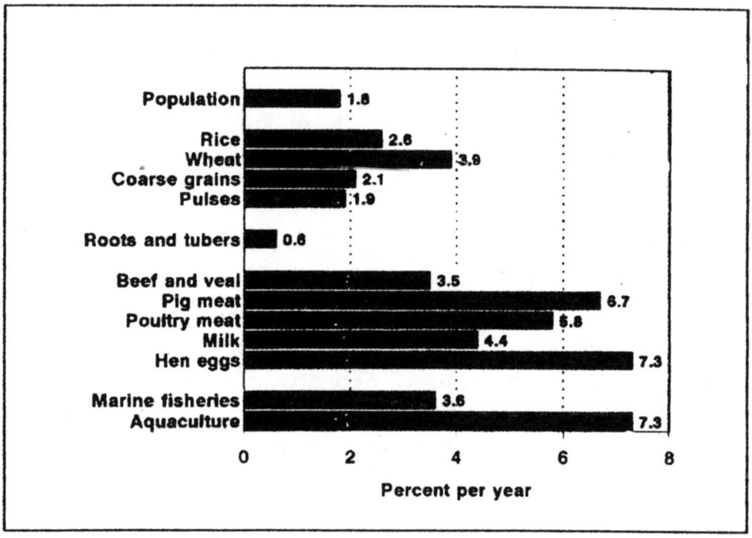

Figure 2. Growth of population and food production in Asia and the Pacific between 1980 and 1990 (FAO/RAPA 1991)

an increase of 5.8 % for poultry meat production and 3.9 % for wheat (Figure 2). Population rose by 1.8 % in that decade. Because of the large quantities of molluscs and seaweeds produced in the region (Figure 3) only about 54 % of the total of 12.9 million t produced by Asian aquaculture in 1990 was finfish and a surprisingly low 4.5% crustaceans. However, whereas Asian production of finfish grew from 1.8 million t in 1975 to 7.0 million t in 1990, crustacean production leapt from 26,000 t to 576,000 t in the same period, nearly six times the rate of increase of finfish (Figures 4 and 5).

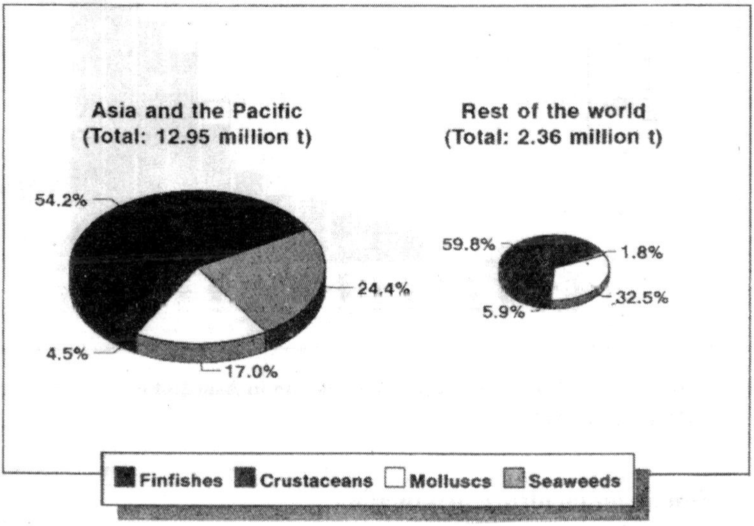

Figure 3. Aquaculture production by major commodity groups in 1990 (FAO 1992)

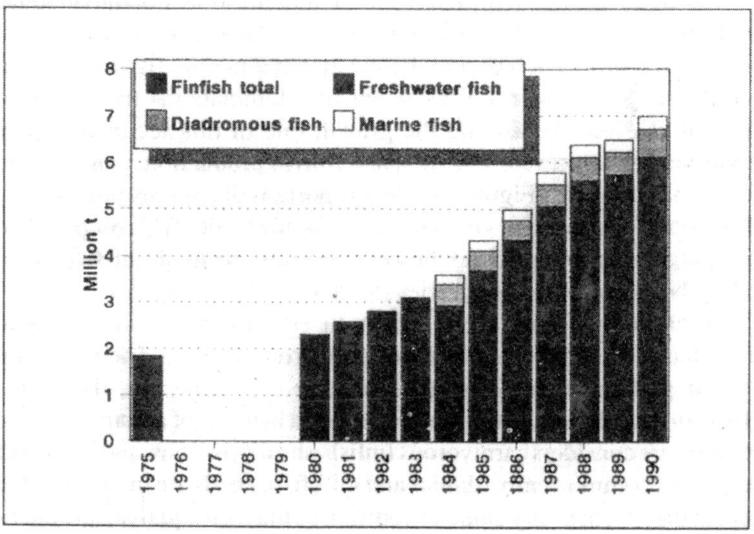

Figure 4. Growth of cultured finfish production in Asia and the Pacific (Csavas 1988; FAO 1992)

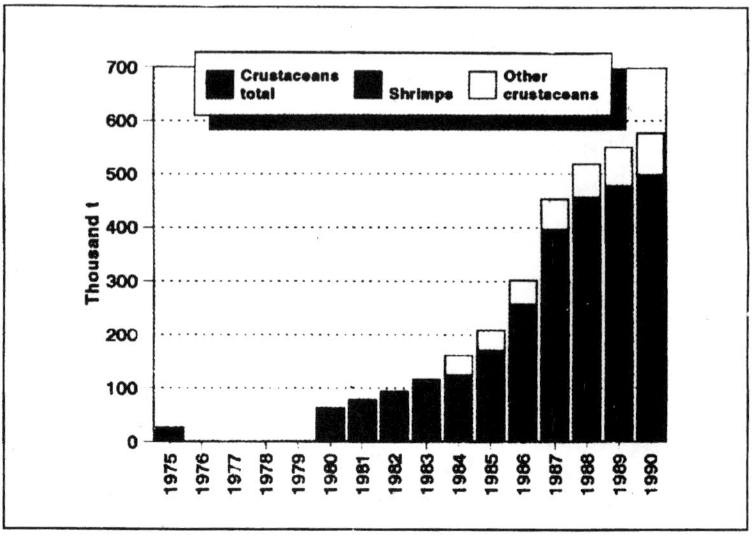

Figure 5. Growth of cultured crustacean production in Asia and the Pacific (Csavas 1998; FAO 1992)

Finfish aquaculture production

While diadromous fish, mainly salmonids, form 38% of finfish aquaculture production in the rest of the world (Figure 6), in Asia the proportion is only 9%. Conversely the proportion of both freshwater and marine fish is greater in Asia than in the rest of the world. Freshwater fish dominate the finfish aquaculture production of Asia (87 %) and the growth rate of this sector is exponential. Carnivorous fish represent 7 % of Asian finfish production, compared to 50 % in the rest of the world (Figure 7). The proportion of carnivorous finfish grown in "developed" countries of the region is high (90 %) compared to their "developing" neighbours (3 %); however most of the total finfish production is reared in the "developing" countries (95 %).

While most finfish feeds are produced for carnivorous fish, aquafeeds are also manufactured for some non-carnivorous species. The most important species or groups in the latter category are common carp, the tilapias and milkfish. In discussing aquaculture production in terms of aquafeed demand this paper therefore considers carnivorous finfish, shrimp, prawns, and other crustacea, together with common carp, tilapias and milkfish. Feeds are made for other types of fish but they form a very small proportion of the total aquafeed production and have not been taken into account.

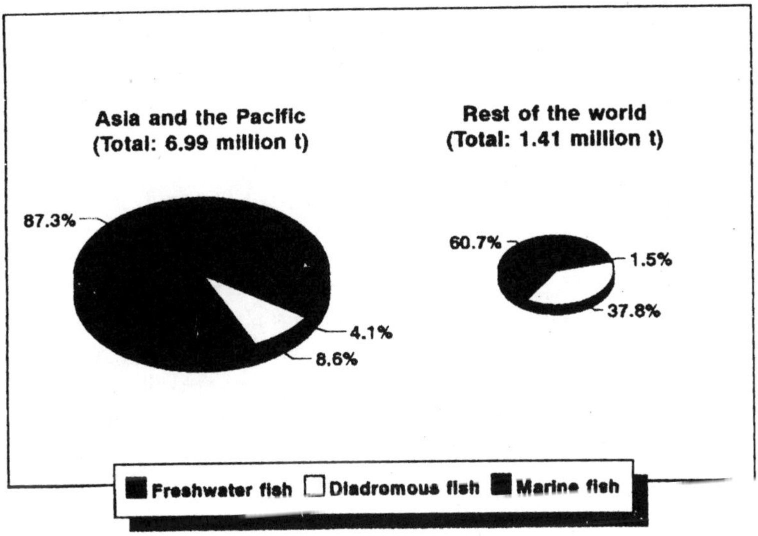

Figure 6. Production of cultured finfish by major species groups in 1990 (FAO 1992)

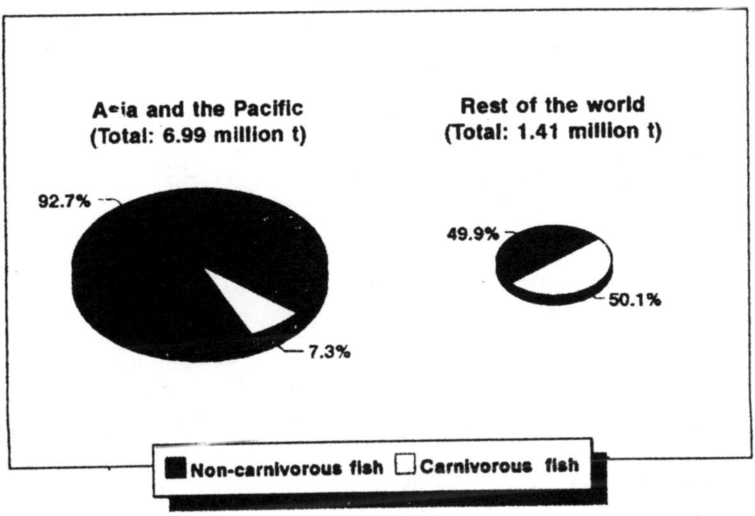

Figure 7. Cultured carnivorous and non-carnivorous finfish production in 1990 (FAO 1992)

In 1990, of the 510,000 t of carnivorous fish produced in Asia (Figure 8), 11 % consisted of salmonids. The other important species groups were catfishes (15 %), yellowtails (32 %), eels (19 %) and seabreams (14 %). Production of the three non-carnivorous fish species groups (Figure 9) totalled 1.6 million t in 1990 of which 53 % was common carp, mostly produced in China.

A glance at the geographical disparities in Asian carnivorous fish production (Figure 10) immediately highlights those countries where commercial aquafeed production is current or has potential - two of the three "developed" countries of the region (Australia and Japan, together with those "developing" countries where the total production of carnivorous fish is more than 2,500 t per year - Taiwan (Province of China), India, China, Thailand, Indonesia, Republic of Korea, Hong Kong, and the Philippines, in descending order of magnitude. At the moment (Figure 8), more carnivorous fish are produced by aquaculture in the rest of the world (nearly 708,000 t) than in Asia (510,000 t).

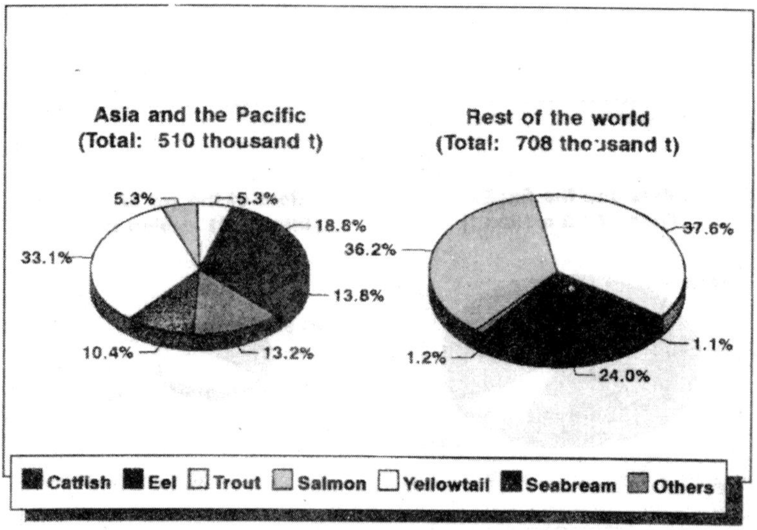

Figure 8. Cultured carnivorous fish by major species groups in 1990 (FAO 1992)

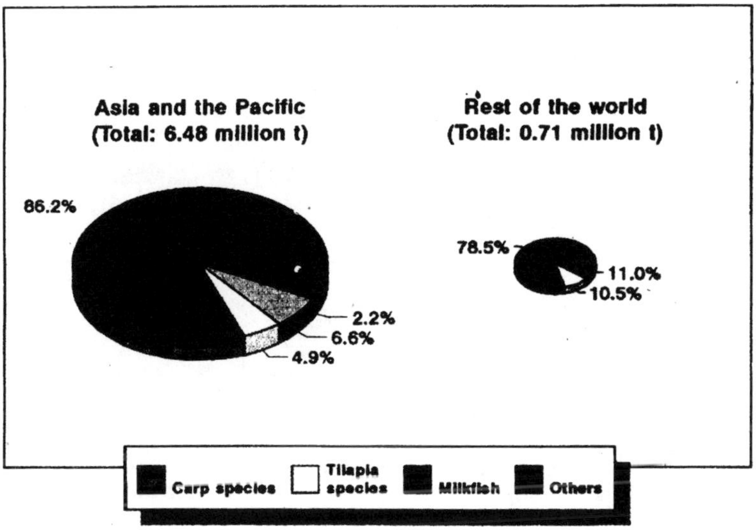

Figure 9. Cultured non-carnivorous fish by major species groups in 1990 (FAO 1992)

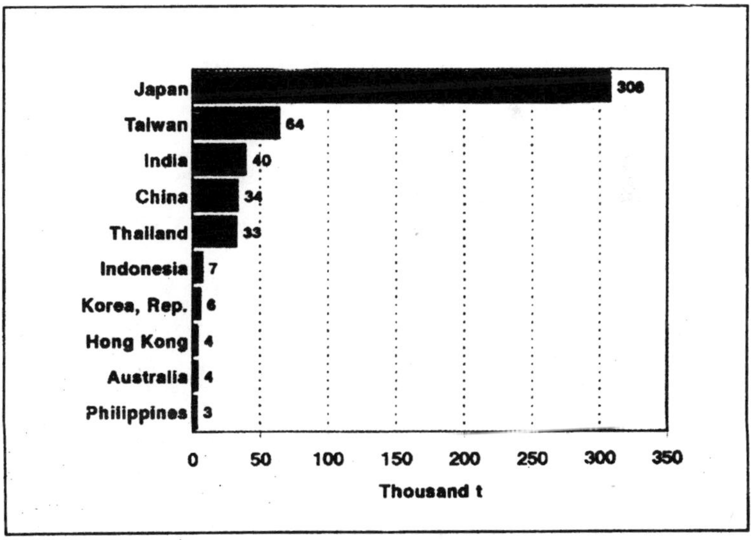

Figure 10. Major producers of cultured carnivorous fish in Asia and the Pacific in 1990 (FAO 1992)

Crustacean aquaculture production

Globally, about 716,000 t of crustacea were produced by aquaculture in 1990 (Figure 11). 80 % of this total was produced in Asia (nearly 576,000 t), almost all of it in the developing countries of the region. China (27 % of the global total) and ASEAN (35 %) mainly Indonesia (14%), the Philippines (8%) and Thailand (14%) were the major producing areas. Vietnam, Taiwan, Bangladesh, India and the Democratic Peoples Republic of Korea were also significant producers.

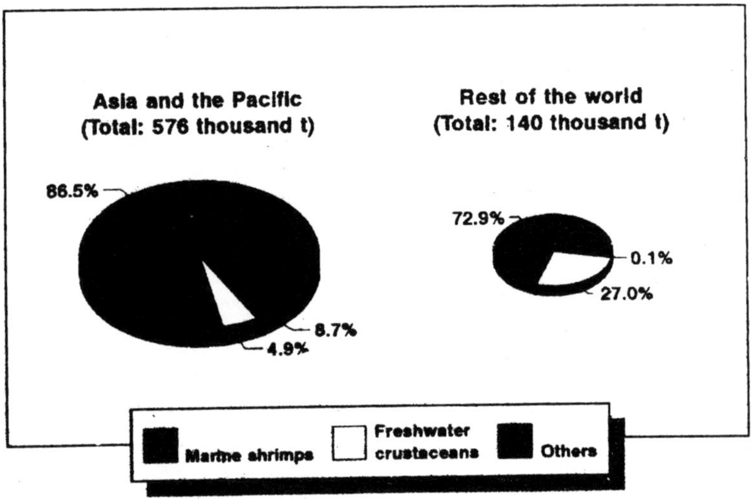

Figure 11. Production of cultured crustaceans by major species groups in 1990 (FAO 1992)

CURRENT AQUAFEED MARKET IN ASIA

Finfish feeds

In Asia, fish feed needs centre round those for yellowtails, eels, seabreams, catfishes, salmonids and the three non-carnivorous fish species (Figure 12). Much of the feed requirements of yellowtails, seabreams, and "other" carnivorous fish are still met by moist feeds consisting of a large proportion of trash fish or low-value marine fish.

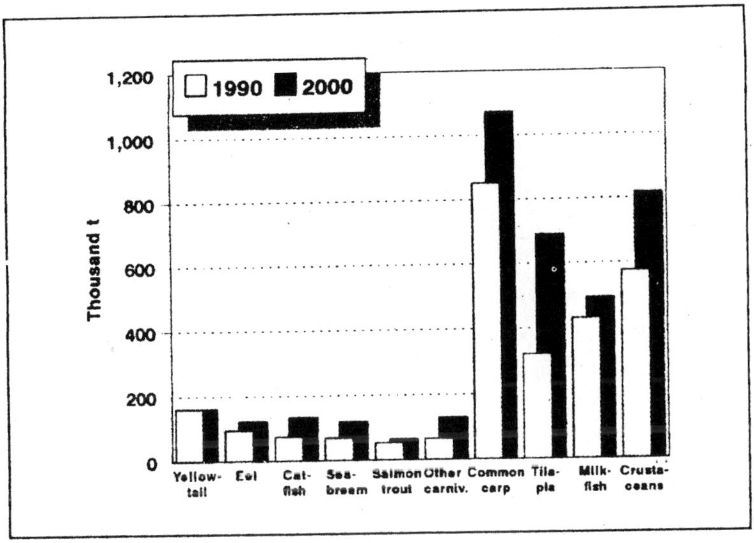

Figure 12. Aquaculture production in Asia and the Pacific in 1990 and estimates for 2000 (commercially fed species only)

We have estimated (Table 1), using a series of assumptions (Table 2), that the Asian market for carnivorous fish feeds in 1990 was about 472,000 t. By far the largest type of carnivorous finfish feed is for eels (37 %), almost entirely in Taiwan and Japan. Salmonid feeds constitute 16 %. Feed for catfish (13 %) and for "other carnivores" (15 %) are the largest species groups, followed by yellowtails (10%).

The remainder is feeds for seabreams (9 %). By country, Japan (54 %) and Taiwan (23 %) were the main users of feeds for carnivorous finfish in 1990. Total consumption for all ASEAN countries was estimated at less than 8%, most of it in Thailand. About 7 % was used in the Indian sub-continent.

Feeds are also made for the non-carnivorous fish listed in Table 1. It is estimated that over 554,000 t of commercial feeds were made for common carp (31 %), tilapia (29 %) and milkfish (40 %) in 1990. Of these most were for the market in Taiwan (49 %). ASEAN (25 %), mainly Indonesia and the Philippines, was also a very significant market for feeds for non-carnivorous finfish feeds.

Table 1. Aquaculture and aquafeed production of selected species groups in 1990 (t)

SPECIES GROUPS	Indonesia	Philippines	Thailand	Other ASEAN	Japan	China	Taiwan	Ind.Subcont.	Others	Total
YELLOWTAILS										
FISH	-	-	-	-	161,106	-	-	-	462	161,568
FEED	-	-	-	-	**48,332**	-	-	-	**139**	**48,471**
EELS										
FISH	290	-	-	-	38,855	-	55,816	-	1,146	96,107
FEED	**522**	-	-	-	**69,939**	-	**100,469**	-	**2,063**	**172,993**
CATFISHES										
FISH	3,880	105	30,480	197	-	-	40	39,260	1,252	75,214
FEED	**3,143**	**85**	**24,689**	**160**	-	-	**32**	**31,801**	**1,014**	**60,924**
SEABREAMS										
FISH	3,200	3,142	1,750	2,425	51,200	-	7,375	-	1,713	70,805
FEED	**1,920**	**1,885**	**1,050**	**1,455**	**30,720**	-	**4,425**	-	**1,028**	**42,483**
TROUTS										
FISH	-	-	-	-	20,072	700	531	1,260	4,668	27,231
FEED	-	-	-	-	**30,108**	**1,050**	**797**	**1,890**	**7,002**	**40,847**
SALMONS										
FISH	-	-	-	-	23,000	-	-	-	4,260	27,260
FEED	-	-	-	-	**27,600**	-	-	-	**6,390**	**33,990**
OTHER CARNIVOROUS*										
FISH	-	-	505	419	26,796	33,000	576	-	3,847	65,143
FEED	-	-	**303**	**251**	**48,233**	**19,800**	**1,037**	-	**2,308**	**71,932**
TOTAL CARNIVOROUS										
FISH	7,370	3,247	32,735	3,041	321,029	33,700	64,338	40,520	17,348	523,328
FEED	**5,585**	**1,970**	**26,042**	**1,866**	**254,932**	**20,850**	**106,760**	**33,691**	**19,944**	**471,640**

* and Ayu

Table 1. (cont.)

SPECIES GROUPS	Indonesia	Philippines	Thailand	Other ASEAN	Japan	China	Taiwan	Ind.Subcont.	Others	Total
COMMON CARP										
FISH	89,810	717	2,000	1,284	16,150	522,369	3,369	181,637	33,194	850,530
FEED	**35,924**	**287**	**800**	**514**	**32,300**	**10,447**	**6,738**	**72,655**	**13,278**	**172,943**
TILAPIAS										
FISH	50,470	76,142	21,900	1,144	5,825	106,071	52,047	5,000	2.142	320,741
FEED	**20,188**	**30,457**	**8,760**	**458**	**9,320**	**2,121**	**83,275**	**2,000**	**857**	**157,436**
MILKFISH										
FISH	128,000	210,882	220	–	–	–	90,716	–	52	429,870
FEED	**16,000**	**26,360**	**28**	**-**	**-**	**-**	**181,432**	**-**	**7**	**223,827**
TOTAL NON-CARNIV.										
FISH	268,280	287,741	24,120	2,428	21,975	628,440	146,132	186,637	35,388	1,601,141
FEED	**72,112**	**57,104**	**9,588**	**972**	**41,620**	**12,568**	**271,445**	**74,655**	**14,142**	**554,206**
FRESHWATER PRAWNS										
PRAWNS	-	-	8,000	137	50	-	11,607	198	8,023	28,015
FEED	**-**	**-**	**10,560**	**181**	**110**	**-**	**25,535**	**109**	**4,413**	**40,908**
MARINE SHRIMP										
SHRIMP	94,960	53,989	89,300	1,512	3,000	186,181	18,126	45,365	42,470	534,903
FEED	**136,742**	**77,744**	**128,592**	**2,177**	**4,320**	**67,025**	**32,627**	**16,331**	**15,289**	**480,847**
OTHER CRUSTACEANS										
CRUST.	2,410	1,000	30	203	30	7,833	1,358	-	109	12,973
FEED	**2,892**	**1,200**	**36**	**244**	**60**	**3,917**	**2,716**	**-**	**55**	**11,120**
TOTAL CRUSTACEANS										
CRUST.	97,370	54,989	97,330	1,852	3,080	194,014	31,091	45,563	50,602	575,891
FEED	**139,634**	**78,944**	**139,188**	**2,602**	**4,490**	**70,942**	**60,878**	**16,440**	**19,757**	**532,875**
TOTAL AQUACULTURE PRODUCTION	373,020	345,977	154,185	7,321	346,084	856,154	241,561	272,720	103,338	2,700,360
TOTAL AQUAFEED MARKET	**217,331**	**138,018**	**174,818**	**5,440**	**301,042**	**104,360**	**439,083**	**124,786**	**53,843**	**1,558,721**

Table 2. Assumptions used in preparing Tables 1 and 3

SPECIES GROUPS	ASSUMED FCR		PRODUCING AREA	ESTIMATED COMMERCIALLY FED (%)		FORECAST GROWTH RATE OF AQ.PROD 1990 - 2000 (% PER DECADE)
	1990	2000		1990	2000	
Yellowtails	2:1	1.8:1	All	15	50	Nil
Eels	2:1	1.8:1	Japan	)	(	Nil
			Taiwan	) 90	90 (	50
			Others	)	(	20
Catfishes	1.8:1	1.5:1	All	45	75	80
Seabreams/ basses	2:1	1.8:1	All	30	40	70
Trouts	1.5:1	1.3:1	Taiwan	)	(	100
				) 100	100 (	
			Others	)	(	10
Salmons	1.5:1	1.3:1	Japan	80	80	25
			Others	100	100	20
Other carnivorous*	2:1	1.8:1	Japan & Taiwan	90	95	100
			Pacific Islands	90	95	100
			Others	30	40	100
Common carp	2:1	1.8:1	Japan & Taiwan	100	100	Nil
			China	1	5	25
			Others	20	25	30
Tilapias	2:1	1.8:1	Japan	80	100	Nil
			Taiwan	80	90	Nil
			Pacific Islands	20	40	100
			China	1	5	200
			Others	20	25	100
Milkfish	2.5:1	2:1	Taiwan	80	90	70
			Others	5	10	Nil

Table 2. (cont.)

SPECIES GROUPS	ASSUMED FCR		PRODUCING AREA	ESTIMATED COMMERCIALLY FED (%)		FORECAST GROWTH RATE OF AQ.PROD 1990 - 2000 (% PER DECADE)
	1990	2000		1990	2000	
Freshwater prawns	2.2:1	2:1	Japan & Taiwan	100	100	200
			ASEAN	60	80	100
			Others	25	50	100
Marine shrimp	1.8:1	1.6:1	Japan	80	100	Nil
			Taiwan	100	100	100
			ASEAN	80	85	25
			China	20	60	35
			Others	20	30	50
Other crustaceans	2:1	1.8:1	Japan & Taiwan	100	100	200
			ASEAN	60	80	100
			Others	25	50	100

* includes Ayu (non-carnivorous but 100% commercially fed)

Crustacean feeds

Using a similar series of assumptions (Table 2) we have estimated that the Asian crustacean feed market in 1990 was about 533,000 t (Figure 12), mostly for marine shrimp (90 %).

Indonesia and Thailand, each with a 26 % share, were by far the largest markets for crustacean feeds in 1990 (Table 1).The ASEAN total (over 360,000 t) constituted 67 % of the Asian total. China (13 %) and Taiwan (11 %) were the other two main centres, with the Indian sub-continent (3 %) and other countries (4 %) in a fledgling state. While Japan was the major user of commercial feeds for carnivorous finfish in 1990 it was a very minor market for crustacean feeds (less than 1 %).

It should be noted that our estimate of the marine shrimp feed market in 1990 (481,000 t) is about 30 % less than that of Akiyama (1992a), which was 691,000 t. His estimate of 250,000 t of shrimp feed production in China is, in our view, far too high. This is also so, to a lesser extent, for Thailand and Taiwan. On the other hand, our estimates for Indonesia and the Philippines are higher than his. Since China, Thailand and Taiwan account for 585,000 t of Akiyama's Asian total of 691,000 t, this represents the major cause of our disparity. Naturally, we believe our estimates are more realistic !

Total Asian aquafeed market in 1990

The current Asian aquafeed market for all species is estimated to total 1.6 million t. ASEAN was the major market for commercial aquafeeds in 1990 (34 %) - Indonesia (14 %), Thailand (11 %) and the Philippines (9 %) being the focal points. The commercial aquafeed market in Taiwan, at 28 % of the Asian total, was by far the largest of any country, being more than Indonesia and Thailand combined. Japan was also a huge market, at 19 % of the Asian total. Aquafeeds in the Indian sub-continent (8 %) and China (7 %) approached the level for the Philippines (9 %). Aquafeeds for carnivorous finfish (30%), non-carnivorous finfish (36 %) and crustaceans (34 %) constituted similar proportions of the total 1.6 million t market in 1990.

It should be noted that our estimate of 1.6 million t of aquafeeds in Asia in 1990 is about 1 million t less than that of Akiyama (1992b), due principally to a gross over-estimate by the latter for China (PRC). Megisson (1990) gave a global estimate of 4 million t for aquafeed production in 1988. In our opinion this also is much greater than reality; we believe that it was less than 3 million t. Gill (1992a) stated that aquaculture accounts for 3 % of total world feed production, stated to be 610 million t in 1992. This would equate to 18.3 million t of aquafeeds, which is clearly wrong; this quantity would be sufficient to feed 100 % of <u>all</u> species of cultured finfish and crustaceans ! We believe that, amongst the several figures available, ours of 1.6 million t in Asia in 1990 is the most reasonable estimate and certainly the only one for which the assumptions have been published.

FUTURE PROSPECTS FOR AQUAFEEDS

What of the future: what will be the scale and nature of the Asian aquafeeds industry in the year 2000 ? Using a series of assumptions on the expansion of aquaculture production, improvements in food conversion efficiency, and the proportion of stock fed with commercial feeds, which vary by species group and location (Table 2), we have derived forecasts of the Asian aquafeed market in the year 2000 (Table 3). Aquaculture production in 1990 and forecasts for 2000 have been compared in Figures 12 and 13, while the potential aquafeed markets in 1990 and 2000 are illustrated in Figures 14 and 15.

Table 3. Aquaculture and aquafeed production of selected species groups in 2000 (t)

SPECIES GROUPS	Indonesia	Philippines	Thailand	Other ASEAN	Japan	China	Taiwan	Ind.Subcont.	Others	Total
YELLOWTAILS										
FISH	-	-	-	-	161,106	-	-	-	462	161,568
FEED	-	-	-	-	**144,995**	-	-	-	**416**	**145,411**
EELS										
FISH	348	-	-	-	38,855	-	83,724	-	1,375	124,302
FEED	**564**	-	-	-	**62,945**	-	**135,633**	-	**2,228**	**201,370**
CATFISHES										
FISH	6,984	189	54,864	355	-	-	72	70,668	2,254	135,386
FEED	**7,857**	**213**	**61,722**	**399**	-	-	**81**	**79,502**	**2,535**	**152,309**
SEABREAMS										
FISH	5,440	5,341	2,975	4,123	87,040	-	12,538	-	2,912	120,369
FEED	**3,917**	**3,846**	**2,142**	**2,969**	**62,669**	-	**9,027**	-	**2,097**	**86,667**
TROUTS										
FISH	-	-	-	-	22,079	770	1,062	1,386	5,135	30,432
FEED	-	-	-	-	**28,703**	**1,001**	**1,381**	**1,802**	**6,676**	**39,563**
SALMONS										
FISH	-	-	-	-	28,750	-	-	-	5,112	33,862
FEED	-	-	-	-	**29,900**	-	-	-	**6,646**	**36,546**
OTHER CARNIVOROUS*										
FISH	-	-	1,010	838	53,592	66,000	1,152	-	7,694	130,286
FEED	-	-	**727**	**603**	**91,642**	**47,520**	**1,970**	-	**13,157**	**155,619**
TOTAL CARNIVOROUS										
FISH	12,772	5,530	58,849	5,316	391,422	66,770	98,548	72,054	24,944	736,205
FEED	**72,338**	**4,059**	**64,591**	**3,971**	**420,854**	**48,521**	**148,092**	**81,304**	**33,755**	**817,485**

* and Ayu

Table 3. (cont.)

SPECIES GROUPS	Indonesia	Philippines	Thailand	Other ASEAN	Japan	China	Taiwan	Ind.Subcont.	Others	Total
COMMON CARP										
FISH	116,753	932	2,600	1,669	16,150	652,961	3,359	236,128	43,152	1,073,704
FEED	**52,539**	**419**	**1,170**	**751**	**29,070**	**58,767**	**6,046**	**106,258**	**19,418**	**274,438**
TILAPIAS										
FISH	100,940	152,284	43,800	2,288	5,825	318,213	52,047	10,000	4,284	689,681
FEED	**45,423**	**68,528**	**19,710**	**1,030**	**10,485**	**28,639**	**84,316**	**4,500**	**3,085**	**265,716**
MILKFISH										
FISH	128,000	210,882	220	-	-	-	154,217	-	52	493,371
FEED	**25,600**	**42,176**	**44**	-	-	-	**277,591**	-	**10**	**345,421**
TOTAL NON-CARNIV.										
FISH	345,693	364,098	46,620	3,957	21,975	971,174	209,623	246,128	47,488	2,256,756
FEED	**123,562**	**111,123**	**20,924**	**1,781**	**39,555**	**87,406**	**367,953**	**110,758**	**22,513**	**885,575**
FRESHWATER PRAWNS										
PRAWN	-	-	16,000	274	150	-	34,821	396	16,046	67,687
FEED	-	-	**25,600**	**438**	**300**	-	**69,642**	**396**	**16,046**	**112,422**
MARINE SHRIMP										
SHRIMP	118,700	67,486	111,625	1,890	3,000	251,344	36,252	68,047	63,705	722,049
FEED	**161,432**	**91,781**	**151,810**	**2,570**	**4,800**	**242,290**	**58,003**	**32,663**	**30,578**	**775,927**
OTHER CRUSTACEANS										
CRUST.	4,820	2,000	60	406	90	15,666	4,074	-	218	27,334
FEED	**6,941**	**2,880**	**86**	**585**	**162**	**14,099**	**6,518**	-	**174**	**31,445**
TOTAL CRUSTACEANS										
CRUST	123,520	69,486	127,685	2,570	3,240	267,010	75,147	68,443	79,969	817,070
FEED	**168,373**	**94,661**	**177,496**	**3,593**	**5,262**	**256,389**	**134,163**	**33,059**	**46,798**	**919,794**
TOTAL AQUACULTURE PRODUCTION	481,985	493,114	233,154	11,843	416,637	1,304,954	383,318	386,625	152,401	3,810,031
TOTAL AQUAFEED MARKET	**304,273**	**209,843**	**263,011**	**9,345**	**465,671**	**392,316**	**650,208**	**225,121**	**103,066**	**2,622,854**

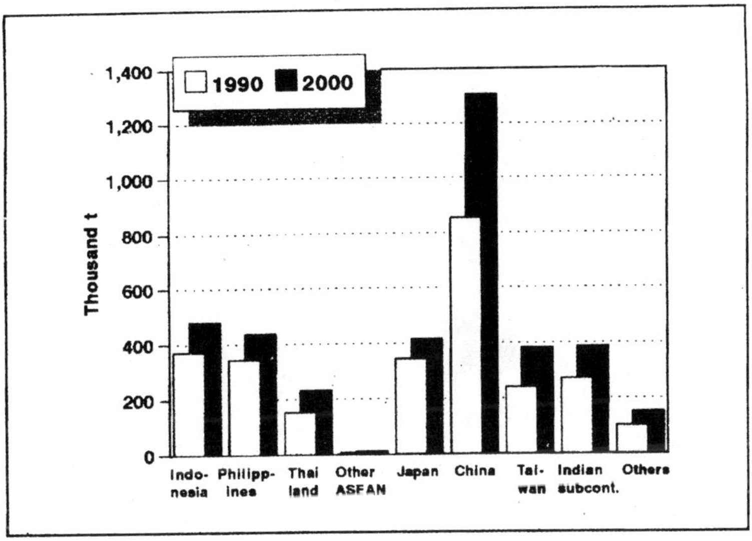

Figure 13. Aquaculture production in Asia and the Pacific in 1990 and estimates for 2000 (commercially fed species only)

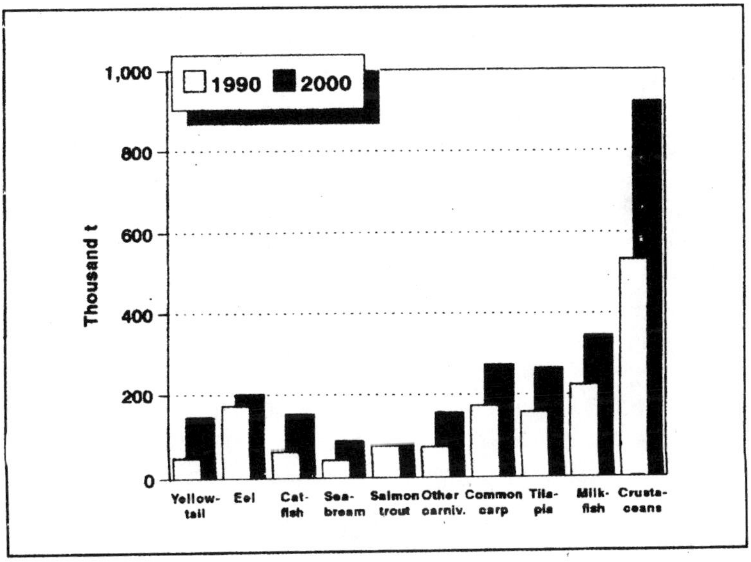

Figure 14. Aquafeed markets in Asia and the Pacific in 1990 and forecasts for 2000 by species

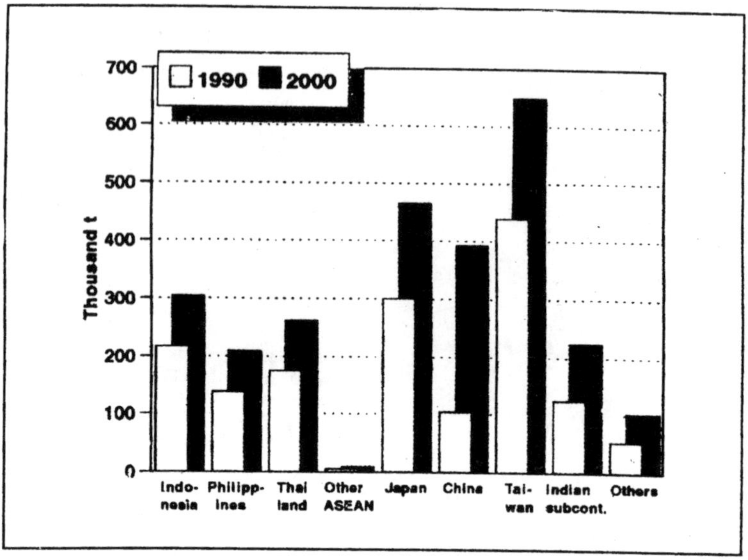

Figure 15. Aquafeed markets for Asia and the Pacific in 1990 and forecasts for 2000 by country

Finfish feeds in the year 2000

Growth rates for farmed carnivorous fish in Asia are extremely difficult to predict because, unlike crustacea, there are so many species which are or will be produced, in varying locations, each with entirely different markets: some local, some global.

Asian production of carnivorous fish is expected to grow by over 40 % above 1990 levels to 736,000 t by 2000, while production of the three major non-carnivorous species which are partially fed by commercial feeds is forecast to increase to 2.26 million t in 2000, 41 % over 1990. The Asian aquafeed market in 2000 is expected to include 817,000 t of feeds for carnivorous fish and 886,000 t for non-carnivorous finfish. This represents increases of 73 % and 60 % respectively over 1990.

In the year 2000 the principal markets for carnivorous finfish feeds in Asia will still be Japan (51 %), and Taiwan (18 %). However, while expansion in these countries will be significant, the market for carnivorous finfish feed by 2000 will grow faster still, and is expected to more than double over 1990, in China, the Indian sub-continent and ASEAN. Eel feed as a proportion of the total feeds for carnivorous finfish is expected to decrease from 37 % in 1990 to 25 % in 2000, the demand being expected to increase by only 28,000 t in the decade. Demand for salmonid feeds will be almost static, increases in aquaculture production being compensated by improvements in food conversion ratio (FCR). Conversely, the market for commercial feeds in Asia is expected to treble by 2000 for yellowtails, to about 145,000 t, to increase by a factor of 2.5 to 152,000 t for catfishes and to double for seabreams and other carnivorous fishes.

The market for feeds for the three non-carnivorous finfish groups in Asia will be around 886,000 t by 2000. This represents a 60 % increase over 1990. The proportions of the increased market for aquafeeds for these three species groups, common carp (31 %), the tilapias (30 %) and milkfish (39 %) will remain almost identical to 1990. However, the share destined for Taiwan (42 %) will decrease. ASEAN's share is expected to increase from 25% in 1990 to nearly 29% in 2000. The market for non-carnivorous finfish feeds in China in the year 2000, though still a small proportion of the Asian total (10 %) is expected to be about seven times greater by weight than in 1990.

Crustacean feeds in the year 2000

This Asian market is forecast to expand by 387,000 t (73 %) over 1990. The market for fresh water prawn feeds will nearly treble to 112,000 t. 444,000 t of crustacean feeds are expected to be used in ASEAN countries in 2000, a significantly smaller share (48 %) of the total than in 1990. The Chinese market, which is expected to be over 250,000 t by 2000, will constitute a 28 % share of Asian crustacean feed production. Taiwan will retain a sizeable share (15%) but less than Thailand (19%) or Indonesia (18%). The market in the Indian sub-continent will still be small (5 %).

The estimates of the crustacean feed market in the year 2000 have been based on an expansion of aquaculture production from the official statistics of 576,000 t in 1990 (Table 1) to 817,000 t in 2000 (Table 3).

Total Asian aquafeed market in the year 2000

The total market for commercial aquafeeds in Asia is expected to grow by 1.0 million t (68 %) to about 2.6 million t by the year 2000 (Table 3). Of this total, about 35 % will consist of feeds for crustacea and 34 % for non-carnivorous finfish (Figures 14 and 15). Taiwan is expected to remain the largest single aquafeed market, at over 650,000 t; however, at 25%, this is a decreasing proportion of the Asian total. ASEAN (over 786,000 t) is expected to constitute 30 % of the total, also a significantly lower proportion than in 1990, while Japan's share remains static. China is expected to consume 15 % of Asian commercial aquafeeds by 2000 (7 % in 1990), increasing its feed production by a factor of 3.8 to nearly 400,000 t.

FARM-MADE AQUAFEEDS

As we have shown above, it is possible to estimate commercial aquafeed production in 1990 and to attempt to predict how this market may change by the year 2000 in response to anticipated increases in aquaculture production, improvements in feed conversion efficiency and increasing reliance on compounded feed. However, what we have not shown is the amount of farm-made feed made, either now or in the future.

We do not propose to attempt to do so now because there is insufficient data on which to make such estimates and predictions. We hope that the expert consultation, to which our paper is an introduction, will provide sufficient information on the utilization of farm-made aquafeeds in Asia and the Pacific to make such estimates possible.

The estimates we have made for commercial aquafeed production are based upon the current production of carnivorous finfish, crustacea and three species groups of non-carnivorous finfish, and upon our forecasts for increased aquaculture production of these groups by the year 2000. Using similar assumptions on FCR to those used in deriving our estimates for commercial aquafeed production it can be shown that, in 1990, most aquaculture production, even of these types of fish and crustaceans, was <u>not</u> produced through the use of commercial aquafeeds (Figure 16).

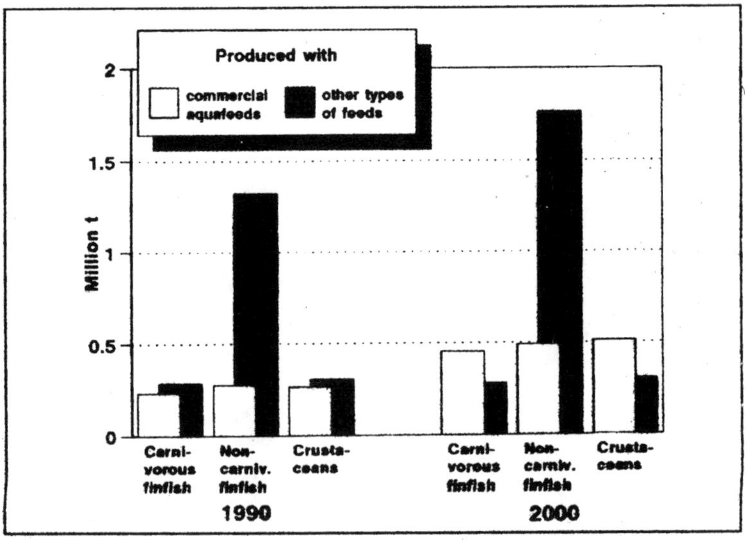

Figure 16. Volume of selected cultured species produced through the use of commercial aquafeeds and other types of feeds in Asia and the Pacific

By 2000 this may have changed significantly for carnivorous finfish and crustaceans, so that more than 50 % will be cultured through the use of commercial aquafeeds. However, even then, most common carp and tilapias will still be produced through other nutrient inputs. The same will apply to milkfish outside Taiwan.

Our calculation (Table 4) highlights our belief that, both now and in the year 2000, around 2.0-2.3 million t of carnivorous finfish, crustacea and common carp, tilapias and milkfish will be produced without the use of commercial feeds. A significant proportion is reared through the use of dry and moist farm-made feeds, ranging from simple mixtures, sometimes made into a doughballs or other forms by hand, to relatively complex pelleted feeds. The scope for greater use of farm-made feeds is immense. The need for improvements in formulation, and in production and storage technology for the small-scale farmer is similarly great and provides the challenge which we hope this meeting will emphasize.

Table 4. Estimate of the proportion of aquaculture production * produced in Asia through the use of commercial aquafeeds in 1990 and 2000 (million t)

	CARNIVOROUS FINFISH		NON-CARNIVOROUS FINFISH		CRUSTACEANS		TOTALS	
	1990	2000	1990	2000	1990	2000	1990	2000
ESTIMATED COMMERCIAL AQUAFEED PRODUCTION (see Tables 1 and 3)	0.472	0.817	0.554	0.886	0.533	0.920	1.559	2.623
AQUACULTURE PRODUCTION (see Tables 1 and 3)	0.523	0.736	1.601	2.257	0.576	0.817	2.700	3.810
ESTIMATED AQUACULTURE PRODUCTION ACHIEVED BY COMMERCIAL AQUA-FEEDS (2)**	0.236	0.454	0.277	0.492	0.267	0.511	0.780	1.457
ESTIMATED AQUACULTURE PRODUCTION ACHIEVED BY OTHER MEANS (3)***	0.287 (55%)	0.282 (38%)	1.324 (83%)	1.765 (78%)	0.309 (54%)	0.306 (37%)	1.920 (71%)	2.353 (62%)

* carnivorous finfish, selected non-carnivorous finfish (common carp, tilapias and milkfish) and crustaceans

** assuming average FCR of 2:1 for 1990 and 1.8:1 for 2000

*** natural productivity; fertilization; integrated aquaculture; farm-made feeds

LIST OF REFERENCES

Akiyama, D.M. 1992a. "Guesstimated" world production of shrimp feed. World Shrimp Farming 17.11.37.

Akiyama, D.M. 1992b. Future considerations for the aquaculture feed industry, p. 5-9. *In* D.M. Akiyama and R.K.H. Tan (eds.) Proceedings of the Aquaculture Feed Processing and Nutrition Workshop. Thailand and Indonesia, 19-25 September 1991. American Soybean Association, Singapore.

Anon. 1990a. Feed boom forecast. Fish Farming International 17(3):1.

Anon. 1990b. CP Feed mill regains lost popularity with local stock investors. Bangkok Post, Bangkok. April 30, 1990:18.

Anon. 1990c. World feed panorama: Census 1990. Feed International, January 1990: 7-9.

Csavas, I. 1988. Shrimp farming developments in Asia, p. 63- 92. *In* INFOFISH (eds.) Shrimp '88 Conference Proceedings. 26-28 January 1988, Bangkok, Thailand. INFOFISH, Kuala Lumpur, Malaysia.

Csavas, I. 1990a. Shrimp aquaculture development in Asia, p. 207-222. *In* M.B. New, H. de Saram and T. Singh (eds.) Technical and Economic Aspects of Shrimp Farming: Proceedings of the Aquatech '90 Conference, 11-14 June 1990, Kuala Lumpur, Malaysia. INFOFISH, Kuala Lumpur, Malaysia.

Csavas, I. 1990b. New developments in Asian aquaculture, p.11-57. *In* Aquaculture International Congress Proceedings, September 4-7 1990, Vancouver B.C., Canada.

FAO. 1992. Aquaculture production (1985-1990). FAO Fisheries Circular No. 815, Revision 4, FIDI/C 815 Rev. 4, Statistical Tables. 206 p.

FAO/RAPA. 1991. Selected indicators of food and agriculture development in Asia-Pacific region, 1980-90. RAPA Publication 1991/18. 211 p.

Gill, C. 1992a. Focus on feed: Global perspective. Feed International, January 1992:4-6.

Gill, C. 1992b.More feed, fewer feedmillers.Feed International,January 1992:10-16.

Megisson, P.A. 1990. Future world of aquaculture. Paper presented at the Rovithai International Shrimp Seminar, July 1990, Bangkok, Thailand.

New, M.B. 1989. Formulated aquaculture feeds in Asia : some thoughts on comparative economics,industrial potential, problems, and research needs in relation to the small-scale farmer, p.19-30. *In* Report of the Workshop on shrimp and finfish development, Johore Bahru, Malaysia, 25-29 October 1988, Report No. ASEAN/SF/ 89/GEN/11. ASEAN/UNDP/FAO Regional Small-Scale Coastal Fisheries Development Programme, Manila, Philippines.

New, M.B. 1990. Compound feedstuffs for shrimp culture, p.79-118. *In* M.B. New, H. de Saram and T. Singh (eds.) Technical and Economic Aspects of Shrimp Farming: Proceedings of the Aquatech '90 Conference, 11-14 June 1990, Kuala Lumpur, Malaysia. INFOFISH, Kuala Lumpur, Malaysia.

New, M.B. 1991. Turn of the millennium aquaculture: navigating troubled waters or riding the crest of the wave? World Aquaculture 22(3):28-49.

New, M.B. and U.N.Wijkstrom. 1990. Feed for thought: some observations on aquaculture feed production in Asia. World Aquaculture 21(1):17-19 and 22-23.

Rosenberry, R. (ed.). 1990. World Shrimp Farming 1989. Aquaculture Digest 28 p.

Rosenberry, R. (ed.). 1991. World Shrimp Farming 1990. Aquaculture Digest 40 p.

Rosenberry, R. (ed.). 1992. World Shrimp Farming 1991. Aquaculture Digest 44 p.

Shiau, S.Y. and I.C.Liao. 1989. Structure and outlook for the fish feed industry in Taiwan, p.142-159. *In* Proceedings of the 29th Annual Conference of the International Association of Fish Meal Manufacturers, 20-24 November 1989, Hong Kong.

Wijkstrom, U.N. and M.B. New. 1989. Fish for feed: a help or a hindrance to aquaculture in 2000 ? Infofish International 6/89:48-52.

Yamamoto, T. 1992. Development of aquaculture in Japan within her fisheries. Resource Paper SEM-18-92 presented at the APO Seminar on Aquaculture, 25 August to 4 September 1992, Tokyo, Japan.

SUPPLEMENTARY FEEDING IN SEMI-INTENSIVE AQUACULTURE SYSTEMS

Sena S. De Silva

Faculty of Aquatic Sciences, Deakin University, P.O. Box 423, Warrnambool, Victoria 3280, Australia

DE SILVA, S.S. 1993. Supplementary feeding in semi-intensive aquaculture systems, p. 24-60. *In* M.B. New, A.G.J. Tacon and I. Csavas (eds.) Farm-made aquafeeds. Proceedings of the FAO/AADCP Regional Expert Consultation on Farm-Made Aquafeeds, 14-18 December 1992, Bangkok, Thailand. FAO-RAPA/AADCP, Bangkok, Thailand, 434 p.

INTRODUCTION

A 'supplement' is defined as a "thing added to supply deficiencies". In an aquacultural nutritional context it will be tantamount to the supply of feed to meet one or more nutrient deficiencies of the system for the well being of the stock. It is a truism that the supply of deficiencies in relation to nutrition of the stock in semi-intensive aquaculture practices is one of the least addressed aspects of the industry. Therefore, it continues to remain a rather grey area in our knowledge. The aquaculture industry in Asia, in particular, is embarking on an increasing phase of intensification, driven by a multitude of economic forces amongst which are an increasing need for even the small-scale farmer to transform from subsistence aquaculture to an income generating enterprise. In the light of increasing competition for primary resources, such as water and land, such economic needs become even more demanding. This increasing intensification of the industry at large however, has to take place within the framework of certain potential global constraints on the industry such as the "fishmeal trap" (Wijkstrom and New 1989). In such a scenario issues related to supplementary feeding are likely to come to the forefront.

Supplementary feeding in semi-intensive aquaculture has to be considered in the perspective of the past - and future trends of the industry. Moreover, the treatment will be meaningful only if supplementary feeding is considered in the context of other strategies that are prevalent in Asian aquaculture which directly or indirectly reduce nutritional deficiencies. In this paper an endeavour is made to address these issues and in particular in the context of the envisaged growth of the industry.

GENERAL TRENDS IN ASIAN AQUACULTURE

The development of semi-intensive aquaculture (to be defined later) in Asia and the Pacific, and in tropical Asia were recently reviewed by Csavas (1989) and Bailey and Skladany (1991) respectively. A majority of the recent reviews however, have tended to highlight the growth of shrimp and carnivorous finfish culture, particularly from the point-of-view of the aquafeed industry and its potential problems (Wijkstrom and New 1989; New 1991). In this presentation an attempt will be made to focus on other aspects of the aquaculture industry in Asia which are of relevance to supplemental feeding.

The backbone of Asian aquaculture is finfish. Finfish contribute over 50% to the total aquaculture production in Asia (Table 1; Figure 1), and has been steadily increasing over the last five years. In addition, finfish production in Asia accounts for more than 80% of the world's total production of cultured finfish (Figure 2), and it ranks second in relative terms only to the cultured seaweed production of the world. In the latter case the cultured aquatic plant production of Asia contributes nearly 98% to the world's total.

The mainstay of Asian aquaculture historically, as well as at present and in the foreseeable future, is non-carnivorous finfish (Table 2; Figure 3). The non-carnivorous finfish production in Asia has been increasing steadily, and presently (1989) contributes 90% to the total, cultured non-carnivorous finfish production of the world (Figure 4). On the other hand, the relative contribution of cultured carnivorous finfish production in Asia has been on the decline. The reason for this decline was addressed by Csavas (1990).

Csavas (1990) also focused on some of the key features of the Asian aquaculture industry, some of which are directly relevant to the present topic. Those concerning finfish and shrimp culture, the two practices in which supplementary feed and feeding are likely to have a direct bearing, are as follows:

- very little hard data exist regarding the level of intensity of Asian aquaculture. Nevertheless, the typical freshwater culture systems are beyond doubt extensive or semi-intensive polycultural systems with some fertilization and supplementary feeding. This is reflected in the major species cultured and the average value of the product. For example in 1988 it was US$ 1.18/kg in developing countries of Asia, as opposed to US$ 3.21/kg elsewhere;
- by and large freshwater aquaculture is a small-scale activity in Asia. A "small-scale activity" is not easily definable either. Perhaps in the present context small-scale can be considered as non-corporate culture activities which are single family owned and managed;
- the development of non-carnivorous finfish production still shows an exponential growth, and has continued to exceed that of carnivorous species;

Table 1. Aquaculture production 1985-1990 (million t)

Production*	1985	1986	1987	1988	1989	1990
Total Aquaculture						
World	11.171	12.210	13.180	14.556	14.420	15.323
Asia	9.317	10.234	11.056	12.307	12.070	12.906
% Asia	83.40%	83.82%	83.89%	84.55%	83.70%	84.23%
Finfish						
World	5.250	5.998	6.875	7.578	7.784	8.411
Asia	4.335	4.993	5.780	6.376	6.474	6.989
% Asia[1]	82.57%	83.24%	84.07%	84.14%	83.17%	83.09%
% Asia[2]	46.53%	48.79%	52.28%	51.81%	53.64%	54.15%
Crustacea						
World	0.279	0.388	0.588	0.645	0.675	0.715
Asia	0.209	0.302	0.453	0.517	0.548	0.575
% Asia[1]	74.91%	77.84%	77.04%	80.16%	81.19%	80.42%
% Asia[2]	2.24%	2.95%	4.10%	4.20%	4.54%	4.46%
Molluscs						
World	2.180	2.394	2.666	2.949	2.914	2.965
Asia	1.320	1.520	1.785	2.059	2.042	2.155
% Asia[1]	60.55%	63.49%	66.95%	69.82%	70.08%	72.68%
% Asia[2]	14.17%	14.85%	16.15%	16.73%	16.92%	16.70%
Aquatic Plants						
World	3.434	3.400	3.023	3.350	2.998	3.188
Asia	3.425	3.389	3.010	3.321	2.958	3.144
% Asia[1]	99.74%	99.68%	99.57%	99.13%	98.67%	98.62%
% Asia[2]	36.76%	33.12%	27.22%	26.98%	24.51%	24.36%
Other Animals						
World	0.028	0.030	0.028	0.034	0.049	0.044
Asia	0.028	0.030	0.028	0.034	0.048	0.043
% Asia[1]	100.00%	100.00%	100.00%	100.00%	97.96%	97.73%
% Asia[2]	0.30%	0.29%	0.25%	0.28%	0.39%	0.33%

* % Asia[1] = Asian contribution of the particular commodity to the world production of that commodity; % Asia[2] = contribution of the commodity to total Asian aquaculture production

Source: FAO (1992)

- in the developing countries of the region traditional extensive or semi-intensive pond culture dominates finfish and shrimp aquaculture. The general tendency over the past decade was to intensify these to shift from an uncontrolled "polyculture" to monoculture of shrimp by replacing trapping of wild juveniles with stocking;

- in coastal areas the dominant function of aquaculture is income generation, the production of cash-crops sold on distant markets;

- in many cases even semi-intensive shrimp farms have shifted to use factory made feeds, and this trend is likely to increase in the ensuing years.

Table 2. Trends in finfish aquaculture production 1985-1990 (million t)

Production*	**1985**	**1986**	**1987**	**1988**	**1989**	**1990**
Total Finfish	5.250	5.998	6.875	7.578	7.784	8.411
Non-carnivorous						
World	4.515	5.189	5.965	6.552	6.670	7.192
Asia	3.953	4.595	5.343	5.897	6.008	6.485
% Asia[1]	87.6%	88.6%	89.6%	90.0%	90.1%	90.2%
% Asia[2]	75.3%	76.6%	77.7%	77.8%	77.2%	77.1%
Carnivorous						
World	0.735	0.809	0.910	1.026	1.114	1.219
Asia	0.382	0.398	0.437	0.479	0.466	0.504
% Asia[1]	52.0%	49.2%	48.0%	46.7%	41.8%	41.4%
% Asia[2]	7.3%	6.6%	6.4%	6.3%	6.0%	6.0%

* % Asia[1] - Asian contribution to the particular group

* % Asia[2] - Asian contribution of the commodity to total world finfish production

Source: FAO (1992)

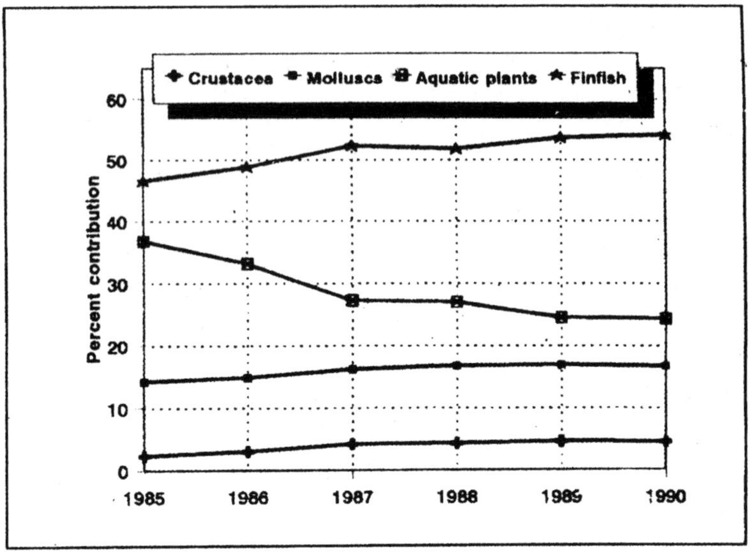

Figure 1. Share of major commodity groups in the aquaculture production of Asia (based on FAO 1992)

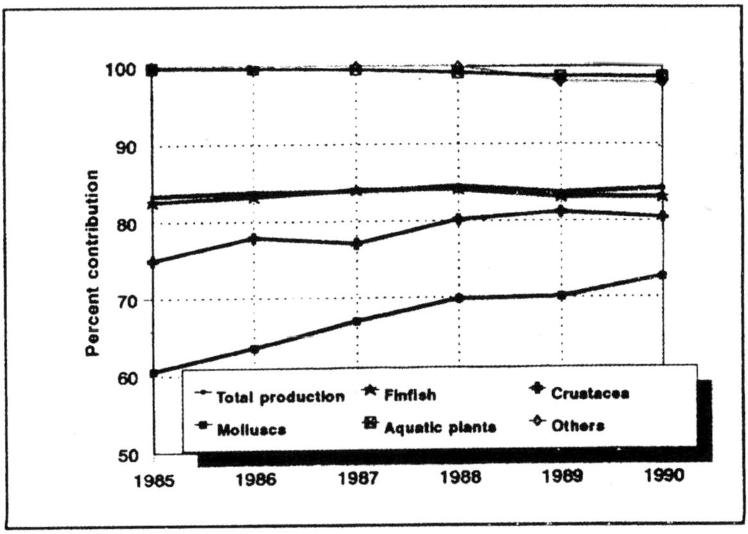

Figure 2. Contribution of Asia to the overall aquaculture production of the world (based on FAO 1992)

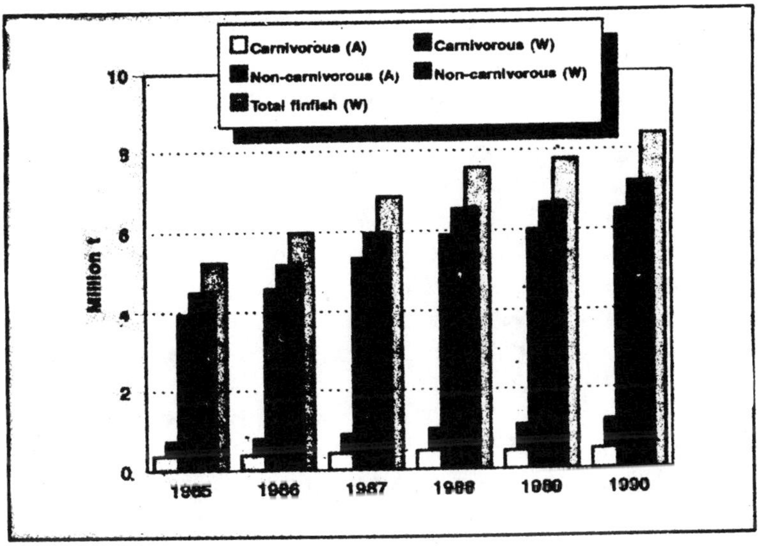

Figure 3. Trends in carnivorous and non-carnivorous finfish production in Asia (A) and the world (W) (based on FAO 1992)

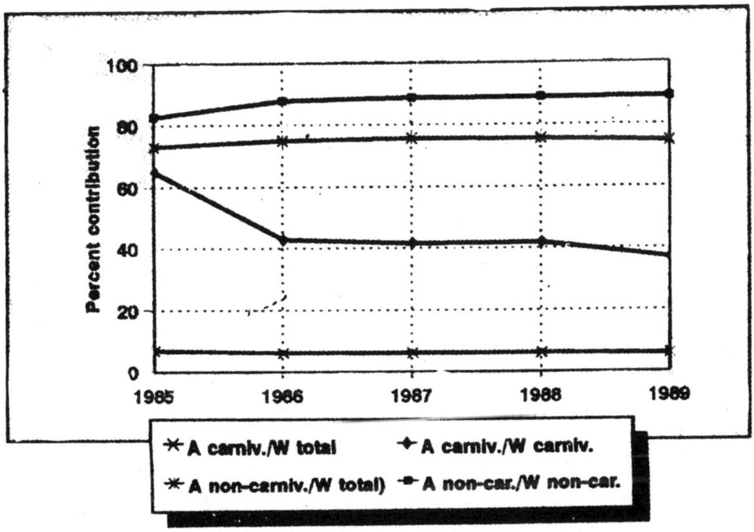

Figure 4. Contribution of Asia (A) to the world (W) production of carnivorous and non-carnivorous finfish and to the world's total finfish production

SEMI-INTENSIVE AQUACULTURE: A SEARCH FOR A DEFINITION

Perhaps semi-intensive aquaculture will continue to remain a term and a practice that will never be defined precisely. On the other hand, some might argue that such a definition is not strictly needed, and that perhaps broadly accepted guidelines are sufficient.

One of the best available guidelines depicting semi-aquaculture practice was presented schematically (Figure 5) by Tacon (1988). Accordingly, semi-intensive aquaculture would constitute a significant quantum in the continuum between extensive and intensive aquaculture. Within this broad spectrum the semi-intensive practices could differ from each other in the degree of management, provision of feed, the stocking density and yields. For example Csavas (1990) suggested that in respect of Asian shrimp culture such practices could result in yields between 1,500-5,000 kg/ha/yr. He related the scale of intensity with improving external inputs and increasing experience in production techniques and farm management. Yields ranging from 5,890 - 14,000 kg/ha/yr have been reported recently from the semi-intensive composite culture practices of Indian major carps from Andhra Pradesh, India (Veerina *et al.* in preparation). The upper range of this production level far exceeds that from most intensive culture practices in temperate waters.

On the other hand, Schroeder (1978) reported yields of up to 30 kg fish/ha/day (common carp and tilapia) through intensive manuring, and Zhu *et al.* (1990) reported an increased net yield of 10.2 kg/ha/day in Chinese integrated farm ponds as a result of manure application. The purpose of focusing on these data is not to emphasise the importance of fertilization in semi-intensive aquaculture but to draw attention to the fact that perhaps yields are one of the weakest criteria in determining the degree of intensity of a culture practice. Secondly, and more importantly, it raises the question whether fertilization needs to be considered as a feed supplement.

The use of manure in fish farming is an old Asian practice, and has been adequately reviewed by Wohlfarth and Schroeder (1979) and Wohlfarth and Hulata (1987). The influence of manure on aquaculture is schematically depicted in Figure 6. Manure is thought to act in supplying a deficiency; in this case primarily nutrients in the culture system, supplementation of which enhances the growth of natural food organisms. Hence, I am inclined to suggest that manure be considered as an indirect feed supplementation; perhaps a supplementation that is least competitive with the other forms of direct or indirect nutrient supplementation that are used in semi-intensive aquaculture. On the other hand, Jana and Chakraborty (1990) suggested that a better approach to carp culture would be to introduce live plankton rather than direct manuring. Here again, the question of wide-scale adoption of such a practice by the small-scale farmer arises.

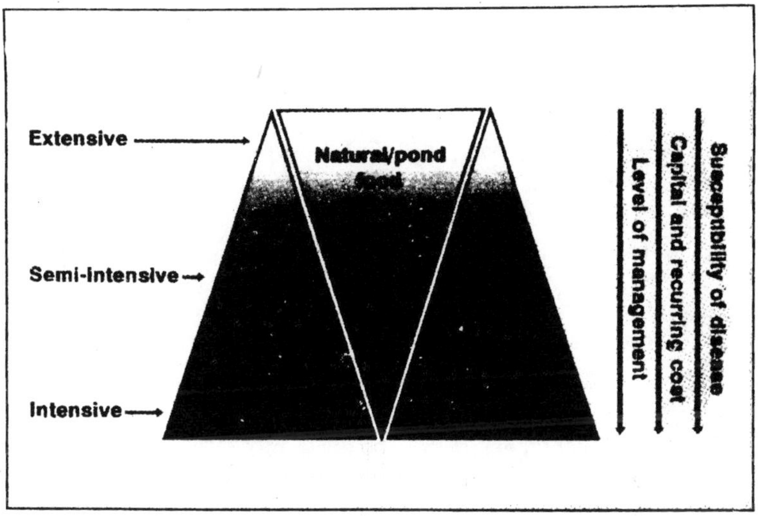

Figure 5. Schematic representation of the range of aquaculture practices in relation to the inputs (modified after Tacon 1988)

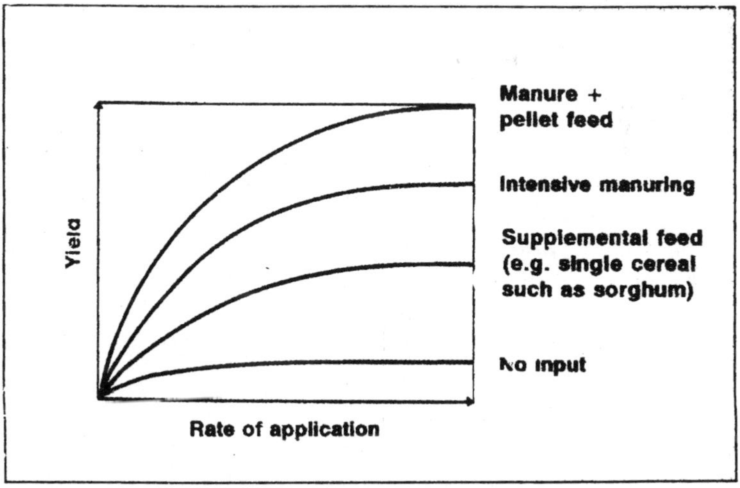

Figure 6. Schematic depiction of the influence of application of manure in comparison to other inputs on yields in semi-intensive aquaculture

SUPPLEMENTARY FEEDS

Interactions with natural food supplies

Supplementing the naturally available food in a culture system is the most simplistic functional interpretation of supplementary feeds. In reality, the role of supplementary feeds in the nutrition of the cultured organism is much more intricate and is closely interlinked to the dynamics of the pond/system. The intricacy is analogous to the interpretation of interactions between species in a simple aquatic community explained in terms of a food chain as opposed to a food web.

The natural food organisms in a pond are a rich source of protein, often the mix of organisms containing 50-60% of protein (dry matter basis). According to Albrecht and Breitsprecher (1969) the mean protein, carbohydrate and lipid content of fish food organisms in ponds were 52.1, 27.3 and 7.7% respectively, and the calorific value ranged from 1.6-5.7 kcal/g dry matter, with a mean of 3.9 kcal/g. Proximate composition of some of the common individual food organisms is given in Table 3. Thus it is evident that the nutritional value of natural food organisms is adequate to meet the nutritional requirements of most of the semi-intensively cultured organisms in Asia, which according to Prus (1970) range from 28-35% of protein and 96 mg/kcal protein to energy ratio. It has also been shown, for example, that the food of tilapia (*Oreochromis mossambicus*), from quasi-natural impoundments, even when it contains a high proportion of detritus, is well digested (De Silva *et al.* 1984).

In semi-intensive culture there are complex interactions between the natural food organisms and supplementary feeding practice(s). The possible qualitative interactions in a well prepared pond which is to be subjected to supplementary feeding is schematically depicted, in Figure 7. The above simplification does not take into account the possible changes in the feeding habits of the cultured organisms. The trophic niches of fish species are known to change with natural food productivity, the standing crop of fish and the plasticity of the feeding behaviour. Such changes are documented for a number of species (see Werner and Hall 1976; De Silva *et al.* 1984; Odum 1970). Furthermore, in polyculture systems fish species are more capable of changing their diet readily (Weatherly 1963). Also Spataru *et al.* (1980), in their studies on common carp, have shown that supplementary feeding can affect feeding habits and food selection, the fish tending to select a narrower range of food items when on supplementary feeding regimes.

There is some experimental evidence to show that, as the difference between standing crop (SC) of the cultured organism(s) and the critical standing crop (CSC) of natural food organisms increases, the deficit in natural protein supply also increases. Under such conditions supplementation with "protein rich" diets becomes necessary to maintain growth and production.

Table 3. Proximate analyses of organisms serving as food for pond fishes

		Dry matter composition*				
Organism/group	**Dry matter (%)**	**Protein (%)**	**Carbohydrate (%)**	**Lipid (%)**	**Ash (%)**	**Energy (kcal/kg)**
Bacteria					5.4	4,710
Algae						
Cyanophyta (blue greens)		31.3			46.7	2,213
Chlorophyta (greens)	16.8	17.6		3.7	26.9	3,773
Phaeophyta (brown algae)	14.1				32.3	3,056
Bacillariacaea (diatoms)		30.7		9.9	38.3	3,654
Rhodophyta (red algae)	21.7				32.1	3,170
Aquatic mácro-vegetation	15.8	14.6		4.5	13.9	3,906
Protozoa						5,938
Rotifers	11.2	64.3		20.3	6.2	4,866
Oligochaetes	7.3	49.3		19.0	5.8	5,569
Leeches	24.0	61.0			5.1	5,432
Crustacea						
Anostraca *(Artemia)*	11.0	61.6		19.5	10.1	5,835
Cladocera	9.8	56.5	28.2	19.3	7.7	4,800
Copepoda	10.3	52.3	9.2	26.4	7.1	5,445
Ostracoda	35.0	41.5				5,683
Malacostraca (higher Crustacea)	24.6	49.9	18.4	20.3	19.6	5,537
Insects	23.2	55.9	20.1	18.6	4.9	5,075
Plecoptera (stone flies)						4,900
Ephemeridae (mayflies)	17.6	50.2			3.7	5,646
Odonata (dragon flies)	21.1	51.9			5.8	4,985
Hemiptera (water bugs)	26.0	68.8				5,150
Trichoptera (caddisflies)	14.8	34.7			11.8	5,019
Diptera	16.0	55.3			6.9	5,177
Chironomids (larvae)	19.1	59.0	22.5	4.9	5.8	5,034
Molluscs	32.2	39.5	7.5	7.8	32.9	3,889
Aquatic detritus	91.5				12.4	4,701

* since the values are averages of figures collected from different sources they do not necessarily add up to 100%

Source: modified from Hepher (1988)

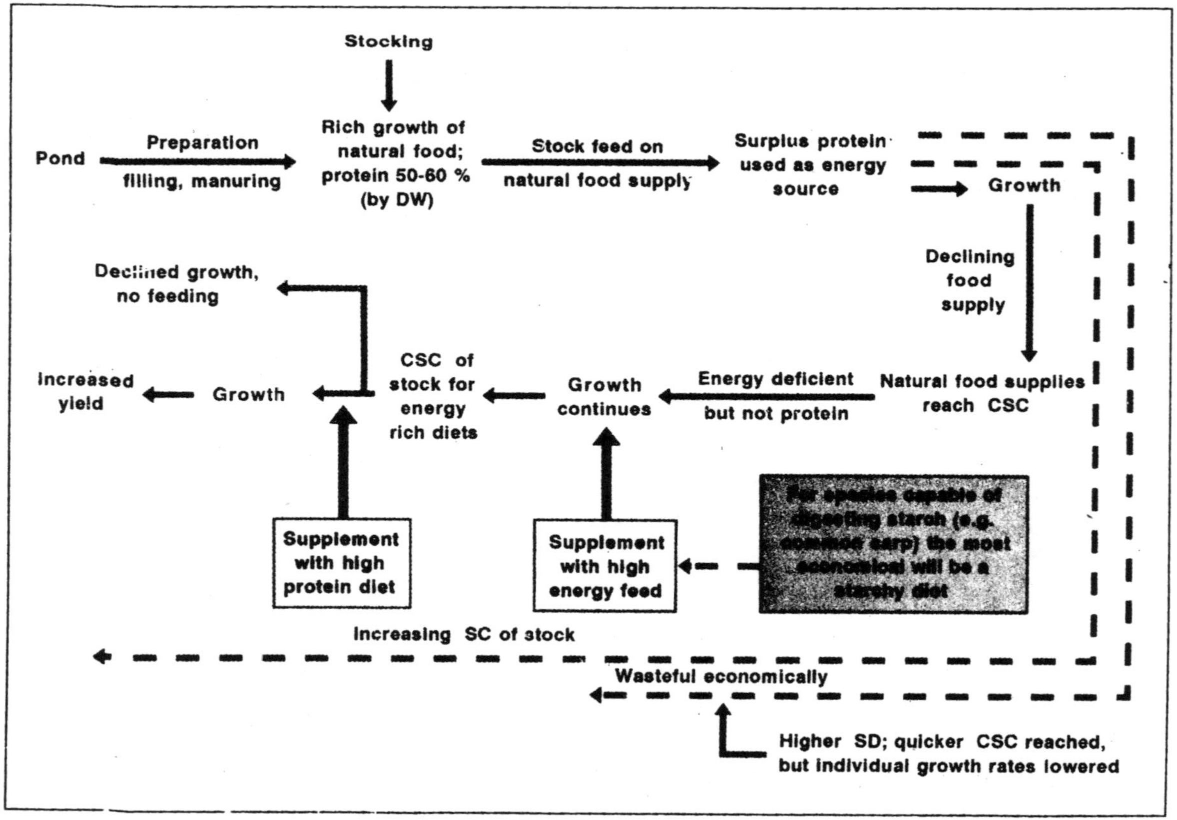

Figure 7. Schematic representation of the qualitative changes that occur in a semi-intensive culture pond. Also indicated is the potential utility of "supplementary feed" strategies.

The changes in protein requirements in supplemental feed(s) in relation to standing crop are depicted in Figure 8. The increase in supplementary protein requirements in relation to increasing SC is gradual. For example Hepher *et al.* (1971) and Hepher (1975) have shown that in common carp under monoculture, a pelleted feed of sorghum (9% protein), or diets of 22.5% and 27.5% protein did not influence the growth until a SC of 800kg/ha was reached. However, when the SC increased further the 9% protein diet was not effective; the 22.5% diet became ineffective when the SC reached 1,400kg/ha. Hanley (1991) on the other hand, reported that in Nile tilapia (*Oreochromis niloticus*) the dietary lipid level did not influence the growth rate or the food conversion ratio (FCR) and recommended that diets for tilapia have to be compounded on the basis of protein rather than energy considerations.

Unlike in an intensive culture system the utilization efficiency of a supplementary feed in semi-intensive culture is dependent on the natural food supply, in addition to other factors such as standing crop and body weight. The latter effects are common to all systems. This phenomenon is best illustrated by the work of Hepher (1978) on semi-intensive culture of common carp (Figure 9). It is apparent from this figure that at a SC of about 500kg/ha, sorghum as a supplementary diet was inadequate and consequently the FCR increased sharply, as opposed to those maintained on a pelleted feed of 25% protein. As pointed out by Hepher (1978) this sharp increase in the FCR is critical from an economic point of view.

Hepher (1988) suggested that in such circumstances one of three options is open to the farmer:

- improve the diet by adding the missing nutrients. This results in an increase in the cost, which may not necessarily be paid off by corresponding increase in yield;
- reduce the SC. This results in a decrease in the yield; and
- increase productivity by manuring.

The above observations indicate that supplementary feeding in semi-intensive aquaculture has many facets; to be economical the farmer has to be alert about the type and quantity of feed to be used. It would be apparent therefore, that feed supplementation in semi-intensive aquaculture, if properly carried out to gain the maximum return, is not an *ad hoc* process. In effect, because of the interactions between the natural food supply and supplementation it is more complex than providing the stock with a nutritionally wholesome diet under intensive culture conditions.

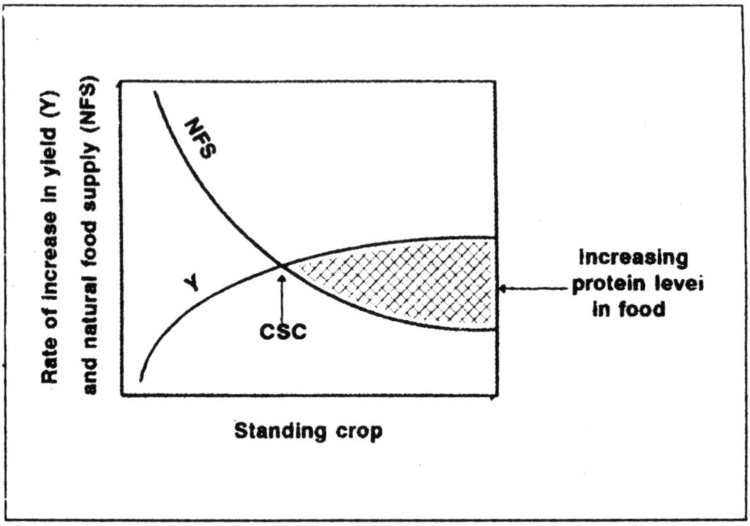

Figure 8. Schematic depiction of changes in the natural food organisms and fish yields, in relation to standing crop of the cultured organism and the ensuing protein needs of the supplemental feed (s)

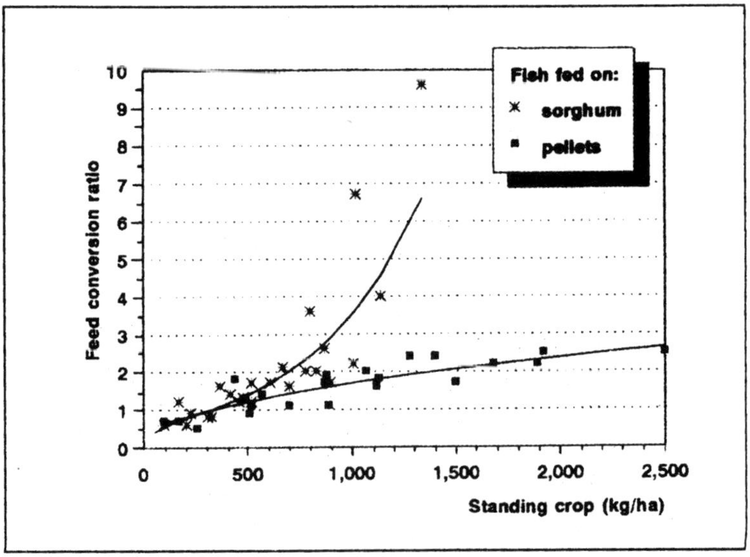

Figure 9. Supplementary feed conversion ratio at different standing crops of common carp in ponds (after Hepher 1978)

Supplementary feeds

Supplementary feeds used in semi-intensive aquaculture in Asia range from kitchen wastes on the one extreme to almost nutritionally complete compounded feeds, such as those used typically in shrimp culture operations. Supplementary feeds can be in the form of single ingredients, simple mixtures of powdered ingredients or ingredients compounded into a dough or pellet form.

In Asia, where aquaculture is mostly a rural occupation, the selection and utilization of supplementary feeds is linked to other agricultural activities in that particular region. This is so because the bulk of the ingredients used as supplementary feeds consists of agricultural by-products, or by-products of the animal husbandry industry. Fish are not capable of efficiently utilizing most of these ingredients, in particular the grasses Table 4 gives a list of commonly used feed types and the FCR values reported in respect of each. The relatively inefficient utilization of grasses and other cellulose-rich materials is to be expected because fish do not possess the enzyme cellulase.

Information available in the literature on the types and extent of usage of different supplementary feeds on farms is scanty. Over the last decade an almost explosive growth of a semi-intensive, polyculture practice of Indian major carps has been witnessed in the south eastern state of Andhra Pradesh in India. Yields ranging from 5,890-14,000 kg/ha/yr have been reported (Veerina *et al.* in preparation). In a survey covering 189 farms in four districts of the State, it was found that 9 major ingredients and 7 feed types are used by the farmers. On average 27,000 kg/ha/yr of feed was utilized as supplementary feed (Tables 5 and 6; Figure 10). This data provides an indication of the diversity of supplemental feeds that are utilized in semi-intensive finfish culture, even within a single region, and also variation in farmer preference for particular ingredients and or ingredient mixes. The scientific rationale behind the adoption of some of the supplemental feed practices is not immediately obvious, however; for example the use of common salt in the feed mix by some Andhra Pradesh farmers. Our knowledge on use of supplementary feeds at farmer level *per se* is very scanty. Until the recent survey by Veerina *et al.* (in preparation) there had been no documentation that common salt was used as an additive to supplementary feed(s). The above is just one example; perhaps there are many more which the scientific community is unaware of and the scientific basis of usage unexplained. This further exemplifies the need for a deviation from the traditional nutritional research based on dose-response approach, as advocated by De Silva and Davy (1992).

Over the last decade or so there had been a shift in emphasis in respect of supplementary feeds; the term supplementary feeds and supplementary

Table 4. FCR reported for some of the commonly used supplementary feeds*

Feeds of animal origin		Feeds of plant origin	
Daphnid	4-6.4	Soybean	3-5
Mysis	2-3.9	Wheat flour	7.2
Gammarus	3.9-6.6	Barley flour	2.6
Prawns & shrimp	4-6	Corn	4-6.0
Earthworm (fresh)	8-10	Wheat bran	6.1-7.3
Clams (flesh)	1.3	Barley bran	7.0
Snail flesh-fresh	22.0	Irish potato	20-30
Snail flesh-dried	10.2	Cereals	4-6
Chironomids	2.3-4.4	Groundnut cake	2-4
Housefly maggots	7.1	Ground maize	3.5
Locust - fresh	10.7	Ground rice	4.5
Locust - dried	5	Oilpalm cake	6-12
Silkworm pupae - fresh	3-5	Manioc leaves	10-20
Silkworm pupae - dried	1.2-2.1	Manioc rind	50.7
		Manioc flakes	17.6
		Manioc flour	49.4
		Banana leaves	25.0
Freshwater fish	4-8	Napier grass	48.0
Fresh sea fish (trash)	6-9	Rye grass**	17-23
Fish flour	1.5-3.0	Sudan grass**	19-28
Fresh meat	5-8	Elephant grass**	30-40
Meat flour	2	Hybrid grass**	25-30
Dried blood powder	1.5-1.7	Lucerne**	25-30
Liver, spleen	5.5-8.0	Clover**	25-30
		Chinese cabbage**	35-40
		Water hyacinth**	50
		Water lettuce**	50

Mixed diets	
Fresh sardine, mackerel, scad, dried silkworm pupae	5.5
Liver, sardine, silkworm pupae	4.5
Silkworm pupae, silkworm faeces, grass, soybean cake, pig manure, night soil	4-8
Raw silkworm pupae, pressed barley, *Lemna* and *Gammarus*	2.6-4
2/3 groundnut cake, 1/3 manioc leaves	3.5
1/2 manioc leaves, 1/2 ground rice	11.0
Manioc leaves and fresh manioc root	26.8
Fish flour, rice flour	2.5-3.0
Meat flour, potato	3.5-4.0
Fresh silkworm pupae, wheat flour	10.4
Fish flour, soybean cake, yeast	1.7-2.8
Fish flour, cottonseed meal, yeast	1.6-3.4

* from Ling (1967)

** food quotients for herbivorous fish species in China (Yang *et al.* 1985)

Source: Tacon (1988)

Table 5. Usage of supplemental ingredients for Indian carp in semi-intensive culture in four districts of Andhra Pradesh

Feed ingredients (kg/ha/yr)	Usage by farmers (No.)	(%)	Total input (kg/ha/yr) (Range)	(Mean)
Rice bran	16	8	2,000 - 7,000	4,180
Deoiled bran	188	99	4,000 - 43,000	18,430
Groundnut cake	156	83	600 - 12,000	5,310
Cottonseed meal	120	63	1,000 - 10,000	3,730
Sunflower meal	42	22	1,000 - 15,000	3,730
Soybean meal	8	4	2,000 - 7,000	4,200
Millets	6	3	1,000 - 4,000	2,220
Sorghum	6	3	1,000 - 4,000	2,470
Deoiled cake	13	7	900 - 4,670	4,620
Eggs	3	2	100 - 200	160
Salt (%)	83	44	1 - 5	2.1
Minerals (%)	21	11	0.3 - 3	1.7

Source: Veerina *et al.* (in preparation)

Table 6. Combination of supplemental feed ingredients commonly used in semi-intensive culture of Indian major carps in Andhra Pradesh

Item*	Farmers (No.)	(%)	Total input (kg/ha/yr) (Range)	(Mean)	Other ingredients added*
DOB	15	8	5,000 - 33,000	20,340	NIL
DOB+CSM	7	4	19,000 - 40,000	25,280	GNC
DOB+DOC	6	3	10,000 - 33,000	24,830	CSM/GNC/SFM
DOB+GNC	142	75	5,000 - 50,000	27,650	CSM/SFM/SBM/SOR/MIL
DOB+RB	1	<1	22,000	22,260	GNC/CSM
DOB+SFM	2	1	34,000 - 59,000	46,800	CSM/GNC
RB+DOB	15	8	20,000 - 39,000	26,790	GNC/DOC/CSM/SFM/SOR/MIL
NIL	1	<1	0	0	NIL
TOTAL	189	100	5,000 - 59,000	27,000	

* CSM = cottonseed meal; DOB = deoiled (rice) bran; DOC = deoiled (groundnut) cake; GNC = groundnut cake; MIL = millets; NIL = none; RB = rice bran; SBM = Soybean meal; SFM = Sunflower meal; SOR = sorghum.

Source: Veerina *et al.* (in preparation)

ingredients being almost completely discarded from the fish nutrition jargon. No longer is there an emphasis on supplementary ingredients *per se* but on the utilization of such ingredients in compounded feeds in relation to inclusion levels as a substitute for fishmeal. Such ingredients are now commonly referred to as non-conventional feedstuffs, a term which was originally used in respect of the animal husbandry industry. Comprehensive accounts on such feed resources are given by Devendra (1985), Tacon and Jackson (1985), Pantastico (1988) and Wee (1991). The proximate composition of feed ingredients, their suitability for incorporation into practical compounded feeds and other relevant information are dealt with in these reviews. Table 7 summarises the results of some of the experimental work based on the utilization of feed ingredients in fish feeds for tropical species. It is apparent that most oilseed meals and grains, when incorporated into compounded feeds at levels up to about 35% of dietary protein, resulted in achieving more than 80% of the growth reported for control diets where fishmeal was used as the primary source of protein.

The above mentioned findings are unfortunately generally confined to laboratory experiments or small-scale pilot studies. Due to a multitude of factors, these hardly get translated into practical feeds (De Silva and Davy, 1992). Nevertheless, these findings have given some indirect indications of the effectiveness of use of agricultural by-products in compound feeds and/or in simple mix forms, as are commonly utilized by the rural farmer.

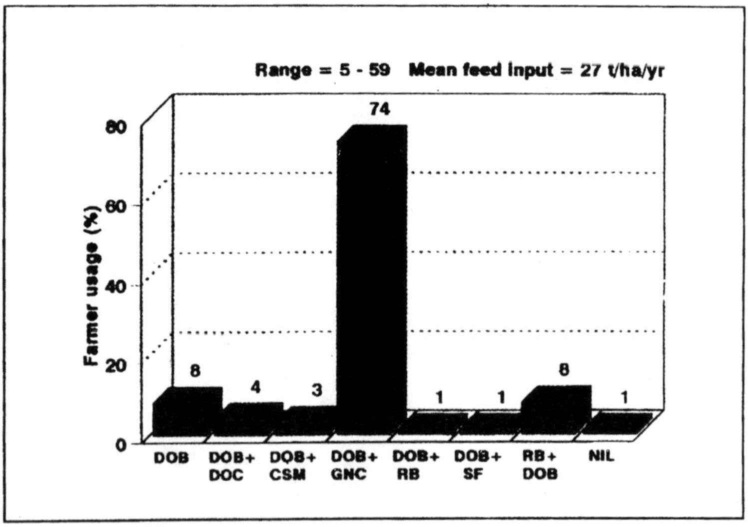

Figure 10. Extent of use of different supplementary feed combinations in semi-intensive carp culture practices in four districts in Andhra Pradesh, India (Veerina *et al.*, in preparation)

Table 7. Summary of results from selected experiments utilizing non-conventional plant feedstuffs (NCPF) in fish feeds for some commonly cultured species

Ingredients	Species of fish	Level of NCPF used	Control diet	Digestibility coefficients	Growth response
Hornwort *Ceratophyllum demersum*	Nile tilapia *Oreochromis niloticus*	40, 30 and 20% of diet replacing fish meal	Chicken feed	Not available (n.a.)	Specific growth rates (SGR) obtained were 69, 99 and 106% of the control diet for 40, 30 and 20% inclusion levels, respectively. SGR of control 7.58%/fish/day
Hornwort *C. demersum*	Nile tilapia *O. niloticus*	50% of diet with 12.5 and 25% fish meal	Commercial pellets	n.a.	Percentage weight gain (PWG) obtained were 52 and 92% of the control diet for diets containing 50% test ingredient with 12.5% and 25% fish meal, respectively. Percentage weight gain for control diet was 114%
Eleocharis ochrostachys	Nile tilapia *O. niloticus*	40, 30 and 20% of diet replacing fish meal	Chicken feed	n.a.	SGR obtained were 65, 93 and 122% of control diet at 40, 30 and 20% inclusion level, respectively. SGR of control was 1.58%/day
Duckweed *Lemna minor*	Common carp *Cyprinus carpio*	40% of diet	60% of rice bran and 40% groundnut oil cake	n.a.	Total weight gain obtained was 83% of control (5.3 kg in 140 days)

Table 7. (cont)

Water hyacinth *Eichhornia crassipes* (dried)	Nile tilapia *O. niloticus*	40, 30 and 20% of diet replacing fish meal	Chicken feed	n.a.	SGR obtained were 64, 93 and 110% of control diet for 40, 30 and 20% inclusion level, resepctively. SGR of control diet was 1.5%/fish/day
Water hyacinth *E. crassipes* (dried)	Rohu *Labeo rohita*	20 and 40% of total dietary protein	Fish meal as 100% of dietary protein	Apparent protein digestibility coefficient (APD) were 71 and 63% for 20 and 40% inclusion level, respectively, and 79% of control	SGR obtained were 79 and 68% of control diet for 20 and 40% inclusion level, respectively. SGR for control was 3.13%/day
Water hyacinth *E. crassipes* (dried)	*C. carpio*	2.5 and 10% of diet	Fish meal remained constant at 35% of diet	n.a.	1.61 and 1.51%/day, respectively
Water hyacinth *E. crassipes* (dried)	*O. mossambicus*	2.5 and 10% of diet	Fish meal remained constant at 35% of diet	n.a.	1.34 and 1.30%/day, respectively
	Puntius javanicus	2.5 and 10% of diet	Fish meal remained constant at 35% of diet	n.a.	0.96 and 1.44%/day, respectively

Table 7. (cont.)

Water hyacinth *E. crassipes* (dried)	*O. niloticus*	100 and 75% of control diet	Commercial diet	n.a.	SGR obtained were 85 and 95% of control diet. SGR from control diet was 1.99%/day
Water hyacinth *E. crassipes* (dried)	*O. niloticus*	50% of dietary protein (37.5% of diet)	Dietary protein by fish meal groundnut meal and rice bran	APD was 49-65%	SGR obtained was 79 and 81% control in recirculating and static water experimental system, respectively. The SGR for control diet was 1.64%/day and 1.58%/day, respectively
Water hyacinth *E. crassipes* (composted)	*C. carpio*	2.5 and 10% of diet	Fish meal remained constant at 35% of diet	n.a.	1.38 and 1.34%/day, respectively
Water hyacinth *E. crassipes* (composted)	*O. mossambicus*	2.5 and 10% of diet	Fish meal remained constant at 35% of diet	n.a.	1.38 and 1.20%/day, respectively
Water hyacinth *E. crassipes* (composted)	*P. javanicus*	2.5 and 10% of diet	Fish meal remained constant at 35% of diet.	n.a.	1.27 and 126%/day, respectively
Water hyacinth *E. crassipes* (composted)	*O. niloticus*	100, 75, 50 and 25% of control diet	Commercial diet	n.a.	SGR obtained were 42, 98, 108 and 98% of the control diet. SGR for control diet was 1.99%/day

Table 7. (cont.)

Leucaena leucocephala leaf meal	*O. niloticus*	20, 40, 80% of dietary protein	Fish meal as 100% dietary protein	n.a.	PWG obtained were 61 and 37%; 37 and 80%, 87 and 8% of the control diet for female and male fish, respectively at levels of 20, 40, 80%, respectively The % weight gains for the control diet were 27 and 72 % for females and males respectively
Leucaena leucocephala leaf meal	*L. rohita*	20, 40% of total protein	Fish meal supplying 100% of dietary protein	APD was 68 and 63%	SGR obtained were 79 and 70% of control diet for 20 and 40% inclusion level, respectively. The SGR for control was 2.34%/day
Leucaena leucocephala leaf meal	*O. niloticus*	25, 50, 100% of dietary protein	Fish meal supplying 100% of dietary protein	APD was 72, 66 and 40, respectively with increasing amounts of leaf meal	SGR obtained were 66, 36 and 18% of control diet for 25, 50 and 100% water inclusion level, respectively. The SGR for the control diet was 3.03%/day
Leucaena leucocephala leaf meal (soaked in water for 48 hours)	*O. niloticus*	25, 50, 100% of dietary protein	Fish meal supplying 100% of dietary protein	APD was 75, 65 and 41% respectively	SGR obtained were 89, 73 and 2.3% of control diet at 25, 50 and 100% inclusion level, respectively. The SGR for the control diet was 3.03%
Leucaena leucocephala leaf meal (soaked in water for 24 hours)	*L. rohita*	20 and 40% of total protein	Fish meal supplying 100% of dietary protein	APD was 71 and 63%, repectively	SGR obtained were 86 and 75% of the control at 20 and 40% inclusion level, respectively. SGR for control diet was 2.34%/day

Table 7. (cont.)

Green gram meal *Phaseolus* sp.	*O. niloticus*	13, 25, 37, 50% of fish meal	Fish meal soybean meal supplying dietary protein (25% content)	n.a.	The percent average daily weight gain (ADG) obtained were 51, 55, 48 and 53% of control with incorporation of plant meal, respectively: The ADG for the control diet was 7.79%/day
Mustard oil cake *Brassica juncea*	*C. carpio*	25 and 50% of dietary protein	Fish meal supplying 100% of protein	APD were 84 and 81%, respectively	SGR obtained were 85 and 67% of control diet for 25 and 50% inclusion level, respectively. SGR for the control diet was 3.58%/day
Linseed meal *Linum usitatissimum*	*C. carpio*	25 and 50% of dietary protein	Fish meal supplying 100% of protein	APD were 85 and 78%, respectively	SGR obtained were 86 and 66% of control diet at 25 and 50% inclusion level, respectively. SGR for the control diet was 3.58%/day
Sesame meal *Sesamum indicum*	*C. carpio*	25, 50 and 70% of dietary protein	Fish meal supplying 100% of protein	APD were 81, 78 and 73% respectively	SGR obtained were 74, 54 and 36% of control at 25, 50 and 70% inclusion level, respectively. SGR for control diet was 3.58%/day

Source: modified from Wee (1991)

DISPENSATION OF SUPPLEMENTARY FEEDS

In a great majority of semi-intensive culture practices in Asia, particularly in small-scale, single-owner rural operations, the most common method of dispensation of supplementary feeds is in powder form. There is very little scientific information on the efficacy of this form of feeding. Swingle (1958) reported that yields of channel catfish fed on a dry mixture of ingredients and in pelleted form were 1,247 kg/ha (FCR 3.3:1) and 2,648 kg/ha (FCR 1.6:1) respectively. To the author's knowledge, comparable studies from Asian aquaculture are lacking and this focuses on the problem of relative dearth of on-farm based nutritional research in Asian aquaculture.

The reasons for the decreased ineffectiveness of this form of feeding are manifold; significant wastage of food, individual fish being unable to ingest sufficient quantities of each of the constituent ingredients and thereby not obtaining a nutritionally favourable diet as well as the digestibility of the individual ingredients being different when ingested on their own. However, farmers continue to use this practice, sometimes in modified form. Farmers continue to provide supplemental feed, in one form or the other, only because they experience an obvious advantage through the practice (Table 8). This applies not only to the practice *per se* but, as pointed out earlier, to the use of additives in feeds of which the scientific rationale is not that obvious. It is conceivable that the direct influence of the feed *per se* on growth and yield are marginal. Nonetheless, feed waste and the undigested material enhances natural food production in the system which results in increased yields over and above what would have been achieved without supplementation.

In semi-intensive culture practice in Andhra Pradesh, India, one aspect of the culture practice remains uniform; the method of feeding in the grow-out ponds. Feed, a mixture of ingredients in powdery form (Figure 10) is presented in perforated polythene bags suspended in the pond at a number of points. Input-output analysis of the farms have shown that the feed ingredients, such as groundnut cake, cottonseed meal, de-oiled rice bran, sunflower seed and mineral mix, apparently provided as direct supplemental feed have a positive but moderate influence on yields (Veerina *et al.* in preparation). A few random samples of fish from these farms in 1989 were found to have only animal and plant material in the stomachs. This information, though grossly inadequate, reinforces the need for more scientific investigation to determine the extent to which supplementary feeds in most aquaculture practices contribute directly to the nutrition of cultured organisms. It is expected that the findings from such studies will lead to the next step of improving such feeds used by farmers, using the simplest of technologies available to them.

In a general sense, the feeding practices in semi-intensive aquaculture in a rural farming scenario are as diverse as the feeds used. In most semi-intensive aquaculture practices, feeding is rather *ad hoc* and feed in a powder form is broadcast manually, once or twice a day. Use of feeding trays or comparable devices, such as use of perforated bags with feed, is becoming increasingly popular. The use of the latter practices, though less labour intensive and/or time saving, is bound to degrade feed quality, through continued leaching of nutrients. There is also the possibility of establishment of prominent hierarchical effects which could result in the prevention of access to the food source to a significant proportion of the stock. Perhaps the greatest challenge lies in the development of efficient ways of dispersing feed in shrimp culture; to make feeding more effective whilst preserving the quality of the feed for a longer period with these rather sluggish feeders.

Table 8. Information on the influence of supplementary feeding on yields in semi-intensive culture practices

Species	Yields (kg/ha)				Feed	Source
	-Fert.	+Fert.	+Suppl. - Fert.	++Suppl. +Fert.		
Chinese & Indian carp	1,053	1,398-2,303	3,314-4,005	4,244-5,506	1:1 bran:oilcake	Şinha (1979)
Milkfish		588	685	-	rice bran	) Sumagaysay
			980	-	22% protein pellet	) *et al.* (1991)
			1,156	-	27.4% protein pellet	)
Nile tilapia, common carp & silver barb		6,000*		8,600*	1% rice bran	) Yakupitiyage
				9,000*	2% rice bran	) *et al.* (1991)
		9,100**		11,600**	1% rice bran	)
				9,800**	2% rice bran	)

* integrated system (No. of ducks stocked = 15)

** integrated system (No. of ducks stocked = 30)

The quantity of feed required per unit body weight (or % body weight) for maintenance and growth decreases as the fish increase in size. However, the relative feed required (g per fish) increases due to increase in weight. Under semi-intensive culture conditions therefore, supplementary feed requirements have to be balanced, taking into account increasing body weight and natural food availability. Based on pond studies, Marek (quoted by Hepher 1988) developed a feeding table for common carp culture in Israel (Table 9). Such carefully planned feeding strategies are rarely used in the rest of Asia, apart from feeding guidelines that have been developed for different species by various researchers. Some of these guidelines are summarised by Tacon (1988).

There is evidence to show that, under experimental conditions or in pilot scale trials, most cultured fish species tend to perform better when the frequency of feeding is increased, there being an optimum feeding frequency (Chiu 1989; Tuang and Shiau 1991). However, there is no data available on the effect of feeding frequency for on-farm semi-intensive culture systems in Asia. Perhaps, this is another area of investigation which should be given priority, based on which possible means of obtaining higher yields can be recommended to small-scale farmers.

More recently the extension of the hypothesis developed by De Silva (1985), that the daily provision of high protein diets is wasteful, has been tested for semi-intensive polyculture of two Indian major carps (catla and rohu) and common carp. Nandeesha *et al* (in preparation) found that 15.3% of protein input in the feeds and 10-20% of overall feed costs can be saved by adopting one of three mixed feeding schedules, based on rice bran and rice bran:mustard oilcake mix, supplemental feeds commonly used in Indian carp culture. Obviously, there is more to be done and understood in adopting mixed feeding schedules in semi-intensive culture. There are also practical implications, such as the storage of two or more types of feeds and how to dispense the correct feed according to strict schedules, which tantamounts to increasing farmers' awareness of the practice. Adoption of mixed feeding schedules, with feeds already in use, as a feed cost saving strategy, is perhaps relatively more easily extendible than new types of feeds.

Generally, there is a reluctance in the scientific community to readily accept new concepts; more so if such concepts originate from the third world (Gaillard 1991). It may be some time, therefore, before mixed feeding schedules become common practice in semi-intensive culture in Asia.

Table 9. Feeding chart for common carp cultured in ponds during summer (water temperature over 20°C) fed diets of differing protein content

Fish weight (g)	Density of fish per hectare						
	2,000-4,000 g (%) D*	**4,000-6,000 g (%) D***	**6,000-8,000 g (%) D***	**8,000-12,000 g (%) D***	**12,000-15,000 g (%) D***	**15,000-20,000 g (%) D***	**20,000-50,000 g (%) D***
20 - 50	1(2.9)I	2(5.7)I	2(5.7)I	2(5.7)I	2(5.7)I	2(5.7)II	2(5.7)II
50 - 100	2(2.7)I	3(4.0)I	4(5.3)I	4(5.3)II	4(5.3)II	4(5.3)II	4(5.3)III
100 - 200	6(4.0)I	6(4.0)II	9(6.0)II	9(6.0)III	9(6.0)III	9(6.0)III	9(6.0)IV
200 - 300	10(4.0)II	11(4.4)II	11(4.4)III	12(4.8)III	12(4.8)III	12(4.8)IV	12(4.8)IV
300 - 400	11(3.1)II	13(3.7)III	14(4.0)III	15(4.2)III	15(4.2)III	15(4.2)IV	15(4.2)IV
400 - 500	14(3.1)II	15(3.3)III	16(3.5)III	17(3.8)III	17(3.8)IV	17(3.8)IV	17(3.8)IV
500 - 600	15(2.7)II	17(3.1)III	18(3.3)III	19(3.5)IV	19(3.5)IV	19(3.5)IV	19(3.5)IV
600 - 700	15(2.5)III	18(2.8)III	19(2.9)III	19(2.9)IV	19(2.9)IV	19(2.9)IV	19(2.9)IV
700 - 800	16(2.1)III	18(2.4)III	19(2.5)IV	19(2.5)IV	19(2.5)IV	19(2.5)IV	19(2.5)IV
800 - 900	17(2.0)III	18(2.1)III	19(2.2)IV	20(2.3)IV	20(2.3)IV	20(2.4)IV	20(2.3)IV
900 - 1,000	17(1.8)III	19(2.0)III	29(2.1)IV	21(2.2)IV	21(2.2)IV	21(2.2)IV	21(2.2)IV
1,000 - 1,100	18(1.7)III	20(1.9)IV	21(2.0)IV	22(2.1)IV	22(2.1)IV	22(2.1)IV	22(2.1)IV
1,100 - 1,200	18(1.6)III	20(1.7)IV	21(1.8)IV	22(1.9)IV	22(1.9)IV	22(1.9)IV	22(1.9)IV

* based on four pelleted diets of increasing protein: 12% (I), 18% (II), 25% (III), and 30% (IV).

Source: Marek (1975), cited in Hepher (1989)

COMPOUNDED FEEDS AS SUPPLEMENTATION

The use of compounded diets as supplementary feeds is becoming increasingly popular in Asian aquaculture (Csavas 1990). For example, the Asian shrimp industry is almost totally dependent on such feeds. In the light of increasing competition for primary resources such as land and water, Asian aquaculture, particularly semi-intensive finfish culture which essentially caters for lower and middle income groups, can only develop (or may only survive) in the ensuing years if such practices become more profitable and productive. One of the main keys to profitability even in semi-intensive aquaculture will be a reduction in feed cost, which is generally estimated to exceed 50% of operating costs.

The basic nutrient requirements of the major species cultured in the world are fairly well known (Wilson 1991); at least to the extent that permits reasonably good diets to be formulated. New (1987) and Tacon (1988) gave some of the compounded feeds used in the culture of tropical species, semi-intensively. However, most of the feed formulations cited were experimental (test diets) and the overall proximate composition of the feeds were not easily obtainable. I will make no attempt to repeat this information, but I do consider conceptually the validity of usage of test diets in semi-intensive aquaculture. Some of the common features recognisable in these diets are:

- most of them contain a significant proportion of fishmeal, as a protein source; and
- the formulations have been made in order to meet the complete nutrient requirements of the species in question.

The need to incorporate micro-nutrients such as vitamins in supplementary feeds is also questionable. Dickson (1987) demonstrated that vitamins incorporated into supplemental diets for tilapia culture had no influence on growth and reproduction. Comparable observations have been reported on survival and growth in juvenile *Penaeus monodon* in modified extensive culture systems (Trino *et al.* 1992) and for *P. vannamei* (Castille and Lawrence 1989). In a similar vein Castell *et al.* (1988) commented that it was unlikely that severe essential fatty acid (EFA) deficiencies would be encountered in aquaculture practice; perhaps the probability is further reduced in semi-intensive culture in tropical Asia.

In semi-intensive aquaculture, where external feed input is expected to supplement natural food production, the use of a nutritionally balanced feed seems to be questionable; a waste of resources, as well as an economically unsound practice. In addition, there is increasing evidence to show that there are significant differences between the dietary protein requirements for optimal growth as opposed to the economically optimal protein level; the concept seems to be applicable at least to two groups of fish which are commonly cultured in Asia (De Silva *et al.* 1989; De Silva and Gunasekera 1991). It follows, therefore that formulation of compounded diets for semi-intensive aquaculture has to focus on the fact that the diets need not necessarily be nutritionally complete and, to be economical, that the protein level in the diets could be reduced even further.

Recently, Green (1992) demonstrated that substitution of organic manure (chicken litter - 1,000 kg total solids for a 0.1 ha earthen pond) for a 23% protein pelleted feed in monosex culture of *Oreochromis niloticus* resulted in better net returns. The total production cost for feed only, manure followed by feed, and manure plus simultaneous feeding, resulted in yields of 5,305, 4,794, and 4,351 kg/ha and costs of US$ 5,336/ha, US$ 4,645/ha and US$ 3,471/ha respectively. This brings me back to the point that I raised earlier; should not manure be considered as a feed input, and accordingly some of the ensuing research priorities be directed to such studies?

Perhaps, for most countries and/or regions the whole emphasis on formulations and manufacture has to be looked at afresh. Rural aquaculture in Asia is such that, in order to make any impact, new strategies may have to be considered. A schematic representation of a possible strategy is given in Figure 11.

A LOOK INTO THE FUTURE

Globally, the potential resource limitation of fishmeal on further expansion of the aquaculture industry has been focused upon by Wijkstrom and New (1989). According to New (1991), Asian aquaculture in 1988 utilized about 53.5% of global aquafeed production which was almost equally distributed between the production of carnivorous and non-carnivorous finfish and shrimp (Table 10). The aquafeed industry has grown very rapidly over the last five years. For example, the feed industry in the ASEAN alone had only 5 feed mills producing 27,000 t of feed in 1985, but by 1988 the number of feed mills had increased to 53, producing 264,000 t of feed (Boonyaratapalin and Akiyama 1989). New (1991) estimated that the Asia-Pacific shrimp industry will require nearly 1.1 million t of feed by the year 2000 (to produce 1.084 million t of shrimp, of which 52% will be produced through intensive culture). Csavas (1989), based on a predicted increase of carnivorous fish production to 800,000 t and shrimp production to 1.2-1.3 million t, estimated that the actual fishmeal requirements of the industry by year 2000 would be 375,000 t, approximately a doubling of the present usage.

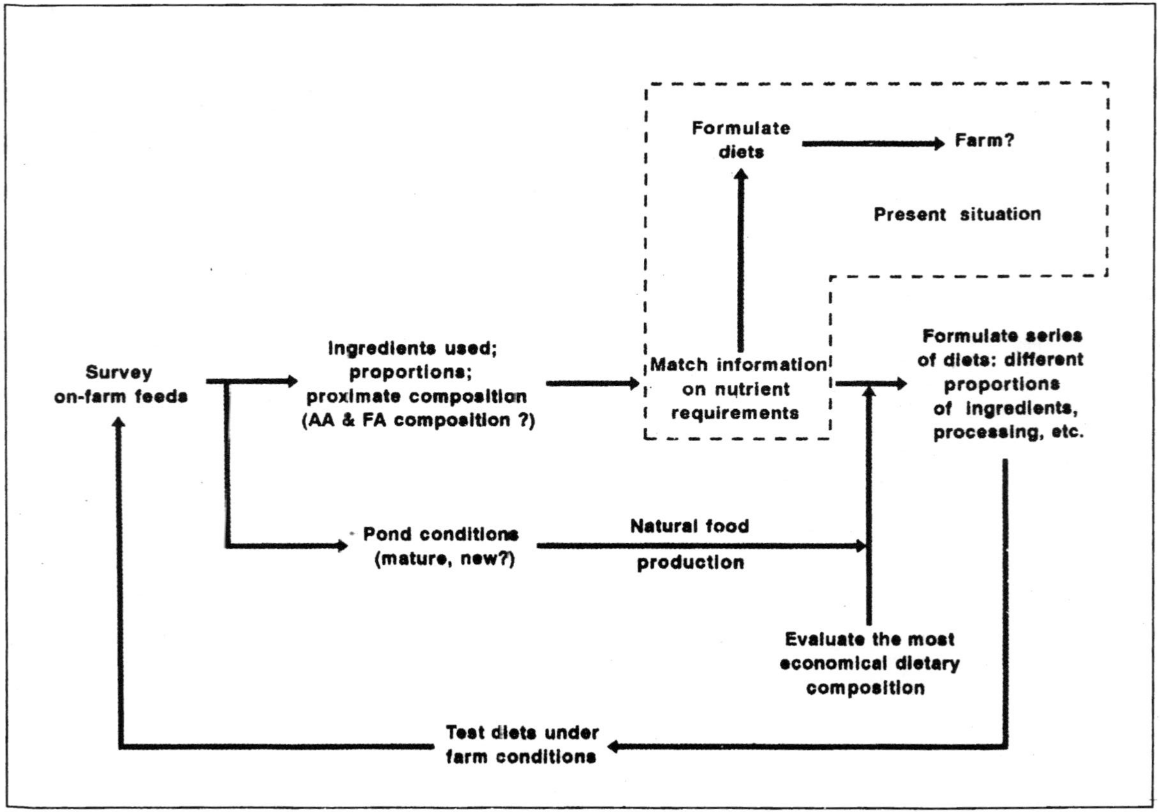

Figure 11. Schematic representation of the steps involved in the formulation of economic compounded diets for semi-intensive aquaculture

Table 10. Aquafeed utilization in the world and in Asia-Pacific (thousand t)

	Carnivorous finfish	Non-carnivorous finfish	Shrimp	Total
World				
1988	1,272	424	513	2,209
1990		653		
Asia-Pacific*				
1988	402	339	441	1,182
(a)	(34.0%)	(28.7%)	(37.3%)	
(b)	(18.2%)	(15.3%)	(20.0%)	(53.5%)
1990			580	
2000			1,132	

* (a) - feed utilization as a percent of Asian total
(b) - feed utilization as a percent of global total

Source: New (1991)

Most predictions on fish meal requirements for the growing aquaculture industry are based on the known FCR values of cultured species. The high feed cost for aquatic organisms is attributed to the relatively high protein requirements of fish. However, this is being increasingly disputed now (Bowen 1987; Lovell 1989). Apart from this, there is increasing controversy regarding the "real requirements" of even some of the most studied species such as the rainbow trout (Murai 1992). It is also interesting that recent work on the optimum dietary protein requirements of the bighead carp *Aristichthys nobilis*, (Santiago and Reyes 1991) has shown that it is closer to the economically optimal dietary protein level as predicted by De Silva and Gunasekera (1991). In effect therefore, there is increasing information being gathered which indirectly shows that we have been wasting dietary protein, wittingly or unwittingly. There is a trend towards more stringent legislation being introduced in developed countries with regard to fish feed specifications which primarily originated in an attempt to curtail environmental degradation from aquaculture activities. Whatever the reasons for the introduction of such legislation, these new feed specifications entail a reduction (Table 11) of the protein level in the feeds (amongst other components) without compromising growth and production.

Table 11. Summary of feed specifications recently introduced in Denmark

	Before 1 January 1990	After 1 January 1990	After 1 January 1992
FCR (max)	1.2	1.1	1.0
Energy (kcal/kg)			
Gross*	5,593	5,712	5,999
% Digestibility (minimum)	70	74	78
% Nitrogen*+	9.0	9.0	8.0
% Protein (of feed)	50.6	50.6	45.0
% Phosphorus*+	1.1	1.1	1.0
% Dust*+	1.0	1.0	1.0

* % dry matter; + maximum

Source: after Kiaerskou (1991)

Few have hitherto attempted to predict the growth of non-carnivorous finfish culture, which constitutes the bulk of aquaculture production in Asia, as well as globally. Similarly, the demand for supplementary feed ingredients which are more commonly used in semi-intensive aquaculture has not been addressed. A case in point is evident from Andhra Pradesh (Table 5) where nearly 18,340 kg/ha/yr of de-oiled rice bran is utilized for supplemental feed. Swaminathan (1992) drew attention to the global problem of environmental and food security. He focused on the urgent need to increase production of certain produce such as, for example, to double the production of rice by year 2000. For the aquaculture industry in rural Asia to conform to the envisaged growth the supply of feed ingredients which can be incorporated into supplemental feeds need to be readily available.

The availability of some of the common ingredients used in supplemental feeds in Asia over the last 6 years has been relatively static (Table 12). If pulses (rarely used by farmers) are not taken into account the total availability (for all uses) of the major ingredients in 1990 would amount to about 80 million t. Assuming that 10% of this is available to the aquaculture industry, the total amount of ingredients available will be 8 million t, an increase of only 12% over

Table 12. Production of selected crops in Far Eastern developing countries of Asia (thousand t)

Crop	1985		1986		1987		1988		1989		1990	
	Total	By-Prod.	Total	By-Prod.	Total	By-Prod.	Total	By-Prod.	Total	By-Prod.	Total	By-Prod.
Rice	416,582	41,658*	414,883	41,488	408,739	40,874	432,411	43,241	458,125	45,812	463,302	46,330
Wheat	144,693	14,469**	153,728	15,373	144,987	14,458	147,111	14,711	162,224	16,222	162,567	16,257
Pulses	21,393		22,523		21,060		21,264		21,577		22,980	
Oil crops (cake equivalent)	35,514	15,093***	35,992	15,297	37,495	15,936	39,049	16,596	38,367	16,306	41,148	17,488

* rice-bran
** wheat bran
*** based on extraction rate for cotton seed

Sources: FAO (1991); Devendra (1977)

the six-year period 1985-1990, or averaging 2% per year. This obviously is much less than the rate at which the aquaculture industry has grown over the last decade, as well as its envisaged growth in the present decade. In such a scenario it is conceivable that, in the ensuing years, the cost of commonly used supplemental feeds and/or ingredients will increase, and accordingly the cost of supplemental feeding will also increase. Apart from direct cost increases, these ingredients are beginning to have competing uses and thereby also influence their ready availability to the farmer. In other words, a more serious "trap" in respect of supplemental feeds is in the offing. Perhaps it is opportune for us to pay more attention to availability and efficient usage of supplementary feed ingredients and not to literally put all the eggs in the "fishmeal basket", before it becomes too late.

New (1990) advocated simple farm-made moist feeds, and/or sun-dried feeds derived from these, for farms capable of producing up to 300 t of shrimp per year. Lippert (1990), on the other hand, pointed out that increasing dependence on farm-made feeds will decrease the incentive for major feed companies to develop better feeds. In shrimp farming, most of which borders on the upper scale of semi-intensive culture, utilization of practical, compounded feed is an accepted practice. Moreover, the profit margins in shrimp culture permit dependence on commercial feeds, or investment in small-scale machinery to make on-farm feeds a practical proposition. This scenario might not necessarily hold true for semi-intensive culture of non-carnivorous finfish, the great bulk of aquaculture produce in Asia. Therefore, the future strategies for this sector may have to be different.

The growth of the shrimp industry has relatively higher physical constraints and therefore higher physical limitations (based on the availability of suitable sites) in comparison to semi-intensive culture of finfish. Whilst both industries face the common problem of availability of fishmeal, admittedly to varying degrees, the sheer volume of production and the higher degree of dependence of the semi-intensive finfish culture industry on agricultural by-products tends to make this industry in some ways more vulnerable in the future.

It is essential that future research efforts on supplemental feed development take a different direction. It is the author's belief that there is an urgent need to move away from the traditional dose-effect studies on nutritional requirements of cultured species in Asia and on the emphasis of development of fishmeal substitutions by agricultural by-products for compounded diets, as has been pointed out earlier by De Silva and Davy (1992). There is an urgent need to work in conjunction with the practioners and for research on supplemental feeds to be based on improvements in ongoing farm practices. The "top-down" approach does not appear to have delivered the goods. Perhaps a more pragmatic "bottom-up" approach is urgently needed.

LIST OF REFERENCES

Albrecht, M.L. and B. Breitsprecher. 1969. Untersuchungen ueber die chemische Zusammensetzung von Fischenahrtieren und Fischfutter-mitteln. Z-fish N.F. 17: 143-163.

Bailey, C. and M. Skladany. 1991. Aquaculture development in tropical Asia. A re-evaluation. p. 66-73. Natural Resources Forum, February 1991.

Boonyaratapalin, M. and D.M. Akiyama. 1989. The aquaculture industry in Southeast Asia, p. 41-56. *In* Proceedings of the Third International Symposium on Feeding and Nutrition in Fish, Toba, Japan.

Bowen, S.H. 1987. Dietary protein requirements in fishes - a reassessment. Canadian J. Fish Aquat. Sci. 44 : 1995-2000.

Castell, J.D., D.E. Conklin, J.S. Carigie, S.P. Lall and K. Norman-Boudreau. 1988. Aquaculture nutrition, p. 291-308. *In* M. Bilio, H. Rosenthal and C.J. Sindermann (eds.) Realisms in Aquaculture: achievements constraints, perspectives. European Aquaculture Society, Belgium.

Castille, F.L. and A.L. Lawrence. 1989. The effects of deleting dietary constituents from pelleted feed on the growth of shrimps in the presence of natural food in ponds. J. World Aquaculture Society 20(1):22A (Abstract).

Chiu, Y.N. 1989. Consideration for feeding experiments to quantify dietary requirements of essential nutrients in fish, p. 46-57. *In* S.S. De Silva (ed.) Fish Nutrition Research in Asia. Asian Fisheries Society, Special Publication 4, Manila, Philippines.

Csavas, I. 1989. Development of semi-intensive and intensive aquaculture in Asia and the Pacific. Paper presented at 29th Annual Conference of the International Association of Fish Meal Manufacturers, November 1989, Hong Kong.

Csavas, I. 1990. Aquaculture development and environmental issues in the developing countries of Asia. Paper presented at the Conference on Environmental Issues in Third World Aquaculture Development, September 1990, Bellagio, Italy.

De Silva, S.S. 1985. Performance of *Oreochromis niloticus* (L.) fry maintained on mixed feeding schedules of different protein content. Aquaculture and Fisheries Management 16:621-633.

De Silva S.S. and F.B. Davy. 1992. Strategies for finfish nutrition research for semi-intensive aquaculture in Asia. Asian Fisheries Science 5:129-144

De Silva, S.S. and R.M.G. Gunasekera. 1991. An evaluation of the growth of Indian and Chinese major carps in relation to the dietary protein content. Aquaculture 92:237-241.

De Silva, S S. and D.E.M. Weerakoon. 1981. Growth food intake and evacuation rates of grass carp, *Ctenopharyngodon idella* fry. Aquaculture 25:67-76.

De Silva, S.S., M.K. Perera and P. Maitipe. 1984. The composition, nutritional status and digestibility of the diets of *Sarotherodon mossambicus* from nine man-made lakes in Sri Lanka. Environmental Biology of Fishes 11:205-19.

De Silva, S.S., R.M.G. Gunasekera and D. Atapattu. 1989. The dietary protein requirements of young tilapia and an evaluation of the least cost dietary protein levels. Aquaculture 80:271-284.

Devendra, C. 1977. The utilization of agro-industrial by-products in Asia and the Far East, p.111-124. *In* New Feed Resources. FAO Animal Production and Health Paper No. 4. FAO, Rome, Italy.

Devendra, C. 1985. Non-conventional feed resources in Asia and the Pacific. 2nd ed. FAO/ APHCA Publication No.6. FAO, Bangkok. 140 p.

Dickson, M.W. 1987. The supply of vitamins in feeds for intensive tilapia farming in Zambia. Aquaculture and Fisheries Management 18:221-230.

FAO. 1991. FAO Yearbook - Production. Vol. 44, (1990). FAO Statistics Service No. 99. FAO, Rome, Italy. 283 p.

FAO. 1992. Aquaculture production (1984-1990). FAO Fisheries Circular No. 815, Revision 4. FAO, Rome, Italy. 206 p.

Gaillard, J. 1991. Scientists in the Third World. The University Press of Kentucky,USA.191 p.

Green, B.W. 1992. Substitution of organic manure for pelleted feed in tilapia production. Aquaculture 101:213-222.

Hanley, F. 1991. Effects of feeding supplementary diets containing varying levels of lipid on growth, food conversion, and body composition of Nile tilapia, *Oreochromis niloticus* (L). Aquaculture 93:323-334.

Hepher, B. 1975. Supplementary feeding in fish culture. Proceedings International Congress on Nutrition, 9(3):183-198.

Hepher, B. 1978. Ecological aspects of warm water fish pond management, p.447-465. *In* S.D. Gerking (ed.) Ecology of Freshwater Fish Production. Blackwell Scientific Publisher, Oxford.

Hepher, B. 1988. Principles of fish nutrition, p.121-142. *In* M. Shilo and S. Sarig (eds.) Fish Culture in Warm Water Systems: Problems & Trends. CRC Press, NY, USA.

Hepher, B., J. Chervinski and H. Tugari.1971. Studies on carp nutrition III. Experiments on the effect on fish yields of dietary protein source and concentration. Bamidgeh 23:11-37.

Jana, B.B. and R. Chakraborty. 1990. Exogenous introduction of live plankton, a better approach to carp growth than direct-manure systems. Prog.Fish-Cult. 52:252-260.

Kiaerskou, J. 1991. Production and economics of "low pollution diets" for the aquaculture industry, p. 65-76. *In* C.B. Cowey and C.Y Cho (eds.) Nutritional Strategies and Aquaculture Wastes. University of Guelph, Canada.

Ling, S.W. 1967. Feeds and feeding of warm-water fishes in ponds, p.291-309. *In* T.V.R. Pillay (ed.) Proceedings of the World Symposium on Warm Water Pond Fish Culture. FAO Fisheries Report No. 44(3).

Lippert, I.E. 1990. Shrimp farming and its markets. Fish Farming International 17(2):26-29.

Lovell, T. 1989. Energy requirements for growth : fish versus farm animals. Aquaculture Magazine 15(4):65-66.

Murai, T. 1992. Protein nutrition of rainbow trout. Aquaculture 100:191-207.

Nandeesha, M.C., K.D. Murthy, S.S. De Silva and K. Dathatri. (in preparation). Evaluation of mixed feeding schedules in carp culture - field trials on catla, rohu and common carp.

New, M.B. 1987. Feed and feeding of fish and shrimp. FAO Field Document ADCP/REP/87/ 26. FAO, Rome, Italy. 275p.

New, M.B. 1990. Compound feedstuffs for shrimp culture, p.79-123. *In* M.B. New, H. de Saram and T. Singh (eds.) Technical and Economic Aspects of Shrimp Farming. INFOFISH, Kuala Lumpur, Malaysia.

New, M.B. 1991. Compound feeds: a world review. Fish Farmer 14(2):39-42.

Odum, W. 1970. Utilization of the direct grazing and plant detritus food chain by *Mugil cephalus*, p. 222-240. *In* H. Steele (ed.) Marine Food Chains. Oliver and Boyd, London.

Pantastico, J.B. 1988. Non-conventional feed resources in aquaculture: An overview of work done in the Philippines, p.71-87. *In* S.S. De Silva (ed.) Finfish Nutrition Research in Asia. Heinemann Publishers Asia Pte. Ltd, Singapore.

Prus, T. 1970. Calorific value of animals as an element of bioenergetical investigations. Polish Archives Hydrobiologie 17:183-199.

Santiago, C.B. and O.S. Reyes. 1991. Optimum dietary protein level for growth of bighead carp (*Aristichthys nobilis*) fry in a static water system. Aquaculture 93:155-165.

Schroeder, G.L. 1978. Autotrophic and heterotrophic production of micro - organisms in intensely-manured fish ponds, and related fish yields. Aquaculture 14:303-325.

Sinha, V.R.P. 1979. Contribution of supplementary feed in increasing fish production through composite fish culture in India, p. 565-574. *In* J.E. Halver and K. Tiews (eds.) Finfish Nutrition and Fish Feed Technology. Vol. 1, Heenemann, Berlin.

Spataru, P., B. Hepher and A. Halevy. 1980. The effect of the method of supplementary feed application on the feeding habits of carp (*Cyprinus carpio* L.) with regard to natural food. Hydrobiologia 72:171-178.

Sumagaysay, N.S., F.E. Marquez and Y.N. Chiu-Chern. 1991. Evaluation of different supplemental feeds for milkfish (*Chanos chanos*) reared in brackish water ponds. Aquaculture 93:177-189.

Swaminathan, M.S. 1992. Environment and global food security. Food Tech. 46(7):89-95.

Swingle, H.S. 1958. Experiments on growing fingerling channel catfish to marketable size in ponds. Proc. Southeast Association of Game Fish Commissioners 12:63-72.

Tacon, A.G.J. 1988. The nutrition and feeding of farmed fish and shrimp - a training manual. 3. Feeding methods. FAO Field Document, Project GCP/RLA/075/ITA. Field Document 7/E, FAO, Brasilia, Brazil. 208 p.

Tacon, A.G.J. and A.J. Jackson. 1985. Utilization of conventional and non-conventional protein sources in practical feeds, p. 119-145. *In* C.B. Cowey, A.M. Mackie and J.G. Bell (eds.) Nutrition and Feeding in Fish. Academic Press, London.

Trino, A.T., V.D. Penaflorida and E. Bolivar. 1992. Growth and survival of *Penaeus monodon* juveniles fed a diet lacking vitamin supplements in a modified extensive culture system. Aquaculture 101:25-32.

Tuang, P-H. and S.-Y. Shiau. 1991. Effect of meal frequency on growth performance of hybrid tilapia, *Oreochromis niloticus* x *O. aureus*, fed different carbohydrate diets. Aquaculture 92:343-350.

Veerina, S.S., M.C. Nandeesha and S.S. De Silva. (in preparation). Analysis of input - output interrelationship in the Indian major carp farming system in Andhra Pradesh, India.

Veerina, S.S., M.C. Nandeesha, K. Rao Gopal and S.S. De Silva. (in preparation). Status and technology of Indian major carp farming in Andhra Pradesh, India. Asian Fisheries Society, Indian Branch, Special Publication 9, Mangalore.

Weatherly, A.H. 1963. Notions of niche and competition among animals with special reference to freshwater fish. Nature (Lond.) 197:14-17.

Wee, K.L. 1991. Use of non-conventional feedstuffs of plant origin as fish feeds: is it practical and economically feasible? p.13-22. *In* S.S. De Silva (ed.) Fish Nutrition in Asia. Asian Fisheries Society Special Publication, 5. Asian Fisheries Society, Manila, Philippines.

Werner, E.E. and D.J. Hall. 1976. Niche shifts in sunfishes, experimental evidence and significance. Science 191:404-406.

Wijkstrom, U.N. and M.B. New. 1989. Fish for feed: a help or a hindrance to aquaculture in 2000? Infofish International 6/89:48-52.

Wilson, R.P. (ed.) 1991. Handbook of nutrient requirements of finfish. CRC Press, Boston, USA. 94 p.

Wohlfarth, G.W. and G. Hulata. 1987. Use of manures in aquaculture, p. 353-367. *In* D.J.W. Moriarty and R.S.V. Pullin (eds.) Detritus and Microbial Ecology in Aquaculture. ICLARM Conference Proc. 14, Manila, Philippines.

Wohlfarth, G.W. and G.L. Schroeder. 1979. Use of manure in fish farming - a review. Agricultural Wastes 1:279-299.

Yakupitiyage, A., P. Edwards and K.L. Wee. 1991. Supplementary feeding of fish in a duck-fish integrated system I. The effect of rice bran, p. 143-157. *In* S.S. De Silva (ed.) Fish Nutrition Research in Asia. Asian Fisheries Society Special Publication, Manila, Philippines.

Yang, H.Z. *et al.* 1985. Chinese integrated fish farming system and their effects. Freshwater Fisheries Research Center, Wuxi, China. [cited in Tacon (1988)]

Zhu, Y , Y.Yang, J. Wan, D.Hua and J.A. Mathias. 1990. The effect of manure application rate and frequency upon fish yield in integrated fish farm ponds. Aquaculture 91:233-251.

FEED FORMULATION AND ON-FARM FEED MANAGEMENT

Albert G.J. Tacon

Inland Water Resources and Aquaculture Service
FAO, Via delle Terme di Caracalla, 00100 Rome, Italy

TACON, A.G.J. 1993. Feed formulation and on-farm feed management, p. 61-74. *In* M.B. New, A.G.J. Tacon and I. Csavas (eds.) Farm-made aquafeeds. Proceedings of the FAO/AADCP Regional Expert Consultation on Farm-Made Aquafeeds, 14-18 December 1992, Bangkok, Thailand. FAO-RAPA/AADCP, Bangkok, Thailand, 434 p.

INTRODUCTION

The nutritional performance and success of an artificial diet for farmed fish or shrimp is dependent upon a number of important factors which can be broadly considered under two major headings: feed formulation and manufacture; and on-farm feed management and aquatic environment (Figure 1). Sadly, all too often it is believed that the role of the nutritionist and aquafeed manufacturer ends once the diet has been formulated and physically produced. It does not; the ultimate success of an artificial diet usually being dependent upon the management of the finished feed on the farm. This paper attempts to provide some general guidelines for the formulation and on-farm management of artificial diets for fish and shrimp, with particular reference to home-made aquafeeds.

FEED FORMULATION

The formulation of a practical diet for fish or shrimp is based upon a series of technical and economical considerations. These have been discussed previously (New 1987; Tacon 1988; Lall 1991) and can be summarized as follows:

- the market value of the species to be fed - the farm gate value of the cultured species will set the upper limit which can be spent on the formulation and finished diet (*i.e.* as a general rule total dry diet cost should not exceed 25% of the farm gate value of the cultured species);

- the financial resources of the farmer (if home-made aquafeeds are to be produced) - capital available for the purchase of feed ingredients and feed

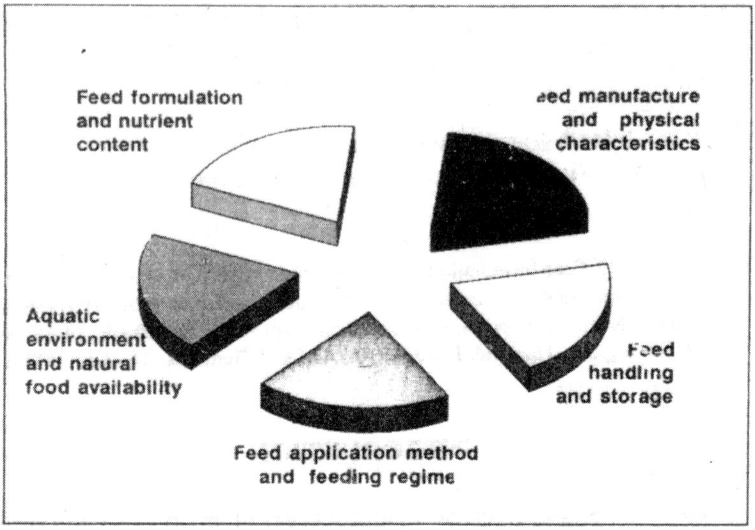

Figure 1. Major factors determining the nutritional performance and success of an artificial diet feeding regime

manufacturing equipment (*i.e.* the formulation will depend upon the money in the farmer's pocket and the availability of services such as electricity or fuel);

- **dietary nutrient requirements of the species to be fed (or closest relative)** - including dietary protein, amino acid, fatty acid, mineral, vitamin and energy requirements for each phase of the culture cycle (*i.e.* for larvae/fry, juveniles, production and broodstock, if known);

- **natural feeding habits of the species in question** - in the absence of detailed information on the dietary nutrient requirements of the species an analysis of its natural feeding habits in the wild will indicate its position in the aquatic food chain (*i.e.* herbivore, detritivore, omnivore or carnivore), preferred food items and feed size (*i.e.* plant and animal feedstuffs), feeding station (*i.e.* surface, mid-water or benthic feeder), and feeding behaviour (*i.e.* daylight or nocturnal feeder, visual or olfactory feeder, rapid or slow feeder, continuous or intermittent feeder). Apart from determining dietary food preferences, an analysis of the natural feeding behaviour of the target species will also determine the physical characteristics of the artificial diet to be produced (*i.e.* feed size and shape, texture, palatability, buoyancy and water stability);

- **available feed ingredient sources, composition and cost** - including seasonal availability, current usage, proximate composition, quality control, digestibility and nutrient availability, and additional ingredient handling and processing cost prior to mixing or pelleting;

- intended feed manufacturing process to be used - grinding, mixing, cold pelleting, steam pelleting, expansion pelleting, flaking, and/or microencapsulation (*i.e.* the formulation will depend upon the manufacturing process employed and whether the diet is to be produced as a mash, crumble, paste, ball, moist pellet or dry pellet); and

- intended stocking density and farm production unit - extensive, semi-intensive or intensive fish/shrimp production systems within tanks, raceways, concrete lined ponds, earthen ponds, pens or floating cages (*i.e.* the intended stocking density and production unit will dictate the relative availability of natural food organisms per unit biomass of cultured fish/shrimp over the production cycle and the consequent necessity for a complete or supplementary diet formulation).

Despite the obvious importance of all the above mentioned factors, sadly little or no consideration is usually given by nutritionists and feed manufacturers concerning the intended fish/shrimp stocking density and consequent availability of natural food organisms within the proposed production unit. For example, at present almost all commercial feeds produced for semi-intensive pond farming systems are formulated as nutritionally complete diets irrespective of the intended stocking density employed and natural food availability (Akiyama 1991; Tacon 1991a; Chamberlain 1992; De Silva and Davy 1992). To a large extent this has been due to the almost total absence of published information concerning the dietary nutrient requirements of fish and shrimp within semi-intensive pond farming systems and the apparent difficulty of predicting the nutritional contribution of natural food organisms within the overall nutritional budget of the pond raised species (for review see Tacon 1987, 1988; Hepher 1988).

It follows from the above that the formulation of a supplementary diet is dependent upon the density and total biomass or standing crop of the cultured fish or shrimp species present, and the fertility and consequent availability of natural food organisms within the water body. As a general rule the contribution of natural pond biota toward the overall nutritional budget of the cultured fish or shrimp will be highest at low stocking densities or at the start of the culture cycle when the total biomass or standing crop is lowest. The consequent availability and relative contribution of natural pond biota thereafter decreases over the course of the production cycle with increasing fish or shrimp size and standing crop. For example, Figure 2 shows the influence of shrimp (*Penaeus monodon*) market size and pond stocking density on the reported performance (as indicated by the recorded apparent food conversion ratio - AFCR) of a pelleted shrimp feed within commercial shrimp farms in Thailand (ponds stocked with *ca.* 30-50 post-larvae/m^2; Charoen Pokphand Aquaculture Co., Ltd., personal communication). From the data it is clear that the lower AFCRs observed at the lowest shrimp market sizes and stocking densities were almost certainly due to the higher contribution of natural pond biota toward the overall nutritional budget of the farmed shrimp at these lower densities and market sizes.

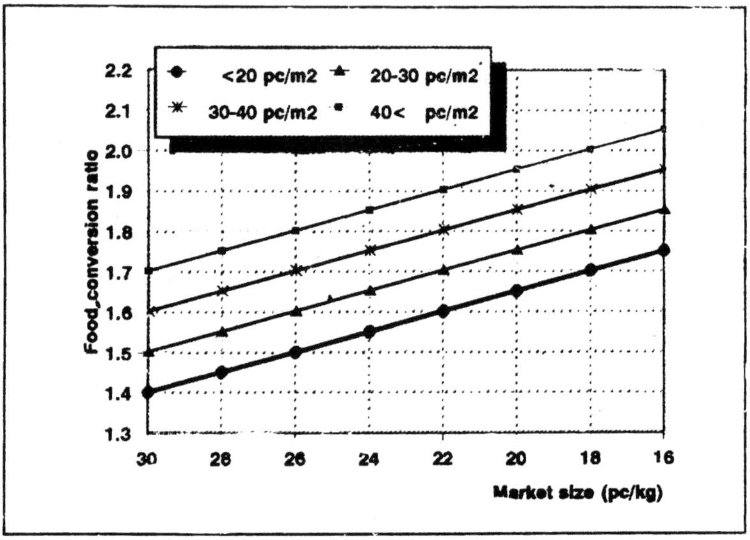

Figure 2. Influence of market size and stocking density on food conversion ratio of pond cultured black tiger shrimp fed CP feed in Thailand

The nutritional and economic importance of natural food organisms within semi-intensive pond farming systems cannot be over emphasized. For example, the tracer studies of Lilyestrom *et al.* (1987) and Anderson *et al.* (1987) with pond reared prawns (*Macrobrachium rosenbergii*; 0.5-1.5 animals/m^2) and shrimp (*Penaeus vannamei*; 20 animals/m^2) showed that natural pond organisms accounted for 18-75% and 53-77% of the growth of the cultured species, respectively; in both studies animals were fed complete pelleted feeds. In view of these results it is perhaps not surprising therefore that no difference was observed in the growth of prawns (*Macrobrachium rosenbergii*) in outdoor concrete ponds in Thailand when fed a 35% protein diet, a 15% protein diet or a broiler starter diet (stocking density 5 animals/m^2, pond water exchanged every three weeks; Boonyaratpalin and New 1982), or in the growth and survival of shrimp (*Penaeus vannamei*) in earthen ponds in Hawaii receiving no fertilizer and feed input, fertilization only, or fed a commercial pelleted shrimp diet (stocking density 7.1-9.4 animals/m^2; Lee and Shleser 1984). Similarly, Bostock (1991) reported that there was no difference in the growth of pond reared shrimp (*Penaeus monodon*; 10/m^2) in India fed a high-nutrient pelleted diet or a locally produced dough-ball costing one third of the price. In addition, numerous studies have been conducted concerning the necessity for dietary vitamin supplements within artificial feeds. In general the results have shown dietary vitamins to be dispensable under practical farming conditions for tilapias and shrimp/prawns when reared under semi-intensive pond farming conditions (Boonyaratpalin and New 1982; Castille and Lawrence 1989; Viola 1989; Tacon 1991b; Trino *et al.* 1992).

On the basis of the studies of Hepher and co-workers in Israel it has been found that dietary energy is generally the first limiting nutrient within earthen ponds at low carp (*Cyprinus carpio*) stocking densities and that, as the density and standing crop increases, so the requirement for supplemental dietary nutrients such as proteins, minerals and vitamins becomes greater (Figure 3). It follows therefore that the dietary protein and essential micronutrient content of supplementary feeds intended for use within semi-intensive pond farming systems should be low at the start of the culture operation and then progressively increased with increasing fish size and standing crop (Tacon 1988). This is the complete reverse of an intensive clear-water farming system where dietary nutrient levels usually decrease with increasing fish or shrimp size and age. Sadly, the majority of researchers, feed manufacturers and farmers alike still employ gradually decreasing dietary nutrient levels with increasing fish/shrimp age groups for semi-intensive fish and shrimp production systems. Clearly, this situation will have to be rectified if feed and production costs are to be reduced and maximum benefit is to be gained from semi-intensive farming systems. However, it is encouraging to note that the tide may soon be turning. For example, a commercial shrimp feed manufacturer in the Philippines has recently introduced two new feed lines on to the market; a conventional shrimp grower/ finisher diet containing 37% crude protein and a new diet designed for extensive to semi-intensive farming systems (for ponds larger than one hectare with a shrimp stocking density of not more than 10/m^2) containing 30% crude protein.

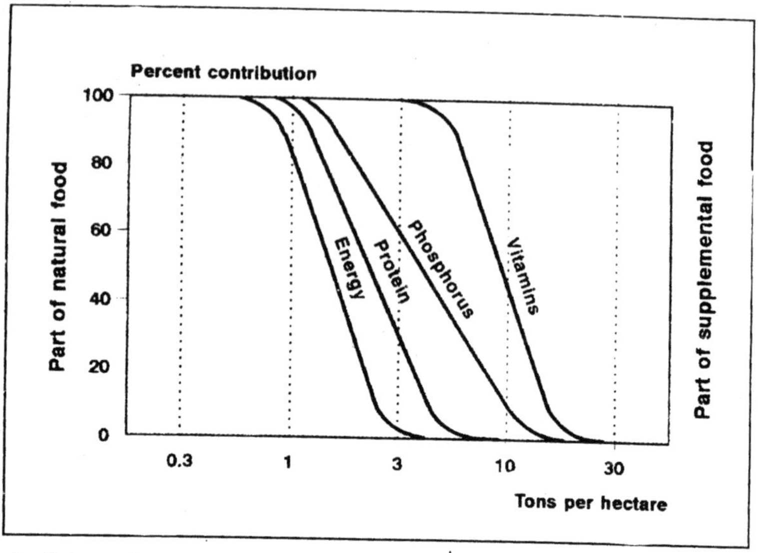

Figure 3. **Schematic model of the contribution of nutrients from natural and supplemental food for increasingly intensive carp culture in Israel (Viola 1989)**

ON-FARM FEED MANAGEMENT

As mentioned before, the success of an artificial diet is dependent not only upon the formulation and manufacturing process used to produce the diet but also upon the management of the diet on the farm.

Feed handling and storage

Since aquafeeds are composed of highly perishable nutrients it is essential that adequate handling and storage procedures are employed to protect the finished feed from the natural elements (*i.e.* light, heat, humidity, air and water) from the time the feed emerges from the die plate to the time the feed is consumed by the fish or shrimp. The economic and nutritional consequences of poor feed handling and storage can be both profound and immediate, with a reduction in the nutritional performance of the feed (as indicated by increased FCR and reduced fish or shrimp growth), either through mechanical damage (*i.e.* increased "fines" content), oxidative damage (*i.e.* lipid peroxidation and vitamin destruction), or microbial/pest infestation (*i.e.* nutrient destruction and mycotoxin production). For a review of feed handling and storage techniques and information concerning the nutritional consequences of poor feed storage, including the pathological effects of oxidized lipids and mycotoxin contamination, see Chow (1980), New (1987), Akiyama (1989), De la Cruz *et al.* (1989), Tsai (1989a), Coelho (1991), Lovell (1992) and Tacon (1988, 1992).

Although the negative effects of feed handling and storage may be minimised by the use of dietary anti-fungal compounds and higher dietary antioxidant/vitamin fortification levels it is important to note that many dietary vitamins can be lost from finished feeds in a matter of minutes, just by exposing the surface of feed bags to the warming and destructive ultra violet rays of the sun. The latter is a common phenomenon during feed transportation and when bags of feed are left by the side of a tank or pond for feeding later the same day. Clearly, farmers must be made aware of the serious consequences of poor feed handling/storage and provided with adequate guidelines concerning proper on-farm feed management procedures. For example, Annex 1 shows the recommended feed handling and storage guidelines given to farmers by a commercial shrimp feed manufacturer in the Philippines. In addition to the guidelines listed in Annex 1, it should also be stated that feeds should not be stored in direct sunlight.

Feed application method and feeding regime

Maximum benefit from an artificial diet will only be achieved if the diet is ingested in its entirety and supplied to the fish or shrimp at a sufficient rate to satisfy its requirements for optimal growth and feed efficiency. Sadly, dietary feeding regimes are more often than not related to the convenience of the feeding technician or farm worker during his or her working day rather than to the behavioral feeding requirements of the cultured fish or shrimp species. For example, despite the nocturnal feeding habit of most pond reared shrimp and the rapid loss of water soluble nutrients from rations on prolonged immersion in water (Cuzon *et al.* 1982; Brown *et al.* 1989; Lim and Cuzon 1992; Table 1), the majority of commercial shrimp farmers in Latin and North America still only feed their animals once or twice daily during the morning or late afternoon (Hirono 1989; Jaenike 1989; Villalon 1991). Clearly, farmers must be made aware of the nutritional and economic losses they will be incurring through dietary leaching for every additional moment their diet remains uneaten in water; a feed pellet behaving in much the same way as a dissolving sugar cube in a hot cup of tea! It follows from the above that pond reared shrimp should be fed on a "little and often" basis, with the bulk of the feed fed during the night (*i.e.* from sun-set to sun-rise) when their feeding activity is highest (Tacon 1988; Akiyama

1989; Villalon 1991). The benefits of increasing the frequency of feed application are immediate and include reduced leaching and feed loss, and improved growth and feed efficiency (Villalon 1991; Sumudra 1992).

Table 1. Reported nutrient loss from a pelleted shrimp diet after a one hour seawater immersion period* (Cuzon *et al.* 1982)

Nutrient	Level at start	Level after 1 hour	Percent loss (%)
Dry matter (%)	100	81	19
Crude protein (%)	52	41	21
Carbohydrate (%)	16	8	50
Vitamin C (mg/kg)	3,089	332	89
Thiamine (mg/kg)	29.5	0.7	98
Riboflavin (mg/kg)	55	7.5	86
Pyridoxine (mg/kg)	14	1	93
Pantothenate (mg/kg)	100	5.9	94
Niacin (mg/kg)	120	17	86
Inositol (mg/kg)	4,000	1,928	52
Choline (mg/kg)	3,368	1,835	45

* diet contained 15% wheat gluten as a binder to maintain pellet shape in water for at least 48 hours

It is also important to mention here that each diet has an ideal dietary feeding rate range where both growth and feed efficiency are optimal for each fish/shrimp age class; the optimal feeding level in turn varying with body weight, dietary nutrient level, water quality, stocking density and natural food availability (for review see Hepher 1988, and Tacon 1988). Sadly, the majority of feeding tables recommended by feed manufacturers for use within semi-intensive farming systems are generally hypothetical and arbitrarily set, irrespective of dietary composition, pond fertilization rate, natural food availability, stocking density and standing crop. Furthermore, since natural food organisms play a gradually diminishing role in the overall nutritional budget of pond reared shrimp and fish as the body size and standing crop increases with time, it follows that the feeding rate of the artificial diet should be gradually increased over the course of the culture cycle. However, at present almost all feed manufacturers are recommending semi-intensive farmers to use a decreasing dietary feeding rate with increasing fish/shrimp size and body weight (Akiyama 1989; Villalon 1991; Sumudra 1992). Clearly, optimal feeding rates and frequency of feed presentation must be determined for individual feeds and farms by carefully monitoring feed consumption, growth and feed efficiency over several growing seasons (for good pond feed monitoring methods see Akiyama 1989). For example, Figure 4 shows the observed variation in reported FCR of a commercial pelleted shrimp diet within 174 shrimp farms in Thailand (Charoen Pokphand Aquaculture Co., Ltd., personal communication). The fact that the reported FCR ranged from 1 to 2.6 clearly shows the wide variation in farm management practices employed by the shrimp farmers using the same diet (see also Figure 2). From an economic viewpoint, feed costs alone ranged from as low as US$ 1.20/kg to as high as US$ 2.60/kg of shrimp harvested, depending upon the management procedure employed. The results clearly show that there is considerable room for improvement concerning on-farm feed management programmes; the difference between good and poor feed management being the difference between economic success and failure!

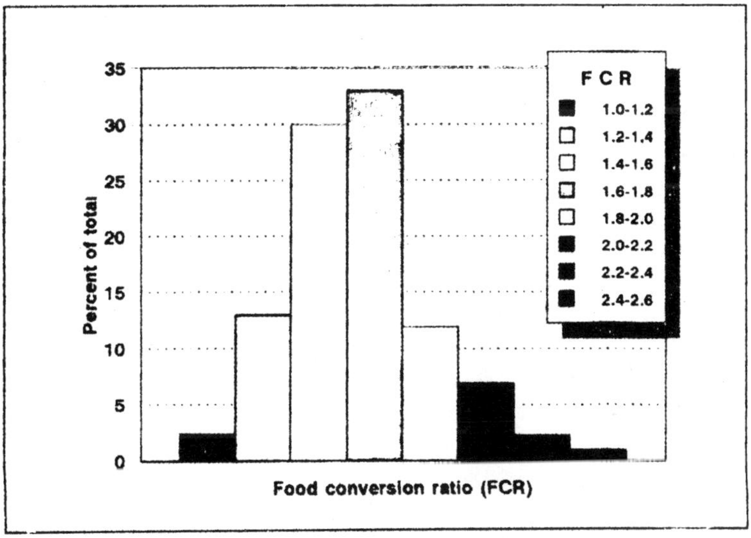

Figure 4. Variation in food conversion ratio in 174 black tiger shrimp farms using CP shrimp feed in Thailand

Aquatic environment and natural food availability

Finally, but not least, the performance of a diet is dependent upon the aquatic environment in which the fish or shrimp is cultured, including:

- **intended fish or shrimp stocking density and natural food availability** (see previous sections and Figures 2-3);

- **intended farm production unit** - tank, raceway, concrete lined pond, earthen pond, pen or floating cage. For example, aquafeed distribution and management within large 5-10 ha ponds is much more difficult (from a practical point of view) and consequently far less effective (from a nutritional point of view in terms of fish/shrimp growth and feed performance or FCR) than within smaller 0.5-1 ha ponds; and

- **water quality and water/pond management** - water temperature, dissolved oxygen concentration, salinity, turbidity, mineral concentration, pH, water depth, water exchange rate, pond preparation and fertilization, water circulation pattern and artificial aeration. For example, Featherstone (1981) found that the performance of salmon starter diets was related to their physical characteristics (*i.e.* feed size and density) and consequent spatial behaviour with respect to the water circulation pattern within the culture tanks used. Similarly, high pond water exchange rates have a negative effect on the action of fertilizers and on natural pond productivity. Furthermore, under certain circumstances the action of fertilizers and performance of aquafeeds can be improved by using artificial pond aeration devices. For a review of the importance of water quality, pond preparation and pond aeration to the success of a semi-intensive pond feeding strategy see Tacon (1988), Boyd (1989), Chiang *et al.* (1989) and Tsai (1989b).

CONCLUSIONS

On the basis of the above discussion it is clear that the nutritional and economic success of an aquafeed is based upon a variety of inter-linked factors and that no one factor can be considered in isolation (Figure 1). It is also apparent that the development of dietary feeding strategies for semi-intensive pond farming systems has proceeded without due regard to fish or shrimp stocking density and natural food availability. This is particularly surprising since the majority of world fish and shrimp aquaculture production is realised within semi-intensive pond farming systems. In fact, one of the major stumbling blocks to aquaculture development within many developing countries has been due to the direct transfer and application of intensive complete diet feeding strategies to semi-intensive pond farming systems. Furthermore, increasing raw ingredient and feed manufacturing costs, and a generally static and/or decreasing market value for the majority of cultured aquaculture species, necessitates that farmers reduce production costs to maintain profitability. Clearly, the road is open for improvements to be made in the area of feed formulation and feed management practices. These should be designed to reduce production costs and maximize the role played by natural food organisms in the overall nutritional budget of the cultured species. The special and unique needs of the semi-intensive fish or shrimp farmer should also be addressed. It is hoped that this paper will help to sow a seed along this road.

LIST OF REFERENCES

Akiyama, D.M. 1989. Shrimp feed requirements and feed management, p. 75-82. *In* D.M. Akiyama (ed.) Proceedings of the Southeast Asian Shrimp Farm Management Workshop, Philippines, Indonesia, Thailand, 26 July-11 August 1989. American Soybean Association, Singapore.

Akiyama, D.M. 1991. Future considerations for the aquaculture feed industry, p. 5-9. *In* D.M. Akiyama and R.K.H. Tan (eds.) Proceedings of the Aquaculture Processing and Nutrition Workshop, Thailand and Indonesia, September 19-25, 1991. American Soybean Association, Singapore.

·rson, R.K., P.L. Parker and A.L. Lawrence. 1987. A 13C/12C tracer study of the utilization of presented feed by a commercially important shrimp *Penaeus vannamei* in a pond growout system. J.World Aquaculture Soc. 18(3): 148-155.

Boonyaratpalin, M. and M.B. New. 1982. Evaluation of diets for *Macrobrachium rosenbergii* reared in concrete ponds, p. 249-256. *In* M.B. New (ed.) Giant Prawn Farming. Elsevier Science Publ.Corp., Amsterdam, Holland.

Bostock, T. 1991. Better feeds for small-scale shrimp farmers. Bay of Bengal News 42:22-26.

Boyd, C.E. 1989. Aeration of shrimp ponds, p. 134-140. *In* D.M. Akiyama (ed.) Proceedings of the Southeast Asian Shrimp Farm Management Workshop, Philippines, Indonesia, Thailand, 26 July - 11 August 1989. American Soybean Association, Singapore.

Brown, P.B., E.H. Robinson and G. Finne. 1989. Nutrient concentrations of catfish feces and practical diets after immersion in water. J. World Mariculture Soc. 20(4):245-249.

Castille, F.L. and A.L. Lawrence. 1989. The effects of deleting dietary constituents from pelleted feeds on the growth of shrimp in the presence of natural foods in ponds. J.World Aquaculture Soc. 20(1): 22A (Abstract).

Chamberlain, G. 1992. Shrimp culture in Indonesia - 4. Feeds for the shrimp industry. World Aquaculture 23(2):38-40.

Chiang, P.D-R., C-M. Kuo and C-F. Liu. 1989. Pond preparation for shrimp growout, p. 48-55. *In* D.M. Akiyama (ed.) Proceedings of the Southeast Asian Shrimp Farm Management Workshop, Philippines, Indonesia, Thailand, 26 July - 11 August 1989. American Soybean Association, Singapore.

Chow, K.W. 1980. Storage problems of feedstuffs, p. 215-224. *In* Fish Feed Technology. FAO Field Document, FAO/UNDP Report ADCP/REP/80/11, Rome, Italy.

Coelho, M.B. 1991. Effects of processing and storage on vitamin stability. Feed International 12(12):39-45.

Cuzon, G., M. Hew, D. Cognie and P. Soletchnik. 1982. Time lag effect of feeding on growth of juvenile shrimp *Penaeus japonicus* Bate. Aquaculture 29:33-44.

De la Cruz, M.C., G. Erazo and M.N. Bautista. 1989. Effect of storage temperature on the quality of diets for the prawn, *Penaeus monodon* Fabricus. Aquaculture 80:87-95.

De Silva, S.S. and F.B. Davy. 1992. Fish nutrition research for semi-intensive culture systems in Asia. Asian Fisheries Science 5:129-144.

Featherstone, P.B. 1981. A comparison of various fish feeding diets for the Atlantic salmon (*Salmo salar* L.). M.Sc thesis. University of Stirling, Stirling, Scotland, 75 p.

Hepher, B. 1988. Nutrition of pond fishes. Cambridge University Press, England. 388 p.

Hirono, Y. 1989. Shrimp farm management in Ecuador, p. 2-10. *In* D.M. Akiyama (ed.) Proceedings of the Southeast Asian Shrimp Farm Management Workshop, Philippines, Indonesia, Thailand, 26 July - 11 August 1989. American Soybean Association, Singapore.

Jaenike, F. 1989. Management of a shrimp farm in Texas, p. 11-21. *In* D.M. Akiyama (ed.) Proceedings of the Southeast Asian Shrimp Farm Management Workshop, Philippines, Indonesia, Thailand, 26 July - 11 August 1989. American Soybean Association, Singapore.

Lall, S.P. 1991. Concepts in the formulation and preparation of a complete fish diet, p. 1-12. *In* S.S. De Silva (ed.) Fish Nutrition Research in Asia. Proceedings of the Fourth Asian Fish Nutrition Workshop. Asian Fisheries Society Special Publication No.5, Asian Fisheries Society, Manila, Philippines.

Lee, C-S. and R.A. Shleser. 1984. Production of *Penaeus vannamei* in cattle manure-enriched ecosystems in Hawaii. J.World Maricul.Soc. 15:52-60.

Lilyestrom, C.G., R.P. Romaire and P. Aharon. 1987. Diet and food assimilation by channel catfish and Malaysian prawns in polyculture as determined by stomach content analysis and stable carbon isotope ratios. J. World Aquaculture Soc. 18(4): 278-288.

Lim, C. and G. Cuzon. 1992. Water stability of shrimp pellets. Paper presented at the Third Asian Fisheries Forum: Fisheries Towards 2000, 26-30 October 1992. Asian Fisheries Society, Singapore.

Lovell, R.T. 1992. Mycotoxins: hazardous to farmed fish. Feed International 13(3):24-28.

New, M.B. 1987. Feed and feeding of fish and shrimp. FAO Field Document, FAO/UNDP Report ADCP/REP/87/26:275 p., Rome, Italy.

Samudra, H.D. 1992. Management of a shrimp farm in Indonesia. ASA Technical Bulletin, AQ35 1992/8: 18 p., American Soybean Association, Singapore.

Tacon, A.G.J. 1987. The nutrition and feeding of farmed fish and shrimp - A training manual. 1. The essential nutrients. FAO Field Document, Project GCP/RLA/075/ITA, Field Document No.2, 117 p. Brasilia, Brazil.

Tacon, A.G.J. 1988. The nutrition and feeding of farmed fish and shrimp - A training manual. 3. Feeding methods. FAO Field Document, Project GCP/RLA/075/ITA, Field Document No.7, 208 p. Brasilia, Brazil.

Tacon, A.G.J. 1991a. Aquaculture nutrition and feeding in developing countries : A practical approach to research and development. Paper presented at the IV International Symposium on Fish Nutrition and Feeding, 24-27 June 1991. Biarritz, France.

Tacon, A.G.J. 1991b. Vitamin nutrition in shrimp and fish, p. 10-41. *In* D.M.Akiyama and R.K.H.Tan (eds.) Proceedings of the Aquaculture Processing and Nutrition Workshop, Thailand and Indonesia, September 19-25, 1991. American Soybean Association, Singapore.

Tacon, A.G.J. 1992. Nutritional fish pathology. Morphological signs of nutrient deficiency and toxicity in farmed fish. FAO Fish. Tech. Paper No.330: 75 p. Rome, Italy.

Trino, A.T., V.D. Penaflorida and E.C. Bolivar. 1992. Growth and survival of *Penaeus monodon* juveniles fed a diet lacking vitamin supplements in a modified extensive culture system. Aquaculture 100:25-32.

Tsai, C.K. 1989a. Quality control in formulated feeds, p. 270-287. *In* D.M. Akiyama (ed.) Proceedings of the People's Republic of China Aquaculture and Feed Workshop, September 17-30, 1989. American Soybean Association, Singapore.

Tsai, C.K. 1989b. Water quality management, p. 56-63. *In* D.M. Akiyama (ed.) Proceedings of the People's Republic of China Aquaculture and Feed Workshop, September 17-30, 1989. American Soybean Association, Singapore.

Villalon, J.R. 1991. Practical manual for semi-intensive commercial production of marine shrimp. Texas A & M Sea Grant College Program, TAMU-SG-91-501, Galveston, Texas. 103 p.

Viola, S. 1989. Production of commercial feeds for warm water fish, p. 143-162. *In* M.Shilo and S.Sarig (editors) Fish Culture in Warm Water Systems: Problems and Trends. CRC Press Inc., Boca Raton, Florida.

Annex 1. Example of feed handling and storage guidelines for a commercial shrimp feed in the Philippines

---

When loading and unloading feeds, always see to it that the bags are carried and put down gently

During transit, make sure that the floor of the vehicle is clean and dry

When transporting feeds, cover the bags on all sides with plastic or any waterproof blanket

Store the feeds in a cool dry place

Keep the feeds away from rats and other animals

Do not store the feeds in the same room with pesticides and other poisonous substances

Keep the feeds at least one foot away from the wall of the storeroom

Do not stack the bags more than 10 bags high. In a large stack of several piles, keep each pile one foot away from the next one. Raise the bottommost layer of each pile four inches above the ground with a grilled and hollow pallet. This will ensure that every stack is ventilated on all sides

When a new batch of feeds arrives in the storeroom, separate it from the older batch and make sure it is used only after the older feeds are consumed (first in, first out)

To ensure that your shrimps receive the feeds in the best possible condition, do not stock the feeds for more than one month. To avoid having excess feeds, purchase smaller quantities of feeds at a time, but more frequently during the grow-out cycle.

Source: San Miguel Foods, Inc, Manila, Philippines

FEED INGREDIENTS AND QUALITY CONTROL

Prasert Sitasit

National Inland Fisheries Institute
Kasetsart University Campus, Bangkok 10900, Thailand

SITASIT, P. 1993. Feed ingredients and quality control, p.75-86. *In* M.B. New, A.G.J. Tacon and I. Csavas (eds.) Farm-made aquafeeds. Proceedings of the FAO/AADCP Regional Expert Consultation on Farm-Made Aquafeeds, 14-18 December 1992, Bangkok, Thailand. FAO-RAPA/AADCP, Bangkok, Thailand, 434 p.

INTRODUCTION

Feed ingredients and finished feeds have played an important role in poultry and livestock production in Thailand for many decades. Many raw materials and finished feeds have been imported and many feed manufacturers were established to produce feeds to satisfy local demand. To protect farmers from the use of poor quality raw materials and finished feeds the Thai government promulgated the first law on animal feed quality control in 1963. Due to advances in animal nutrition and rapid progress in feed manufacturing, this law was revised in 1982.

CURRENT THAI FEEDSTUFF LEGISLATION

According to the 1982 law (Annex 1), animal feeds produced or imported for commercial use have to be licensed by the Department of Livestock Development. Species covered by the law were chicken, ducks, pigs, cattle and buffalo; no aquafeeds were included. This was because, when the law was promulgated, aquaculture was a small enterprise, and most aquafeeds were farm-made or consisted of kitchen wastes. Since 1986, aquaculture production in Thailand has been booming and there has been an increased demand for aquafeeds. Domestic aquafeed production in 1991 was approximately 300,000 t and was expected to reach 410,000 t in 1992. The rapid increase in aquafeed production has caused shortages of feed ingredients and resulted in inconsistent feed quality. Poorly formulated feeds and/or improper manufacturing processes could yield feeds which are poorly digested by aquatic species and have poor water stability. Such low-quality feed will pollute the aquatic environment. Furthermore, feed additives such as growth promoters, hormones, probiotics, and antibiotics, if improperly used in feed formulae, could be a hazard to aquatic species and to public health.

In 1991, to prevent such problems, the Department of Fisheries, together with the Department of Livestock through the support of the Minister of Agriculture and Cooperatives, issued a law controlling the quality of aquafeeds. According to this law aquafeed producers and distributors have to follow the rules, procedures and conditions prescribed (Agriculture and Cooperatives Ministerial Regulation 1991).

According to the regulations, many kinds of feed ingredients, premixes, feed additives and commercial aquatic feeds have to be licensed. At present, commercial feeds and premixes for marine shrimp, freshwater prawns, *Clarias* spp. and herbivorous fish feeds have to be registered with the Department of Fisheries. Some details are given below.

Raw materials

Feed ingredients sold in the market must have guaranteed proximate analyses as shown in Table 1.

Table 1. Composition of some raw materials under legislative control in Thailand

Ingredient	Protein	Fat	Fibre	Moisture	Ash	NaCl
Fish meal	>50	<10	< 2	< 10	<30	< 3
Fish and bone meal	>36	<14	< 2	< 10	<33	< 3
Feather meal	>80	< 4	<1.5	< 11	< 4	-
Soybean meal	>42	< 7	< 8	< 13	< 8	-
Peanut meal	>40	<10	< 8	< 11	<13	-
Fine rice bran	>12	<16	< 8	< 11	<10	-
Rough rice bran	> 5	< 2	<28	< 11	<18	-
Rice bran extract	>14	< 3	<12	< 13	<14	-
Corn meal	> 8	< 3	< 2	< 13	< 2	-

Aquatic commercial feeds

Finished feeds sold in the market are not required to have defined nutritional characteristics but the nutritional value and the quality of the feed stated on the container labels have to comply with those registered with the Department of Fisheries.

Probiotics

Many kinds of probiotics such as *Lactobacillus* and *Pediococcus* spp. are approved for use in aquafeeds.

Hormones

Iodinated casein or thyroprotein, which are used to increase growth rate and feed efficiency in pigs, can be used in fish feeds upon approval of the Department of Fisheries.

Antibiotics

Many antibiotics are currently registered in Thailand for subtherapeutic use as growth promoters in poultry and livestock. However, for aquafeeds, all antibiotics are prohibited in finished feeds.

Containers

Aquafeed containers for feed distribution have to have the following characteristics:

- all containers have to be new, dry and have the capability to protect feeds from moisture;
- metal surfaces inside containers have to be coated with antirust materials which are not dangerous to aquatic animal health;
- vehicles transporting aquafeeds have to be dry and uncontaminated.

Feed preservatives and feed additives

Feed preservatives and additives that are permitted in finished feeds are listed in Table 2.

Table 2. Permitted inclusion levels of major feed preservatives and additives in aquafeeds in Thailand*

Additive/preservative	Permitted inclusion level
Propionic acid	not more than 0.1%
Benzoic acid	not more than 0.1%
Ascorbic acid	not more than 0.1%
Erythorbic acid	not more than 0.1%
Sorbic acid	not more than 0.1%
Thiodipropionic acid	not more than 0.1%
Formic acid	not more than 0.1%
Acetic acid	not more than 0.1%
Lactic acid	not more than 0.1%
Citric acid	not more than 0.1%
Tartaric acid	not more than 0.1%
Malic acid	not more than 0.1%
Fumaric acid	not more than 0.1%
Orthophosphoric acid	not more than 0.1%
Hydrochloric acid	not more than 0.1%
Sulphuric acid	not more than 0.1%
Butylated Hydroxytoluene (BHT)	not more than 0.05%
Butylated Hydroxyanisol (BHA)	not more than 0.05%
Ethoxyquin	not more than 0.015%
Ethylene diamine tetra acetate	not more than 0.024%
5,6-Diacetyl-L-ascorbic acid	as specified
6-Palmitoyl-L-ascorbic acid	as specified
Tocopherols	as specified
Ascorbyl palmitate	as specified
Guar gum	as specified
Methylparaben	as specified
Propionate salts	not more than 0.3%
Benzoate salts	not more than 0.3%
Ascorbate salts	not more than 0.3%
Sorbate salts	not more than 0.3%
Erythorbate salts	not more than 0.3%
Sulphur dioxide	not more than 0.3%
Sulphite salts	not more than 0.3%
Bisulphite salts	not more than 0.3%
Metabisulphite salts	not more than 0.3%
Formate salts	not more than 0.3%
Acetate salts	not more than 0.3%
Diacetate salts	not more than 0.3%
Lactate salts	not more than 0.3%
Citrate salts	not more than 0.3%
Tartrate salts	not more than 0.3%
Dilauryl thiodipropionate	not more than 0.3%

Table 2 (cont.)

Propane 1,2 diol or propylene glycol with mono and diester	as specified
Ethyl 4-hydroxybenzoate	as specified
Sodium ethyl 4-hydroxybenzoate	as specified
Propyl 4-hydroxybenzoate	as specified
Sodium propyl 4-hydroxybenzoate	as specified
Methyl 4-hydroxybenzoate	as specified
Sodium methyl 4-hydroxybenzoate	as specified
Propyl benzoate	as specified
Propyl acetate	as specified
Formaldehyde	as specified
Sodium nitrite	not more than 0.3%
Propyl gallate	not more than 0.1%
Octyl gallate	not more than 0.3%
Dodecyl gallate	not more than 0.1%
Propyl paraben	as specified

* this list applies equally to aquafeeds and other animal feeds

Non nutrients

The non-nutrient compounds which can be added in finished feed at the appropriate level are shown in Annex 2.

Feed form

No restrictions are placed on the forms of aquafeed, as for poultry and livestock feeds. Aquafeeds can be produced in pellet or extruded form, or as meals, crumbles and flakes. The one exception is that shrimp feeds are specially required to be of the sinking type.

Quality control procedures

The Department of Fisheries is responsible for the quality control of aquafeeds. Routine inspections are carried out as follows:

- aquafeeds distributed in the market or on-farm are sampled and analyzed for protein, fat, fibre and moisture and compared to the levels shown on the label and registration documents held by the Department of Fisheries;
- routine inspections of the cleanliness of feed manufacturing premises and equipment and of the raw materials used in feed manufacturing are carried out;
- the examination of finished products emphasizes a check on the presence of antibiotics.

GENERAL NOTES ON QUALITY CONTROL

In feed manufacturing, the purpose of the quality control of raw materials is to ensure that minimum contract specifications are met. These specifications are usually determined by a committee consisting of the nutritionist, management personnel, and the quality control manager. To meet the specifications, sampling, physical inspection and laboratory analysis have to be conducted before the ingredients are accepted and placed in the store. To minimize losses in nutritive value and deteriorative changes in feed ingredients during storage, they must be kept cool in as dry a place as possible. Physical examination and laboratory analysis is also essential for finished aquafeeds, to ensure that specifications are met.

To assure good feed quality, feed ingredients and finished feeds need to be closely monitored: this is known as quality control (QC). QC involves the verification of the quality standards established for each feed ingredient through the process of feed manufacturing and continues as they finally go into storage as finished feed. At first glance, feed ingredients and finished feed may appear to be nutritious and of high quality; however, unless they are systematically analyzed - through physical, chemical, and/or biological means - there is no way that one can be sure of their true value to aquatic animals.

Using good quality food ingredients in feed manufacturing partially guarantees good quality in final products. The purpose of the QC of feed ingredients is to verify that they meet the specifications of both the feedmill and the supplier. Moreover, QC provides knowledge concerning the exact composition of raw materials and the levels of toxic substances present. Feed ingredients arriving in the mill need visual inspection for evidence of wetting or mould growth, presence of scrap metals, stones, or other non-biological contaminants, and presence of insects. Feed ingredients which pass the visual examination will be sampled for further laboratory analysis. Good sampling is very important in QC. An analysis of feed ingredients is only as good as the sample obtained therefrom. If the sample is not representative of the entire batch, the evaluation is useless, no matter how extensively the sample is analyzed. Sampling requires certain techniques.

Feed ingredients received in bulk are sampled by using a scoop. Two samples per ton are taken for bulk consignments smaller than 10 t. Larger consignments, up to 100 t, require one sample per ton or one sample for every two tons, depending on the size of the consignment. Sampling of bagged ingredients is done with a spear probe. Samples from at least 5% of the bags are required. Samples taken should be pooled, mixed, and ground to the size suitable for chemical analysis. Samples submitted to the quality control laboratory should be stored in tightly sealed containers.

Chemical analysis of feed ingredients is the next step in QC. Proximate analyses are usually conducted to reveal the major nutritional characteristics of the samples. These analyses include moisture, crude protein, crude lipid, crude fibre, and ash. Nitrogen-free extract (NFE) is determined by difference. Calcium and phosphorus determinations are necessary for feed ingredients with high ash content and for mineral supplements, such as fish meal, meat and bone meal, calcium phosphate and calcium carbonate sources.

Feed ingredients liable to have high levels of certain anti-nutrients need to be tested prior to use. Trypsin inhibitor and urease content in soybean products, gossypol in cottonseed, aflatoxin in corn and peanut meal, mimosine in *Leuceana* leaf meal, and rancidity in fats, are natural toxins that, at high enough levels, are growth inhibitory and sometimes fatal to the animal consuming them. Therefore, it is advisable to analyze these feed ingredients for their toxic content.

The quality of finished feeds reflects the quality of the ingredients used. However, feed manufacturing and storage can alter the finished feed quality to a certain extent. Therefore, the purpose of QC of the final products is to ensure whether the feed ingredients were added in the proportions required by the formulation and to examine the homogeneity of the final product.

Finished feeds can be sampled during bagging-off by taking a handful from every fifth bag of 40-50 bags and pooling the individual samplings. The products are first visually inspected for homogeneity of size and colour and the amount of fines. The samples are then placed in tightly sealed containers and submitted for physical and chemical evaluation.

Physical evaluation of finished feeds includes the floating ability test for expanded feeds. The feed will be rejected if less than 85% floats after 15 minutes in water. For shrimp feed, water stability is of utmost importance in addition to nutritional quality. In general, water stability up to 3 hours is considered satisfactory.

Chemical analysis of the finished feeds is important to check the specifications, or claims, established for the finished products. Proximate analysis for moisture, crude protein, crude lipid, crude fibre, ash, and nitrogen-free extract (by difference) of the finished feeds is necessary to confirm nutrient content. This analysis also checks the manufacturing process. However, if raw material quality control is properly conducted and if process control is adequate, then the only chemical tests necessary for finished products on a regular basis are for moisture and crude protein. Periodic analysis of other components should also be scheduled, however.

Totally effective quality control for farm-made aquafeeds is impossible. However, physical inspections are appropriate and feasible and will ensure that the correct feeds are used. Inspection of the external features of the ingredients before use, which requires skill and experience, should include:

- assessing whether the ingredient is the normal colour;
- evaluating the odour of each ingredient, which will reflect its freshness and the effectiveness of its method of preservation. Odours such as "stuffy musk", "rancid", or "burning" indicate potential problems;
- feeling if the ingredient is moist (dry ingredients only!). If so, this may mean that the ingredient is mouldy or otherwise in poor condition;
- looking for the presence of insect and mould growth, indicating deteriorative changes in the feed and poor hygiene. Presence of scrap metals, stones, dirt or other non-biological contaminants indicate its impurity.

LIST OF REFERENCES

Agriculture and Cooperatives Ministerial Regulation. 1991. Name, category, type or description of animal feed. Government Gazette, Vol. 108, Part 218, 12 December 1991, Bangkok, Thailand. 103 p.

Chow, K.W. 1980. Quality control in fish feed manufacturing, p. 369-385. *In* Fish feed technology. ADCP/REP/80/11. FAO, Rome, Italy.

Department of Livestock Development. 1990. Animal Feed Quality Act, B.E. 2525. Department of Livestock Development, Ministry of Agriculture and Cooperatives, Bangkok, Thailand. 177 p.

Somsueb, P. 1993. Aquafeeds and feeding strategies in Thailand, p. 365-385. *In* M.B. New, A.G.J. Tacon and I. Csavas (eds.) Farm-made aquafeeds. Proceedings of the FAO/AADCP Regional Expert Consultation on Farm-Made Aquafeeds, 14-18 December 1992, Bangkok, Thailand. FAO-RAPA/AADCP, Bangkok, Thailand.

Annex 1. Thai feedstuff regulation definitions

Current rules and regulations include the following definitions:

- materials intended for animal feed used as published in the Government Gazette by the Minister upon the advice of the animal feed quality control committee, are termed to be "animal feeds";

- "produce" means manufacture, mix, transform, prepare, change the features of, or repack;

- "sell" means distribute, dispose of, give or exchange for commercial purposes and also includes possession for sale;

- "import" covers ordering and entry of materials into Thailand;

- "container" means any material specifically used for packing or wrapping animal feed;

- "label" includes any picture, imprint or statement shown on an animal feed container.

Further details of the regulations are provided by Somsueb (1993).

Annex 2. Non-nutrient compounds permitted in aquafeeds in Thailand

---

Anise oil
Vanillin
Monosodium glutamate
Saccharin sodium
All natural products and corresponding synthetic products
Pulverized bentonite
Hemicellulose plus lignin
Polymethylolcarbamide
Silicon dioxide or colloidal silica
Citric acid
Sodium, potassium and calcium stearates
Dilicic acid
Diatomaceous earth
Calcium silicate
Sodium aluminosilicate
Kaolin and kaolinitic clays
Natural mixtures of steatite and chlorite
Vermiculite
Lignosulphonates
Bentonite and other montmorillonite clays
Ball clays
Aluminium calcium silicate
Magnesium silicate
Hydrated sodium calcium aluminosilicate
Tricalcium silicate
Lecithins
Alginic acid
Alginate salts
Propylene glycol alginate
Propane-1, 2-diol alginate
Agar
Carrageenan
Furcellaran
Locust bean gum
Tamarind seed flour
Guar gum or guar flour
Tragacanth

Acacia or gum arabic
Xanthan gum
D-glucitol or sorbitol
Mannitol
Glycerol
Pectins
Microcrystalline cellulose
Methylcellulose
Ethylcellulose
Hydroxypropylcellulose
Hydroxypropylmethylcellulose
Ethylmethylcellulose
Chrondrus extract
Gum ghatti
Agar-agar
Carboxymethylcellulose
Monoacyl and diacylglycerols
Monoacyl and diacylglycerols esterified with the following acids: acetic, lactic, citric, tartaric, monoacetyltartaric and diacetyl-tartaric
Sucrose esters of fatty acids
Sucroglycerides
Polyglycerol esters of non-polymerized edible fatty acids
Propylene glycol esters of fatty acids, propane-1, 2-diol esters of fatty acids
Stearoyl 2-lactylic acid, sodium stearoyl-2-lactylate, calcium stearoyl-2-lactylate
Stearoyl tartrate
Glycerol poly (ethylene glycol) ricinoleate
Dextrans
Caramel
Sorbitan monostearate
Sorbitan tristearate
Sorbitan monolaurate
Sorbitan monopalmitate
Sorbitan mono-oleate
Polyoxyethylene (20) sorbitan monostearate
Polyoxyethylene (20) sorbitan tristearate
Polyoxyethylene (20) sorbitan monolaurate
Polyoxyethylene (20) sorbitan mono-oleate
Polyoxyethylene (20) sorbitan monopalmitate
Polyoxyethylene (20) sorbitan trioleate

Polyoxyethylene stearate
Penta sodium triphosphate
Polyethyleneglycol esters of fatty acids from soya oil
Polyoxyethylated glycerides of tallow ratty acids
Esters of polyglycerol and of alcohols obtained by the reduction of oleic and palmitic acids
Polypoly ethyelene glycol 6000
Polyoxypropylene-polyoxythylene polymers (M.W. 6,800-9,000)
Propane-1, 2-diol
Sodium, potassium and calcium salts of edible fatty acids of castor oil (polyglycerol polyricinoleate)
Patente blue V
Acid brilliant green BS or lissamine green
Carotenoids
Capsanthin
Beta-apo-8-carotenol
Ethylester of beta-apo-8 carotenoic acid
Lutein
Xanthophylls
Cryptoxanthin
Violaxanthin
Canthaxanthin
Zeaxanthin
Citranaxanthin

ON-FARM FEED PREPARATION AND FEEDING STRATEGIES FOR CARPS AND TILAPIAS

Amararatne Yakupitiyage

Division of Agricultural and Food Engineering
Asian Institute of Technology
GPO Box 2754, Bangkok, Thailand

YAKUPITIYAGE, A. 1993. On-farm feed preparation and feeding strategies for carps and tilapias, p.87-100. *In* M.B. New, A.G.J. Tacon and I. Csavas (eds.) Farm-made aquafeeds. Proceedings of the FAO/AADCP Regional Expert Consultation on Farm-Made Aquafeeds, 14-18 December 1992, Bangkok, Thailand. FAO-RAPA/AADCP. Bangkok, Thailand, 434 p.

INTRODUCTION

Finfish production dominates the inland aquaculture sector in the Asia and Pacific region. Out of nearly 7 million t of finfish produced by inland fish culture in 1990, fresh water fish feeding low in the food chain contributed 93% - 6.5 million t (Csavas 1992). This included 86% carps (5.6 million t) and 5% tilapias (0.33 million t), showing the importance of these species in human nutrition in the Asian region.

A notable feature of carp and tilapia culture in most developing countries is their relatively low farm gate value compared to marine or freshwater carnivorous fish. They are usually produced for local consumption rather than export. These two factors influence the type of feed used in the culture system, the feed preparation method and the feeding practice. For the same reasons, the use of commercial pelleted feed is prohibitively expensive in many countries. For example, in Thailand the farm gate value of carps and tilapias is about US $1.00/kg but it may require 2 kg of commercial feed costing US $0.50/kg to produce them!

The majority of carp and tilapia culture systems, therefore, are at least partially dependent upon natural food. This paper attempts to delineate possible feed preparation methods and feeding strategies for these fish by reviewing the existing literature.

OVERVIEW OF ON-FARM FEED PREPARATION

Farms are delineated by a boundary across which there is inflow and outflow of materials (Figure 1). Farmers also require knowledge of the nutrient requirements of fish, the physical and chemical properties of ingredients, and feed formulation and processing methods, which depend on past experience or have been contributed by research and development (R&D) institutes. Equipment and raw materials need to be purchased outside the farm boundary. Generally available multipurpose equipment may also be used depending upon the scale of operation. Capital and manpower requirements may also be directly related to the size of production.

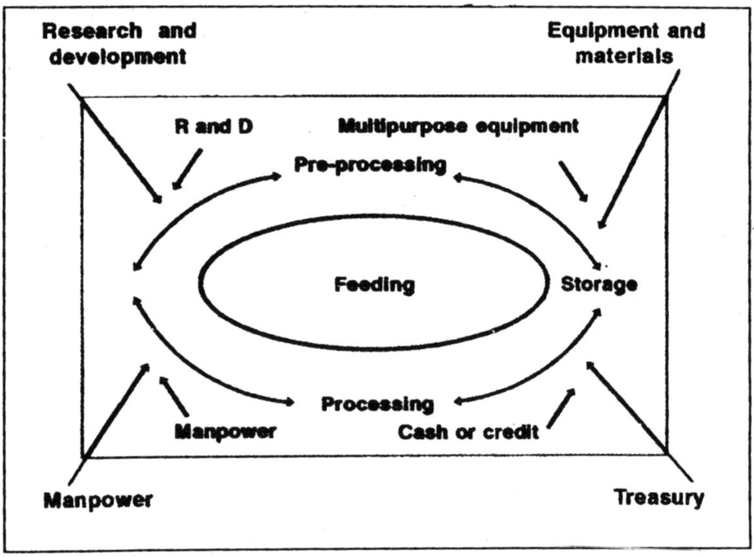

Figure 1. Components of on-farm feed production

The chain of events comprises the acquisition of raw materials, their pre-processing, final processing, storage and/or feeding. Each step adds biological and economic value to the product. Added biological value represents an improvement of palatability (*i.e.* suitable particle size, taste), enhancement of nutritional quality (*i.e.* removing anti-nutritional factors) and an increment of nutrient digestibility. However, the value of improved product biological quality must exceed the accompanying cost of the improvement. Since carps and tilapias fetch a relatively low market price, a cost effective feeding regime is dependent upon the careful selection of feedstuffs and feed ingredients, utilization of an optimum feed formulation, selection of multi-purpose equipment for feed preparation, optimization of feed dispersion and minimization of feed waste. The remaining sections of this paper, therefore, are aimed at elucidating the above factors and at suggesting methods for optimal use of available feed resources for carp and tilapia culture.

SELECTION OF FEEDSTUFFS AND FEED INGREDIENTS

Schaeperclaus (1933) identified two types of farms on the basis of their feed/feeding regime in carp culture in Europe, namely:

- through food or natural food supplied by the pond, equivalent to pasture feeding of terrestrial animals;
- through feeding or feed supplied by the farmer, analogous to stall feeding of terrestrial animals.

Using the same classification, on-farm feeds for carps and tilapias may be divided into two major groups:

- natural food produced by pond fertilization within the culture system;
- addition of pre-prepared feed to the culture system (Figure 2).

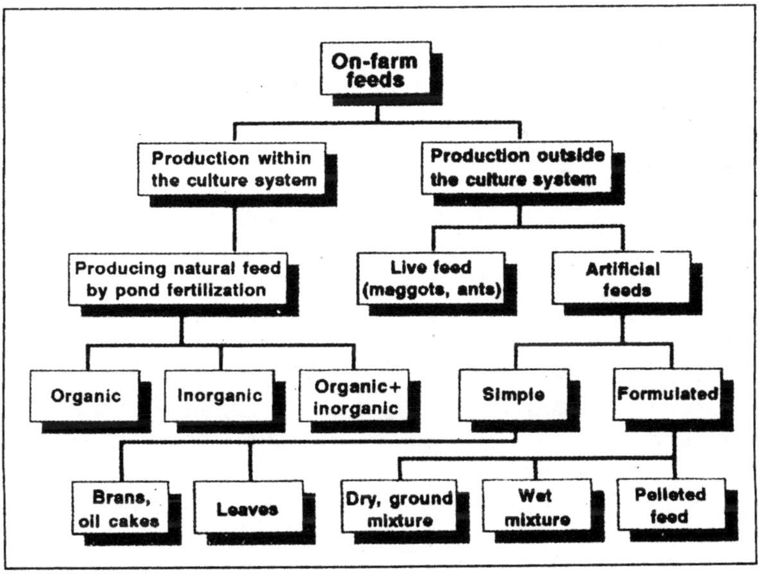

Figure 2. Classification of on-farm feeds

Use of natural feeds through pond fertilization

The main objective of fertilization is to enhance autotrophic and heterotrophic food production. The method of fertilization will be mainly dependent upon the feeding behaviour of fish. For example, enhanced autotrophic production may be sought for species such as tilapias, Indian major carps (rohu, catla) and Chinese carps (silver carp and bighead carp). However, if the target species feed upon benthic organisms or detritus, autotrophic production may indirectly contribute to the fish yield. Hence, a common practice is to grow the bottom feeding carp with filter feeding tilapias in polyculture in order to utilize the different niches of the fish pond ecosystem.

The *in situ* production of natural feed is mainly fulfilled by fertilizing fish ponds with organic manure, inorganic fertilizers or a combination of both. Fertilizing fish ponds is still an art rather than a science. Coleman and Edwards (1987) extensively reviewed different fertilization methods suitable for carp and tilapia culture. Further details on practical fertilization methods used by different authors can be found in Tacon (1988). Edwards (1990) showed that even though the nitrogen (N) and phosphorus (P) conversion efficiency of organic and/or inorganic fertilization of carp/tilapia culture systems are marginally less than feeding fish with complete feed, fertilized systems pollute the environment less than intensive complete feed-based culture systems with water exchange to the surrounding environment. Edwards (1990) also emphasized that a reduction in feed cost can be mainly achieved through fertilizing fish ponds with N and P, rather than attempting to increase yields through the use of complete diets. Potential yields of tilapias and carps under different fertilization regimes are shown in Table 1.

Fertilization rates are dependent upon the micro-ecological conditions including soil type and pond history. It has been demonstrated that the maximum N and P fertilization rates of Asian Institute of Technology (AIT) ponds in acid sulphate soils with considerable P binding capacity are 4 kg N and 2 kg P/ha/day (AIT/CRSP unpublished data). Beyond the above N loading rate, ammonia toxicity develops and the fish standing stock may collapse. However, it should be emphasized that these rates may not be appropriate for other sites where the pond ecology differs from AIT conditions. A number of site specific fertilization rates for carp and tilapia pond culture are summarized by Tacon (1988).

Apart from conventional fertilization techniques, a number of novel fertilization methods have been attempted to produce on-farm natural feeds. For example, the use of waste arising from luxury fish species raised intensively on complete feeds as food for species that feed lower in the aquatic food chain has been attempted, including:

- stock manipulation (*i.e.* the polyculture of catfish and tilapia at a ratio of 10:1 is an existing practice in Thailand);

- luxury species such as walking catfish (*Clarias* spp.) are cultured in a cage utilizing formulated feed with faecal waste passing to the pond functioning as fertilizer for tilapia stocked outside the cage (Lin, unpublished data); and

- feeding large fish with formulated feed within a cage and producing fingerlings through the fertilizing effect of fish excreta in the pond in which the cage is suspended (Lin, unpublished data).

Table 1. Potential fish yield under different fertilization regimes

Fertilization regime	Fish yield (t/ha/yr)	Comments*
Buffalo manure	4 - 5	Polyculture (P) or Monoculture (M)
Muscovy duck manure	8 - 10	P or M
Khaki Campbell manure or chicken manure + inorganic fertilizer	10 - 12	P or M
Inorganic fertilizers (CRSP data) (Urea + TSP)	12 - 15 2 - 4	M - Tilapia M - Silver barb

* polyculture = tilapia + silver barb + common carp

Source: AIT (unpublished data)

Artificial feeds

The phrase "on-farm feeds" creates an instant picture of artificial feed in the mind of the majority of fish nutritionists. Feeding fish either a single feed (*e.g.* rice bran or leaves) or a simple mixture is usually undertaken. These mixtures may be further processed into a wet dough or pellet to minimize dispersion losses. Pellet production provides additional advantages such as an ability to prepare feed in advance, and ease of storage.

Terrestrial based live feed, such as earthworms and maggots, can also be produced on-farm using various organic wastes. Naturally occurring terrestrial live feeds such as termites and snails can also be used depending upon their availability.

Energy supplements

The addition of energy supplements has been mainly used in carp culture. A closer look at the species used in these systems indicates that benthic feeders mainly benefit from energy supplements rather than phytoplankton feeders (Table 2). A plausible reason for this phenomenon is that the relative scarcity of benthic organisms makes a pronounced protein sparing effect of the energy supplement. As phytoplankton is abundant in well fertilized culture systems, the protein sparing action of the energy supplement may not be important to plankton feeding fish.

Table 2. Observed growth of fish fed on an energy supplement

Energy supplement species	Initial weight (g)	Final weight (g) without feed	 with feed	Stocking density (fish/m^2)	Growth period (source)
Rice bran					
Common carp	13	330	360	0.10	5 months
Silver barb	8	80	112	1.50	(1)
Tilapia	13	250	290	1.50	
Rice bran					
Tilapia	31	73	85	2.00	2.5 months (2)
Supplementary feed	<u>Age</u>				
Common carp	0+	227	463	0.91	
Silver carp	1+	457	489	0.15	
Silver carp	2+	1,591	1,620	0.15	Not given
Tilapia	0+	151	144	0.40	(3)
Tilapia	1+	369	417	0.15	

Source: (1) Yakupitiyage *et al.* (1989); (2) Xin (1989); (3) Spataru *et al.* (1980)

Szumeic (1969) and Gurzeda (1969) recommended that for common carp every unit of ingested natural feed should be supplemented with three units of an energy supplement (*i.e.* barley) by weight. Viola *et al.* (1988) reported that at least 2% lipid supplement was necessary within animal protein-free formulated feeds for tilapia. However, if the ponds were well fertilized, added oil had no significant positive effect on tilapia growth (Hung 1989).

Commonly used energy supplements in Asia include agricultural by-products such as rice bran, broken rice and maize. If the particle size is small, special preparation of these ingredients is not necessary. However, cooked carbohydrate (*i.e.* cooked cassava tubers) has certain advantages over non-cooked feed due to its higher palatability and digestibility, and the ability to use it as a wet dough. Feeding rates for rice bran, as reported by Ling (1967), are presented in Table 3.

Table 3. Observed feeding rates of protein (peanut cake) and energy supplement (rice bran) at a stocking density of 1 fish/m²

Month	Rice bran (kg/ha/day)	Peanut cake (kg/ha/day)
1	1.0 - 1.5	-
2	1.5 - 3.0	-
3	3.0 - 5.0	2.0 - 5.0
4 - 5	5.0 - 8.0	5.0 - 10.0
6 - 7	8.0 - 12.0	8.0 - 12.0
8 - 10	12.0 - 16.0	16.0 - 24.0

Source: Ling (1967)

Protein supplements

Empirical data show that approximately 10 g of digestible protein/m² may be available to plankton feeding fish at maximum fertilization rates, leading to maximum autotrophic production (Yakupitiyage, unpublished data). Fish dependent upon secondary production certainly receive less protein than those dependent upon autotrophic production. A protein deficit may develop if the absolute protein requirement of the fish population per unit area is higher than the available protein from natural food.

There are two important factors relating to fish which should be considered: stocking density and body weight. For example, it has been estimated that a protein deficit usually develops at an individual body weight of fish higher than 200 g, at a stocking density of 2 fish/m^2. However, if the stocking density is 0.25 fish/m^2, a protein supplement is required only after fish attain 1 kg body weight (Yakupitiyage, unpublished data). The latter stocking density is widely used in Indian carp polyculture systems.

Several examples of protein supplements used in different countries are summarized in Table 4. It has been demonstrated by a number of authors that plant protein can be used as the main protein source for carps and tilapias without need for the addition of an animal protein component (Viola *et al.* 1988; Nandeesha *et al.* 1989; Hung 1989). Other possible ingredients are listed by Tacon (1987). However, the presence of anti-nutritional factors should be taken into consideration during the selection of ingredients for feed preparation.

Table 4. Some protein supplements used to feed carps and tilapias and their reported food conversion ratio (FCR)

Protein supplement	Protein content (%)	FCR	Species (source)
Silkworm pupae (SP) + rice bran (RB)	25	3.0	Carp (1)
SP+groundnut cake (GNC)+RB	28	3.7	Catla (2)
Soybean meal (SBM)+GNC+RB	27	3.6	Catla (2)
Fish meal (FM) + GNC + RB	30	2.0	Catla (2)
FM + cassava based	30	1.5	Nile tilapia (3)
SBM + cassava based	30	1.5	

Source: (1) Ling (1967); (2) Nandeesha *et al.* (1988); (3) Hung (1989)

Mineral and vitamin supplementation

Viola *et al.* (1988) demonstrated that the addition of 2% di-calcium phosphate was necessary for both carps and tilapia to use soybean-based diets efficiently. Hung (1989) also demonstrated the positive growth effect of the addition of di-calcium phosphate in a well fertilized culture system for tilapia.

However, it is questionable whether calcium is required in these culture systems. As calcium and phosphorus were supplied together, it is reasonable to conclude that either, or both, should be added to supplementary feeds. No data is available concerning the requirement for other minerals. However, it is a common practice in China and India to use common salt as an ingredient in formulated feeds.

The addition of vitamins to supplementary feeds is controversial. Hepher (1972) showed that there was an interactive effect between dietary vitamin supplementation and carp stocking density (Table 5). For example, although added vitamins had no growth stimulating effect in carp stocked at 0.2 fish/m^2, when stocking density was raised to 0.6 fish/m^2, there was a marginal improvement in food conversion efficiency (FCR) and a 15-20% increase in fish yield. Clearly, further research in this field is warranted.

Table 5. Interactive effect of dietary vitamin supplementation and carp stocking density

Stocking density (fish/m^2)	Without vitamins yield (kg/ha/yr)	Without vitamins FCR	With vitamins yield (kg/ha/yr)	With vitamins FCR
0.2	4.42	1.4	4.31	1.4
0.6	6.32	1.7	7.37	1.6

Source: Hepher (1972)

FEED PREPARATION

Existing feed preparation techniques for carps and tilapias are simple. Terrestrial based live feeds such as termites and silkworm pupae are added directly to fish ponds. Snails are crushed before being offered to fish. Leaves are usually chopped when the fish are small but larger fish (> 50g) are fed directly in Chinese carp culture systems.

If dry ingredients are used, a well ground mixture can be offered in feeding bags or the mixture simply dispersed throughout the pond. A reduction in feed waste can be achieved by cooking the carbohydrate source and mixing other ingredients with the gelatinized starch. Such mixtures can be either directly fed as a wet dough or can be pelleted by a simple extrusion process. Prepared pellets may be sun dried and stored for future use. Examples of several protein supplements used as feed for carp and tilapia, including reported FCR, are given in Table 4.

FEEDING

There are three important factors which should be considered when feeding carps and tilapias, namely:

- fluctuation of feed intake and digestibility at *ad libitum* feeding (De Silva and Perera 1983, 1984; De Silva and Gunasekera 1989; Yakupitiyage 1989);
- total dry matter loading rate into the system (AIT, unpublished data); and
- total nitrogen (N) loading rate into the system (AIT, unpublished data).

Feed intake fluctuation

It has been observed under laboratory conditions that there is a fluctuation of feed intake and protein digestibility in fish. De Silva and Perera (1983) suggested that this problem could be overcome by feeding fish with two different diets: a high protein diet followed by a low protein diet. Since it is cumbersome to prepare two diets of different composition and a low protein diet increases the dry matter loading rate to the system, Yakupitiyage (1989) suggested that feeding fish at different rates (*i.e.* a high feeding rate followed by low feeding rate) may be more suitable than using two different feeds. However, as supplementary feeds are usually not used for *ad libitum* feeding, more on-farm feeding trials should be conducted before recommending any of these feeding practices to farmers.

Total dry matter loading rate

The dry matter loading rate into a static culture system will directly influence the early morning dissolved oxygen concentration. The possible upper dry matter loading rate for a short duration (3-6 months) for tilapia/carp polyculture systems was approximately 100 kg feed DM/ha/day to maintain dissolved oxygen at approximately 1 mg/litre at dawn (AIT unpublished data). However, the exact loading rate may depend upon the pond history, as previous organic matter accumulation in the pond bottom may affect the community respiration rate. Ling (1967) reported that 13 t of green fodder, 7.2 t of agricultural by-products and 2 t of animal products were used in Chinese carp polyculture systems at a stocking density of 250 grass carp, 2,000 silver carp and bighead carp, 3,000 mullet, 2,000 milkfish, 1,000 mud carp and 500 common carp for a 10 month period. Assuming green fodder contains 75% moisture and other feed materials contain 10% moisture, the loading rate was approximately 38 kg DM/ha/day.

Total nitrogen loading rate

Table 6 shows the potential N loading rate with pelleted diets of different dietary protein levels. It was assumed that the maximum possible N loading rate was 4 kg N/ha/day, as above these N loading rates ammonia toxicity affects fish growth (AIT, unpublished data). Maximum feeding rate was assumed to be 100 kg DM/ha/day with 80% protein digestibility and a 50% retention of digested protein. The table also demonstrates that reducing dietary protein level may result in reducing the fish population density which can be supported by the supplementary feed, as the absolute protein supplement to the system decreases with decreasing dietary protein content.

Table 6. Possible nitrogen loading rates within fertilized carp and tilapia ponds*

Dietary protein level (%)	Protein supply from 100kg feed (kg/ha/day)	Total nitrogen loading (kg/ha/day)	Ammonia - N loading (kg/ha/day)	Maximum N from fertilizer (kg/ha/day)
40	40	6.4	2.56	1.34
35	35	5.6	2.24	1.73
30	30	4.8	1.92	2.08
25	25	4.0	1.60	2.40
20	20	3.2	1.28	2.72
15	15	2.4	0.96	3.04

* calculations are based upon AIT conditions (unpublished data)

CONCLUSIONS

There is great scope for producing fish economically by utilizing natural food through proper pond fertilization rather than using more expensive formulated feeds. Unfortunately, many fish nutritionists do not consider pond fertilization as a method of producing fish feed and advocate the use of formulated feeds in carp and tilapia culture. Before attempting to draw up recommendations on possible farm-made formulated feeds, investigations on the potential nutritional limiting factors of these culture systems are required.

Farming systems may be divided into industrial agriculture systems, resource-rich farming systems and resource-poor farming systems. Industrial large scale enterprises have apparently had little interest in carp and tilapia culture and the majority of culture systems in Asia belong to the resource-rich or resource-poor farming systems. Formulated feed may have a role in these farming systems provided that the cost of feed is relatively low and the stocking density or fish biomass has been deliberately manipulated to maximise the utilization of the natural food present within the system (*i.e.* that fish biomass is higher than the critical standing crop). People who are most in need of on-farm made aquafeeds for intensifying their culture operation are generally resource-poor farmers and their use of formulated feed will be mainly dependent upon the availability of feed ingredients. A question that requires immediate attention is how to make best use of the low quality feed ingredients available to resource-poor farmers. There may be several answers to this question but current strategy is to use them as supplementary feed in fertilized systems rather than as an ingredients within complete feeds.

LIST OF REFERENCES

Coleman, J.A. and P. Edwards. 1987. Feeding pathways and environmental constraints in waste-fed aquaculture: balance and optimization, p. 240-279. *In* D.W.J. Moriarity and R.S.V. Pullin (eds.) Detritus and microbial ecology in aquaculture. Proceedings of the Conference on Detrital Systems for Aquaculture, 26-31 August 1985, Bellagio, Como, Italy. ICLARM Conference Proceedings 14. ICLARM, Manila, Philippines.

Csavas, I. 1992. Recent developments and issues in aquaculture in Asia and Pacific. Paper presented at the APO Seminar on Aquaculture, 25 August - 4 September, Tokyo, Japan.

De Silva, S.S. and R.M. Gunasekera. 1989. Effect of dietary protein level and amount of plant ingredient (*Phaseolus aureus*) incorporated into the diets on consumption, growth performance and carcass composition in *Oreochromis niloticus* fry. Aquaculture 80:121-133.

De Silva, S.S. and M.K. Perera. 1983. Digestibility of an aquatic macrophyte by the cichlid *Etroplus suratensis* with observations on the relative merits of three indigenous components as markers and daily changes in protein digestibility. J. Fish Biol. 23:675-684.

De Silva, S.S. and M.K. Perera. 1984. Digestibility in *Sarotherodon niloticus* fry: effect of dietary protein level and salinity with further observations on variability in digestibility. Aquaculture 38:293-306.

De Silva, S.S., R.M. Gunasekera and C. Keembiyahetty. 1986. Optimum ration and feeding frequency in *Oreochromis niloticus* young, p. 559-564. *In* J.L. Maclean, L.B. Dizon and L.V. Hosillos (eds.) The First Asian Fisheries Forum. Asian Fisheries Society, Manila, Philippines.

Edwards, P. 1990. Environmental issues in integrated agriculture-aquaculture and waste water-fed fish culture systems. Paper presented at Bellagio conference on Environmental and Third World Aquaculture Development, September 1990, Bellagio, Italy.

Gurzeda, A. 1969. Relationship between natural and artificial food in carp nutrition, p. 118-130. *In* New developments in carp and trout nutrition. Papers submitted to a symposium held in conjunction with the 5th session of EIFAC, Rome, May 1968. EIFAC Tech. Paper 9. FAO, Rome, Italy.

Hepher, B. 1972. Supplementary feeding in fish culture, p. 183-198. *In* Proceedings of the 9th International Congress of Nutrition, Mexico 1972, Vol 3.

Hung, L.T. 1989. Evaluation of soybean meal as a supplementary feed for red tilapia. M.Sc. Thesis, 79 p. Asian Institute of Technology, Bangkok, Thailand.

Ling, S.W. 1967. Feeds and feeding of warm-water fishes in ponds in Asia and the Far East. FAO Fish. Rep. No. 44 (3):291-309.

Nandeesha, M.C., G.K. Sriknath, T.J. Varghese, P. Keshwanath and H.P.C. Shetty. 1989. Growth performance of Indian major carp, *Catla catla* on fish meal free diet, p.137-142. *In* Fish Nutrition Research in Asia. Proceedings of the Third Asian Fish Nutrition Network Meeting. Asian Fish. Soc. Spec. Publ. 4. Asian Fisheries Society, Manila, Philippines.

Schaeperclaus, W. 1933. Textbook of pond culture. Paul Parely Book Publishing House, Berlin. 261 p.

Spataru, P., B. Hepher and A. Halevy. 1980. The effect of the method of supplementary feed application on the feeding habits of carp (*Cyprinus carpio*. L) with regard to natural food in ponds. Hydrobiologia 72:171-178.

Szumeic, J. 1969. Relationship between natural and artificial food in carp feeding, p. 140-160. *In* New developments in carp and trout nutrition. Papers submitted to a symposium held in conjunction with the 5th session of EIFAC, Rome, May 1968. EIFAC Tech. Paper 9. FAO, Rome, Italy.

Tacon, A.G.J. 1987. The nutrition and feeding of farmed fish and shrimp - A training manual. 2. Nutrient sources and composition. FAO Field Document, Project GCP/RLA/075/ITA, Field Document 5/E, Brasilia, Brazil. 129 p.

Tacon, A.G.J. 1988. The nutrition and feeding of farmed fish and shrimp - A training manual. 3. Feeding methods. FAO Field Document, Project GCP/RLA/075/ITA, Field Document 7/E, Brasilia, Brazil. 208 p.

Viola, S., O. Rappaport and G. Zohar. 1988. Animal protein free feeds for hybrid tilapia (*O. niloticus* x *O. aureus*) in intensive culture. Aquaculture 75:115-125.

Xin, Zhang. 1989. Rice bran as an energy-rich supplementary feed in a fertilized system for nile tilapia. M.Sc. Thesis, 87 p. Asian Institute of Technology, Bangkok, Thailand.

Yakupitiyage, A. 1989. Quantitative and qualitative aspects of energy acquisition of the cichiid fish *Oreochromis niloticus*. Ph.D. Thesis, 237 p. Institute of Aquaculture, University of Stirling.

Yakupitiyage, A., P. Edwards and K.L. Wee. 1989. Supplementary feeding of fish in a duck-fish integrated system. I. The effect of rice-bran, p. 143-157. *In* S.S. De Silva (ed.) Fish Nutrition Research in Asia. Proceedings of the Fourth Asian Fish Nutrition Workshop. Asian Soc. Spec. Publ. 5. 205 p. Asian Fisheries Society, Manila, Philippines.

ON-FARM FEED PREPARATION AND FEEDING STRATEGIES FOR CATFISH AND SNAKEHEAD

Wimol Jantrarotai and Patanai Jantrarotai

National Inland Fisheries Institute
Kasetsart University Campus, Bangkok 10900, Thailand

JANTRAROTAI, W. and P. JANTRAROTAI. 1993. On-farm feed preparation and feeding strategies for catfish and snakehead, p.101-119: *In* M.B. New, A.G.J. Tacon and I.Csavas (eds.) Farm-made aquafeeds. Proceedings of the FAO/AADCP Regional Expert Consultation on Farm-Made Aquafeeds, 14-18 December 1992, Bangkok, Thailand. FAO-RAPA/AADCP, Bangkok, Thailand, 434 p.

INTRODUCTION

Inland aquaculture has been practised in Thailand for over five decades with various degrees of diversity. There are about 27 species of fish being cultivated with numerous types of systems ranging from super-intensive farming for commercial production to extensive, mainly for home consumption. More than 60,000 inland farms with a total cultivated area of 49,752 ha exist. Over 95% of the total area consists of pond and paddy-field type culture systems. The remainder are dammed-up ditches, swampy areas and cage culture systems. Total Thai freshwater aquaculture production for the year 1989 was 92,000 t with the top seven species being *Tilapia nilotica, Pangasius sutchi, Puntius gonionotus, Trichogaster pectoralis, Clarias* spp, *Macrobrachium rosenbergii* and *Channa striatus*.

Out of the 27 species being reared, *Clarias* spp. (catfish) and *Channa striatus* (snakehead) are the most intensively cultured. The popularity of catfish farming has been due to its short culture cycle, rapid growth rate and extreme tolerance to inferior water quality and limited water supply. Snakehead, on the other hand, has drawn farmers' attention due to its high farm-gate value. According to official statistics, the production of catfish and snakehead together represents 18% of the total freshwater aquaculture production, with an annual value of US$ 21 million. Catfish production has merely doubled since 10 years ago and that of snakehead is little changed. The reason for the slow increase in total production is that catfish farmers have been facing difficulties such as disease and limited availability of, and high cost, fish feeds. Over-production also resulted in depressed prices, rendering catfish culture less profitable. For snakehead, a decline in wild seed due to overfishing, destruction of spawning grounds and the unreliable availability and rising price of trash fish (the main feed for snakehead) has hindered expansion.

Despite the problems facing further expansion of catfish and snakehead farming, production is quite considerable; according to official statistics an average of about 12,000 t of catfish and 4,000 t of snakehead have been produced annually over the past five years. 75-95% of the production was from the 25 provinces in the central part of Thailand. Culture techniques for these two species have not changed much in the last 10 years. From the nutritional point of view, most farmers still rely on fresh moist feeds prepared on-farm, the so called "farm-made aquafeeds". However, in the future, some modification in feed type and composition may be possible to suit the economic situation of these farms and to increase the farm-gate value of the fish. The purpose of the present paper is to review the current status of farm-made feed preparation and feeding strategies for catfish and snakehead, including an economic comparison between the use of farm-made feeds and commercial feeds.

CULTURE PRACTICES

Catfish

Clarias catfish belong to the family Clariidae and are found throughout Southeast Asia, the Indian subcontinent, and Africa. They are distinguished by the possession of an accessory air breathing organ, which enables them to exist for a long time out of water, or indefinitely in oxygen-poor waters (Bardach *et. al.* 1975). Culture of catfish is established in Thailand where *Clarias batrachus* and *C. macrocephalus* support a thriving domestic demand. Recently, hybrid catfish *(C. macrocephalus* x *C. gariepinus)* were introduced because of their faster growth rate than either of the local species.

Grow-out catfish ponds are 300-2,000 m^2 in area, and 1-1.5 m deep. Little preparation is required for firming the banks or pond bottom since available soils usually retain water adequately. Water supply is introduced and discharged by a simple canal and weir, or is pumped in and out by long-tail pumps. 2-3 cm fry are stocked in production ponds at rates ranging from 60-300/m^2, depending upon their availability and price as well as the intended production. Most operations achieve two crops per year. Each crop requires 3-5 months to produce marketable sized fish of 120-200 g. The incidence of mortality and the quantity and quality of feed used determines the fish yields obtained, which range from 3-12 kg/m^2 per crop.

Snakehead

Voracious carnivores, snakeheads are intensively cultured by farmers with access to capital. The operations of small farms are restricted by the very high capital requirement, largely due to high feed costs. Snakehead grow-out ponds, which are more or less similar to those used for growing catfish, range from 400-3,200 m^2, with a depth of 1.5-2 m. The rearing system depends on adequate, good quality water which is supplied by gravity or by daily water exchange through pumping. Wild fry, with an average length of 1.5-2.0 cm, are stocked by weight at the rate of 0.1-0.4 kg/m^2. Fry numbers are extremely variable ranging from 80-800 fry/m^2. The reason farmers stock at such a relatively high density is to compensate for high mortality rate caused by diseases and cannibalism. Thus, they expect that enough fry will survive to produce a profitable crop. The survival rate of snakehead in grow-out ponds is fairly low, from 7-24%. Culture time to a marketable size of 500-1,000 g ranges from 7-10 months, depending on stocking density and the feed used. Production from successful operations can be as high as 8-16 kg/m^2/crop.

Many snakehead farmers take less risk by avoiding the high mortality experienced during the first two months by stocking ponds with 10-15 g fingerlings or 50-70 g sub-adults at rates of 30-50 fish/m^2 and 20 fish/m^2, respectively. By stocking larger fish, the survival rate until harvesting may be as high as 85-90%, and the time to reach marketable size is reduced from 9-10 to 6-8 months.

NUTRIENT REQUIREMENTS

Clarias catfish

The success in rearing catfish to date, using a simple 8:2 trash fish: rice bran diet has created little interest in exploring the nutrient requirements of the fish *per se*. Rather, most studies have emphasized comparisons of trash fish based diets with mixed feeds, by assessing their quality and cost effectiveness. These studies have been based on the comparative growth performance of fish fed various ingredient combinations. Meeting specific nutrient requirements has not been an important criteria for feed formulation, since little was known about the nutrient requirements of catfish.

Thongutai (1969) reported the superior growth of catfish fed 9:1 trash fish:rice bran diet to those receiving Auburn No. 2 pelleted feed, in which fish meal and soybean meal were included as protein sources. Sitasit (1970) found that pelleted feed with 28% protein was not as good for catfish as trash fish/rice bran in a 6 month trial. Nutritional imbalance in the pelleted feed was suspected. Srisuwantach *et al.* (1981) reported similar findings when they compared the effects of trash fish and pelleted feed in catfish grow-out operations. However, a huge outbreak of disease in pond-reared clarias catfish in late 1982, which was traced to the use of poor quality trash fish, shifted attention to mixed feeds. Research to determine the basic nutrient requirements of catfish began in early 1980. Using a diet with a 2:1 fish meal:soybean meal ratio as the protein source, fed at graded levels to catfish for 45 days, Chotiyarnwong and Chuapoehuk (1981) found that the optimum protein requirement for grow-out *Clarias batrachus* was 30%. Boonyaratpalin (1988) estimated the protein requirement for catfish to be 35-40%, 25-35% and 28-32% for fry, grow-out and broodstock, respectively. Clarias catfish appear to require the same ten essential amino acids as other fish. However, quantitative amino acid requirements have not yet been determined. Requirements for total lipids and essential fatty acids are also not yet known. However, it has been suggested that the dietary inclusion of 6-8% fish oil or soybean oil would provide sufficient essential fatty acids for catfish (Boonyaratpalin 1988). Tacon (1981) implied that clarias catfish are probably able to equally utilize n-3 or n-6 fatty acids, as are channel catfish.

Little is known about the ability of catfish to utilize carbohydrates for energy Broken rice, a high carbohydrate ingredient, has long been used at high levels in combination with trash fish and rice bran for maturing catfish (Sitasit 1968), however. Based on the analysis of the partition between protein and non-protein energy retention, Luquet and Moreau (1990) showed that clarias catfish efficiently utilize non-protein energy from carbohydrates and hence can improve protein retention. Jantrarotai *et al.* (1992) reported that catfish can well utilize up to 49% carbohydrate in diets without any detrimental effects. However, reduction in growth was observed when dietary carbohydrate was increased to 54%.

The optimum energy to protein ratio for best growth of *Clarias batrachus* was reported by Tanomkiate (1984) to be 5.8 kcal/g of protein. This value was much lower than those reported for channel catfish, which ranged from 7.5-11 kcal/g of protein (Lovell and Prather 1973; Garling and Wilson 1976).

Vitamins and minerals play an important role in clarias catfish nutrition in the sense of increasing fry survivval. Taechajanta and Sitasit (1981) found that fry survival was 79% for a group fed mixed feed with vitamin and mineral premixes, in comparison to 35% survival rate in the group fed a diet with no premix. The individual vitamins required by catfish were later determined by Butthep *et al.* (1983) and Sitasit *et al.* (1984). The requirements and deficiency signs of some vitamins are summarized in Table 1. No study has been done to determine the mineral requirements of clarias catfish but the skull fractures often observed in catfish are believed to result from calcium deficiency (Sitasit 1970). Bone meal supplementation of the feed corrects the problem.

Table 1. Major vitamin deficiency syndromes and requirement for some water soluble vitamins by clarias catfish

Vitamin	Requirement (mg/kg feed)	Deficiency syndrome	References
Thiamine	N*		Sitasit *et al.* (1984)
Riboflavin	5	Cloudy lens	Sitasit *et al.* (1984)
Pyridoxine	5	Irritability, equilibrium loss	Sitasit *et al.* (1984)
Pantothenic acid	RD**	Clubbed gills, oedema, fin erosions	Butthep *et al.* (1983)
Folic acid	RD	Fading body colour, pale gills and liver	Butthep *et al.* (1983)
Nicotinic acid	RD	Spasms, whirling, equilibrium loss	Butthep *et al.* (1983)
Vitamin C	1,000	Scoliosis, darkened skin, fin erosion	Sitasit *et al.* (1984)

* N = not required in the diet
** RD = required but dose was not determined

Snakehead

The nutrient requirements for snakehead are much less well known than those for clarias catfish. Snakehead is a strict carnivore which is believed to have a high protein requirement. Boonyaratpalin (1980) reported that the protein requirement for snakehead fed on isocaloric diets (3.1 kcal/g) was 43% and 36% for snakehead fry and one month old fish, respectively. Effects of lipid levels on growth, survival and feed conversion of young snakehead were also evaluated. The minimum level of lipid, which could be n-3 or n-6, required for maximum growth, survival and feed conversion of the fish, appeared to be 6% (Boonyaratpalin 1981a). In another study, Boonyaratpalin (1981b) reported that pantothenic acid is a particularly important vitamin for snakehead growth and survival.

FEEDS OF CHOICE

Clarias catfish

Feeding of catfish in the past was heavily based on trash fish, rice bran and broken rice. The normal on-growing diet was 8:2 trash fish:rice bran. As a fattening diet, 8:1:1 trash fish:rice bran:broken rice was used. The mixtures were hand mixed and minced in a meat mincer to form a slurry. The slurry was hand distributed into ponds until it was all used up or until the fish no longer responded to the food.

Problems with trash fish based feeds

Although trash fish based diets were regarded as one of the best feeds for catfish, many limitations have forced farmers to modify the diets to some extent. As mentioned earlier, catfish farms in Thailand are mainly situated in the central plain region. Trash fish, therefore, had to be transported from port areas. Without proper storage, trash fish spoils easily and its quality can deteriorate during transportation. This can result in risks of transmitting disease. The national disease outbreak in pond-reared freshwater fish in late 1982 was partially blamed on unsanitary trash fish used for feed.

The rapid expansion of shrimp farming in Thailand in the past seven years has led to a high demand for shrimp feeds. Approximately 160,000 t of shrimp feeds are produced annually. This, in turn, results in high competition for raw trash fish for making fish meal, a major ingredient of shrimp feeds. Raw trash fish has, therefore, become more scarce for use in catfish grow-out. Limited availability of trash fish will further dictate an increase in price.

Demand for catfish is mainly for domestic consumption, which cannot be expanded very rapidly. However, when fast growing *Clarias* hybrids were recently introduced, catfish production increased beyond demand causing farm-gate value to fall. The use of trash fish diets will no longer be feasible as their cost may exceed the value of the fish produced.

Since 1980, many commercial catfish feeds have been available on the market. Although catfish feeds have many advantages over trash fish-rice bran diets they were rarely accepted by fish farmers at first. The high unit cost and low bulk volume of commercial feeds in comparison to moist trash fish-rice bran diets were the main concerns of the fish farmers. However, the situation has now completely changed. More and more catfish farmers, particularly those who have no access to trash fish or other cheap ingredients, rely on commercial feeds.

According to a survey in 1992, feeds of choice for intensive clarias catfish farming can be classified into 3 categories: commercial, fresh feeds and farm-made feeds.

Commercial feeds

Commercial feeds are widely used in catfish operations. According to industry sources, approximately 60,000 t of commercial catfish feeds are produced annually, which would place catfish production, allowing for that also produced through the use of farm-made feeds, at about 45,000 t, much higher than official statistics indicate. Catfish feeds are manufactured in several types according to fish size and their nutrient requirements (Table 2). The declared ingredients in commercial catfish feeds are fish meal, soybean meal, full fat soybean meal, corn, broken rice, fresh rice bran, coconut oil meal, vitamins and minerals. Commercial feeds possess good water stability and floating ability and allow easy management of feeding rate. The average price for commercial grower feeds is US$ 500/t. The feed conversion ratio (FCR) for commercial catfish feeds ranges from 1.5-2.0:1, depending upon stocking density and pond management. Taking an average FCR of 1.7:1, the feed cost is US$ 0.85/kg of catfish produced. Most catfish farmers admit that if the farm-gate value of catfish drops below US$ 1.00, the use of commercial feeds is not viable.

Table 2. Specification for clarias catfish feeds manufactured in Thailand (%)

Fish size/age	Feed form	Protein (min)	Lipid (min)	Fibre (max)	Moisture (max)
1 to 4 cm	powder	40	10	8	12
4 cm to 1 month	expanded	32	4	8	12
1 to 3 months	expanded	30	4	8	12
3 months to market	expanded	25	4	8	12
Maturation	expanded	20	3	8	12

Fresh feed materials

This refers to feed materials that can be used without the need to process them mechanically. These include poultry viscera, fish gills and viscera, kitchen refuse and bread. These feed materials can be directly dumped into ponds upon arrival at the farms. The nutritional values of two fresh feed materials are presented in Table 3. Poultry viscera have been regarded as one of the most effective feeds for grower catfish. On a dry matter basis, this ingredient contains 52.9% protein and 42.4% lipid (New 1987). Besides being nutritious, poultry viscera are highly palatable to catfish and naturally float on the water surface. Catfish fed on poultry viscera are fatty with shiny yellowish-skin, which is in high demand. The demand for poultry viscera by catfish farmers is quite competitive. Larger farms with capital resources gain access to consistent supplies of poultry viscera from chicken slaughterhouses year-round by winning contracts at auction. Small farms, which need smaller quantities of poultry viscera either collect them from poultry stalls in the market or buy from larger farms. The price of poultry viscera ranges from US$ 0.02-0.20/kg, depending upon the bargain struck. The feed conversion ratio for poultry viscera is approximately 4:1 which means that the feed cost could be less than US$ 0.80/kg of catfish produced.

Table 3. List of some fresh feed materials and their nutrient composition used to feed clarias catfish in Thailand

Feed materials	Protein (%)	Lipid (%)	NFE (%)	Ash (%)	Moisture (%)
Chicken viscera	13.9	11.2	-	1.2	73.7
Kitchen refuse	3.1	5.0	6.3	0.9	84.7

Farm-made feeds

New (1992) defined farm-made feeds as small-scale feed manufacture encompassing everything from simple hand-formed doughballs to small feed production units. Farm-made feed is an alternative to commercial feeds. The majority of farm-made feeds are prepared by catfish farmers who began their operations prior to the availability of commercial feeds in 1980. These farmers have good experience in processing trash fish-rice bran diets. There is no standard composition of current farm-made feeds for catfish. They may be composed of a single or many sources of raw materials processed by cooking or grinding to some extent. Table 4 shows the composition and nutritional value of some farm-made feeds once used or still in use by farmers. The diversity of the composition of farm-made feeds is dictated by the availability of the individual raw materials. The formulation in farm-made feeds may vary, depending on the amount of each raw material delivered to the farm each day.

From the farmers' point of view, the quality and proportion of feed is not so important as the amount required to feed the ponds. The success of certain farms in using farm-made feeds is usually determined by the availability and cost of raw materials.

Table 4. Composition and nutritional value of some farm-made feeds once used or still in use for rearing clarias catfish

Farm-made feeds	Protein (%)	Lipid (%)	NFE (%)	Ash (%)	Moisture (%)
Trash fish - rice bran (8:2)	15.6	5.6	10.9	5.4	62.5
Trash fish - broken rice (6:4)	13.8	2.6	30.6	3.3	49.7
Soybean cake - chicken bone - noodle waste	12.2	0.1	n.a.	n.a.	48.7
Compound feed	27.0	8.0	n.a.	n.a.	n.a

Serm farm in Minburi, on the outskirts of Bangkok, is an example of a successful clarias catfish farm which depends mainly on farm - made feeds. The farm is over 30 years old and has 10 ponds for catfish ranging from 0.16-1.6 ha. Moist feed is used in the farm. The ingredients are moist tofu waste, chicken bones, and instant noodle wastes or green mustard waste (Table 5). Chicken bones are bought from local chicken slaughterhouses. Instant noodle wastes and green mustard waste are purchased on a daily basis from the processing plants in the region. The supplies are brought to the farm and dumped on a wooden feed preparation platform (6m x 4m), which has been built over the feed processing machinery. The ingredients are roughly mixed by shovel on the platform and swept down into a first mincer through a hole in the platform. The minced material from the first mincer is then dropped into a second mincer, installed just below the first one, to be minced again. The mincer die is 2.5 cm thick with a 1.9 cm die-hole size. The two mincers are powered through two sprockets connected by a single drive chain to a 40 HP motor. When only one mincer is needed, as the fish grow bigger, the screw, knife and die plate of the lower mincer is dismantled. A polypropylene tube is tied to the mincer at the feed discharge point, thus making an extension which prevents the minced feed falling into the second mincer; it is also used to guide the feed into a container.

Table 5. Approximate formula of moist farm-made feed for clarias catfish* in Serm Farm, Minburi, Thailand

Feed ingredient	Cost (US$/t)	Inclusion rate (%)	Notes
Moist tofu waste	40	variable	Mixture of ingredients was minced in 40 HP mincer to form moist feed with 50% moisture content. Protein and lipid content was 12.2% and 1.4% respectively. Average feeding rate was 1.25 t/ha/day. FCR was 3.6:1
Chicken bone	200	variable	
Instant noodle waste**	120	variable	
Total	120-240	100	

* harvested as 200-300 g fish, 4-6 months after stocking. Total yield was 100 t in a 1.6 ha pond (62.5 t/ha). The fish were sold at a farm-gate value of US$ 0.88-1.00/kg

** sometimes substitued by green mustard waste

The nutritional content of the sampled feed from Serm Farm is presented in Table 5. The proportion of the ingredients in the feeds changes daily, depending on the ingredient supplies. The feeds are usually not processed until soybean cake, chicken bone and instant noodle wastes are all available. Lack of any of these ingredients will result in the postponement of feed processing. This results in deterioration in quality of the ingredient(s) which arrived early. The amount of feed processed is dictated by the amount of ingredients available, not the daily requirement of the fish. For this reason, more feed than is needed is sometimes processed. The excess amount of feed is piled up on a plastic sheet on the pond bank for use on the following day. The feed is distributed into the ponds by shovel. Feeding rate depends on visual observations, grow-out ponds being fed once per day at 15.00-17.00 hours. The fish are harvested by size grading about 4 months after the stocking of 2-3 cm fry at the rate of 31 fry/m^2. One of Mr. Serm's 1.6 ha hybrid clarias catfish ponds had a total yield of 100 t at the latest harvest (62.5 t/ha). According to his data, the calculated FCR for his farm-made feed was 3.6:1. He estimated unit cost of his feed to be US$ 0.12-0.24/kg.

Kantho Farm in Chachoengsao once depended on its own farm-made compound feed for *Clarias batrachus*. During 1985-1989 the 9.6 ha farm, with 10 ponds of catfish, was run by Mr. Um Kantho. With assistance from his son, Dr. Uthai Kantho, who is a professor in Animal Nutrition in Kasetsart University, Mr. Um was able to produce floating catfish pellets from simple farm machinery. The composition of the feed and its nutritional value are given in Table 6.

Table 6. Formula of a farm-made floating catfish feed used in Um farm, Chachoengsao, Thailand

Ingredients	Inclusion (%)
Corn	22
Cassava chips	10
Rice bran	15
Soybean meal	30
Fish meal (60% protein)	15
Leucaena	5
Dicalcium phosphate	1
Freshwater fish oil	2
Total	100*
Protein (%)	27
Cost (US$/t)	302
FCR	1.3:1

*plus vitamin mix, as directed

The ingredients were locally bought. Corn and cassava chips need to be finely ground in a cyclone hammer mill driven by a 20 HP motor. The hammer mill has a grinding capacity of 200 kg/hr. through a 1 mm screen. The other ingredients were claimed to be fine enough, needing no further grinding. Each ingredient was weighed separately and mixed in a horizontal mixer with a capacity of 100 kg/batch. After 15 minutes of mixing, 30% of water and the required amount of oil were added before another 10-15 minutes of mixing. The wet mixture was then extruded through a mincer (No. 52) powered by a 5 HP motor to form noodles. The noodles were immediately cut by a knife attached to the shaft of the mincer outside the machine. The length of the pellets could be selected by adjusting the distance between the die and the knife blade. The pellets were sun dried on a 400 m^2 concrete surface for one sunny day and bagged for further use.

Pellets made in this way were able to float for 5 minutes, which was sufficient for catfish to feed on them. Mr. Um showed that three factors are necessary for the production of floating pellets:

- low density ingredients such as cassava chips, *Leuceana* leaf meal and rice bran must be used in the formula;
- ingredients must be finely ground to produce rigid compaction of the pellets, slowing down the penetration of water;
- a thick die plate is important to allow mixtures to stay in the mincer longer before extrusion, resulting in partial gelatinization of some starchy materials. According to Mr. Um, the thickness of his die plate was 0.6 cm.

Drying the pellets was the process that bothered Mr. Um most, because it was not possible on rainy or cloudy days. To minimize the problem, he suggested that the concrete surface should not be too thick. A thin concrete surface heats faster and accelerates the drying of the pellets. A 0.4 cm thick concrete surface was best.

Ponds were stocked with 2 week-old *Clarias batrachus* at the rate of 63/m^2. The fish were fed 30% protein commercial feeds for the first week as they were too small to be fed on his own feed. Thereafter the fish were given farm-made feeds once a day, in the evening, for 6 months until they averaged 200 g. The feed conversion rate was 1.3:1. Survival rate was only 6% and the production was 0.75 kg/m^2 The poor survival rate was due to low pH in the pond water, resulting from acid soil.

This farm-made feed cost US$ 0.30/kg in 1987. At this price, Mr. Um said that even if the farm-gate value of clarias catfish dropped to US$ 0.80/kg, he would have still made a profit if higher survival rate had been acheved. By 1991, Mr. Um had developed his system into a large feed mill, manufacturing floating catfish feed and marine shrimp feed for commercial purposes. The business is now run by his family, with a great deal of success, while Mr. Um is still running the catfish farm; naturally he uses his own commercial catfish feed.

Snakehead

The quality of feeds for snakehead is more important than for catfish. Unlike clarias catfish, snakehead are fed mainly on fresh trash fish. Minced trash fish is the sole feed for snakehead fry and is fed *ad libitum*, the feeding rate being adjusted after observing fish behaviour during feeding. The fish are fed three times a day from fry to the fingerling stage. The inclusion of trash fish in snakehead feed is reduced to 80% by the addition of 20% of rice bran when the fish reach the fingerling size. In some farms a mixture of trash fish and cooked broken rice is used. Trash fish is normally delivered to the farms but some farmers who want to select better quality trash fish collect it directly from ports.

Trash fish and rice bran or cooked broken rice are mixed and minced through a meat mincer to form strands, which are placed on 2 x 0.5 m wooden feeding platforms suspended along both sides of wooden piers. The number of platforms vary with the size of the ponds. Usually a 0.16 ha pond needs 4-6 platforms for thorough feeding of snakehead. The feed conversion rates of such trash fish based feeds range from 3.1:1 to 8.1:1.

Chockchai Farm, one of the most successful snakehead farms in the Suphanburi province of Thailand has used farm-made feeds for more than 20 years. The feeds consist of trash fish, chicken offal (including heads, bones and legs) and rice bran. Trash fish are collected daily from the landing port. Chicken offal is supplied by a slaughterhouse with which the farm has a contract. Rice bran is locally available. The proportion of trash fish:chicken offal:rice bran is approximately 7:2:1. These ingredients are mixed on a concrete surface by shovel and then shovelled into an elevator leading to the mincers. The farm has two mincers, one powered by a 20 HP electric motor and the other driven by a fuel engine providing a back-up in times of power failure. Each mincer can produce up to 2 t/hr of feed. The mincers are operated for 6 hours daily since the farm also supplies feed to contract farmers in the vicinity.

Mr. Chockchai, the owner of the farm, provided information on one of his successful snakehead grow-out ponds. The 0.32 ha pond was stocked with 4,500 kg of 70 g snakehead (1.4 kg/m^2). The fish were fed twice a day. The average daily feeding rate for the entire growing period in this pond, which was 180 days, was approximately 748 kg. One critical factor mentioned by the farmer is that the trash fish used for feeding snakehead must be "daily fresh" otherwise the fish will not accept the mixed feed. The feed conversion rate was approximately 6:1 and the yield of this pond was nearly 27 t (84.2 t/ha).

Production and feed data for the catfish and snakehead grow-out operations described above is summarized in Table 7.

Table 7. Approximate formula of moist farm-made feed for snakehead* in Chockchai Farm, Suphanburi, Thailand.

Ingredients	Cost (US$/t)	Inclusion (%)	Note
Trash fish	220	70	FCR was 6:1
Chicken offal (heads, bones and legs)	120	20	
Rice bran	160	10	
Total	194	100	

* two rai (0.32 ha) pond was stocked with 4,500 kg of 70 g snakehead (14.1 t/ha), and reared to marketable size of 700 g in 6 months. The total yield was 26,950 kg (84.2 t/ha/crop)

ECONOMICS OF FARM-MADE FEEDS

The ultimate goal of aquaculture enterprises is to make profits. Profitability is determined by the difference between total revenue and total cost. Therefore, cost minimization is one way to achieve more profit. In many aquaculture operations today, feed costs account for over half of the total variable operating costs. Since commercial fish feeds are quite expensive, making good quality farm-made feeds might be more economical and result in more profit. The economic viability of farm-made feeds is place and time specific (New 1987). Farm-made feeds may be suitable for certain farms but not for others. Tacon (1990) mentioned the factors to be considered before choosing to make farm-made feeds as the following: market value of the farmed species, financial resources, farming traditions, time available for the farming activities, labour availability, availability of services, feed ingredient availability and cost, feeding habit, feeding behaviour and nutrient requirements, water quality requirements of fish, and food and feeding cost per unit of production per unit time.

The nutritional and physical qualities of farm-made aquafeeds may not match those of commercial feeds but the former can utilize a wide range of locally available ingredients. However, the needs of the species to be farmed require careful consideration. Clarias catfish, by its nature, consumes almost any feed material offered. This allows great opportunities for making feeds on-farm if the farmer has access to the raw materials. Serm Farm is a good example of the economic feasibility of farm-made catfish feeds. Its success is notable (Tables 8 and 9) and is due to the low cost of its moist farm-made feed which has a reasonably good nutritional value and gives a high fish yield. The feed cost in 1992 was US$ 749/t. At this feed price, Serm Farm would still make a gross profit from the operation (all other factors remaining the same as in Table 8) even if the farm-gate value of the fish produced fell to US$ 0.79/kg.

Making compound feeds generally requires more capital than making moist feeds, as a hammer mill and a mixer are required, in addition to a mincer. If conventional raw materials such as fish meal, soybean meal, rice bran, *etc.* are used, the unit cost of these ingredients and their inclusion rate in formulae will dictate the feed cost. Making compound feeds for clarias catfish is economically feasible in some areas, where swine and poultry rearing enterprises are clustered, because ingredients can be selected and formulation modified according to time specific availability and price. In Thailand, these areas include Nakorn Pathom, Chachoengsao, and their adjacent provinces. The Um Farm once produced compound feed for clarias catfish at US$ 302/t. As shown in Table 9, Um Farm was not economically viable in 1992. However if it was not for its acid soil, which resulted in very poor survival rates, the farm would have made a profit by using farm-made feed.

Snakehead farms still rely on trash fish in farm-made feeds, although the cost of trash fish is rising. This is because the farm-gate value of snakehead is reasonably high (US$ 1.92-2.56/kg), well above feeding costs. Cost of snakehead feed *per se* is, therefore, not so important as for clarias catfish, which have a low farm-gate value. Taking the Chockchai Farm as an example (Table 9) feed cost for snakehead is US$ 1.51/kg of fish produced. The farm thus spends a huge amount of money on feed alone, However the farm would still break even (under the conditions shown in Table 8) if the farm-gate value of snakehead did not fall below about US$ 1.53/kg.

It is interesting to note that the Mungkorn clarias farm, which relied solely on commercial feed, failed to make a gross profit (Table 9).

Table 8. Production data of some typical clarias catfish and snakehead farms in Thailand (1992)

Farm	Serm	Um	Mungkorn*	Chockchai
Feed type	moist	compound	commercial	moist
Cultured species	Clarias hybrid	*C. batrachus*	Clarias hybrid	Snakehead
Stocking density (pc/m^2)	31	63	42	1.4 (kg/m^3)
Stocking size (cm)	2-3	2-3	10	66.7(g)
Growing season (days)	180	180	90	180
Average production** (t/ha/day)	0.347	0.042***	0.389	0.39
Average feed input (t/ha/day)	1.25	0.54	0.55	2.34
Feed conversion rate	3.6:1	1.3:1	1.4:1	6:1

* the farm totally depended on commercial catfish feed

** average production $= \frac{\text{yield (harvest) - stocking weight}}{\text{number of days of rearing}}$

*** low production was due to acid soil

Table 9. Economic analysis of typical clarias catfish and snakehead farms in Thailand (1992)

Farm	**Serm**	**Um**	**Mungkorn**	**Chockchai**
Species	**Clarias hybrid**	***C.batrachus***	**Clarias hybrid**	**Snakehead**
Input costs				
Feed				
(US$/t. feed*)	208	302	490	252
(US$/t. fish)	749	393	701	1,512
(US$/ha)**	46,766	2,949	24,526	106,074
Seed				
(US$/ha)	2,480	5,040	8,400	22,400
Total costs				
(US$/ha)	49,246	7,989	32,926	128,474
Yield				
Harvest**				
(t/ha)	62.46	7.56	36.7	84.2
Farm-gate value***				
(US$/kg)	0.88	0.92	0.88	2.40
Revenue				
(US$/ha)	54,965	6,955	32,296	202,080
Gross profit (loss)				
(US$/ha)	5,719	(1,034)	(630)	73,606
Farm-gate value at break-even point				
(US$/kg)	0.79	1.06	0.90	1.53

* cost of farm-made feeds includes 30% mark-up on ingredient cost as suggested by New (1987). Feed costs are mid-1992 except for the Um Farm (1987)

** based on production (yield - stocking weight)

*** farm-gate values based on prices in July 1992, except the Um Farm (1987)

CONCLUSIONS

The majority of farm-made feeds for clarias catfish are moist, comprising a variety of unconventional raw materials. Their composition and nutritional value is therefore highly variable from place to place. Although making moist farm-made feeds is practical and cost-effective for certain farms, it may not be so for others since the raw materials used are unconventional. As long as snakehead rearing depends on trash fish it will be difficult to increase the overall production, since farming would not be possible where trash fish was not available. Production data for snakehead over the past decade support this prediction.

There is no doubt that farm-made feeds are economically viable for farms with access to inexpensive feed materials. However, the quality and quantity of moist farm-made feed cannot be easily controlled, This results in unpredictable and unmanageable fish production. Moist feeds disintegrate easily and therefore pollute the aquatic environment. The quality and quantity of farm-made feeds can be improved by processing conventional feed ingredients into a compound form. However, most of the compound feeds made are of the sinking-type, which are not attractive to farmers. Commercial feeds may be another alternative for catfish and snakehead farmers. However, the cost and palatability of each type of feed, the production management system in use, and the farm-gate value of fish need to be considered prior to making any decision.

LIST OF REFERENCES

Bardach J.E., J.H. Ryther, and W.O. McLarney. 1975. Aquaculture, the farming and husbandry of freshwater and marine organisms. Wiley-Interscience, New York, USA. 868 p.

Boonyaratpalin, M. 1980. Protein requirement of pla chon *(Ophicephalus striatus)*, p. 37-38. *In* National Inland Fisheries Institute Annual Report (in Thai). Department of Fisheries, Bangkok, Thailand.

Boonyaratpalin, M. 1981a. Lipid requirements of snakehead fingerlings. Progress report of the regional project RAS/76/003, Network of Aquaculture Centres in Asia, Bangkok, Thailand. 30 p.

Boonyaratalin, M. 1981 b. Vitamin requirements in snakehead diets. Progress report of the regional project RAS/76/003, Network of Aquaculture Centres in Asia, Bangkok, Thailand. 18 p.

Boonyaratpalin, M. 1988. Catfish feed. National Inland Fisheries Institute. Extension Paper No.528 (in Thai), Department of Fisheries, Bangkok, Thailand. 17 p.

Butthep, C., P. Sitasit, and M. Boonyaratpalin. 1983. Water-soluble vitamins essential for the growth of *Clarias*, p. 118-129. *In* C.Y. Cho, C.B. Cowey and T. Watanabe (eds.) Finfish nutrition in Asia, IDRC, Ottawa, Canada.

Chotiyarnwong, A., and W. Chuapoehuk. 1981. Protein requirements of *Clarias batrachus*. (in Thai). Faculty of Fisheries, Kasetsart University, Bangkok, Thailand. 5 p.

Garling, D.L. and R.P. Wilson. 1976. Optimal dietary protein to energy ratio for channel catfish fingerlings, *Ictalurus punctatus*. J. Nutr. 106:1368-1375.

Jantrarotai, W., P. Sitasit and S.Raichapakdee. 1992. Optimum dietary level of broken rice for growth and performance of hybrid walking catfish. National Inland Fisheries Institute, Technical Paper (in press). Department of Fisheries, Bangkok, Thailand.

Lovell, R.T. and E.E. Prather. 1973. Response of intensively fed catfish to diets containing various protein to energy rations. Proc. Ann. Conf. Southeast Assoc. Game and Fish Comm. 17:455-459.

Luquet, P. and Y. Moreau. 1990. Energy-protein management by some warm water finfishes, p. 751-756. *In* Advances in tropical aquaculture. Actes Colloq. No. 9. IFREMER, Paris, France.

New, M.B. 1987. Feed and feeding of fish and shrimp. ADCP/REP/87/26. FAO,Rome, Italy. 275 p.

New, M.B. 1992. Farm-made feeds. Lecture presented at the AADCP Component 4 Training Course, 19-30 October 1992. AADCP Working Paper WP/20. ASEAN-EEC Aquaculture Development and Coordination Programme, Bangkok, Thailand. 23 p.

Sitasit, P. 1968. Study on the chemical condition of water in pla duk pond at Samutprakarn Province, p. 107-139. *In* Annual Report of the Pond and Experimental Culture Section (in Thai). Department of Fisheries, Bangkok, Thailand.

Sitasit, P. 1970. The experiment of culture of *Clarias batrachus* in cages, p. 160-176. *In* Annual Report of the Pond and Experimental Culture Section (in Thai). Department of Fisheries, Bangkok, Thailand.

Sitasit, P., N. Unpresert and W. Jantrarotai. 1984. Vitamin requirement for growth and survival rate of *Clarias macrocephalus* fry. National Inland Fisheries Institute Paper No. 36. Department of Fisheries, Bangkok, Thailand. 31p.

Srisuwantach, V., R. Soungchomphan and P.Sitasit. 1981. Comparison of the effects of trash fish and pelleted diets in *Clarias* grow-out operations. Report on Programme for the Development of Pond Management Techniques and Disease Control (DOF-UNDP/FAO THA/75/012), Thailand. 21 p.

Tacon, A.J. 1990. Standard methods for the nutrition and feeding of farmed fish and shrimp. Argent Laboratories Press, Redmond, Washington, USA. 208 p.

Tacon, A.J. and M. Beveridge. 1981. Analysis of NIFI *Clarias* Diet No. 12. Report on Programme for the Development of Pond Mangement Techniques and Disease Control (DOFUNDP/FAO THA/75/012), Thailand, 2 p.

Taechajanta, K. and P. Sitasit. 1981. Assesment of a vitamin and mineral premix in an artificial feed for pla duk oui *(Clarias macrocephalus)* fry. Report on Programme for the Development of Pond Management Techniques and Disease Control (DOF-UNDP/ FAO THA/75/012), Thailand, 6 p.

Tanomkiate, K. 1984. Effect of feed with various protein to energy ratios on growth and survival of *Clarias batrachus*. Masters Thesis, Kasetsart University, Bangkok,Thailand. 60 p.

Thongutai, K. 1969. Study on the growth rate of pla duk in different stocking rate, p. 114-128. *In* Annual Report of the Pond and Experimental Culture Section (in Thai). Department of Fisheries, Bangkok, Thailand.

ON-FARM FEED PREPARATION AND FEEDING STRATEGIES FOR MARINE SHRIMP AND FRESHWATER PRAWNS

M. Boonyaratpalin[1] and M.B. New[2]

Feed Quality Control and Development Division
National Inland Fisheries Institute
Kasetsart University Campus, Bangkok 10900, Thailand

[2] ASEAN-EEC Aquaculture Development and Coordination Programme (AADCP), P.O. Box 1006, Kasetsart Post Office
Bangkok 10903, Thailand

BOONYARATPALIN, M. and M.B. NEW. 1993. On-farm feed preparation and feeding strategies for marine shrimp and freshwater prawns, p. 120-134. *In* M.B. New, A.G.J. Tacon and I. Csavas (eds.) Farm-made aquafeeds. Proceedings of the FAO/AADCP Regional Expert Consultation on Farm-Made Aquafeeds, 14-18 December 1992, Bangkok, Thailand. FAO-RAPA/AADCP, Bangkok, Thailand, 434 p.

INTRODUCTION

Our paper describes some examples of farm-made aquafeeds for marine shrimp and freshwater prawns currently in use in Thailand and shares some farmers' experiences on the topic. It is hoped that this information may be of use to other countries in the region.

In Thailand, aquafeeds were made for shrimp and prawns long before commercial feeds became available. Previously chicken feeds were used for freshwater prawn culture but these were unsatisfactory, not only nutritionally but also because of their poor water stability. Furthermore, neither the ingredients nor the compounded feed were of an appropriate particle size. Farm-made feed production for *Macrobrachium* started around 1980 during a period of rapid expansion initiated through the Chacheongsao Fisheries Station and assisted by an UNDP/FAO project. To this date, most freshwater prawn farms in Thailand use farm-made feeds.

By contrast, farm-made feed manufacture for marine shrimp in Thailand is very rare; complicated formulae, the necessity for careful sourcing and quality control of ingredients, and the need for sophisticated feed technology generally being stated as the reasons for this. However, it should be borne in mind that most Thai marine shrimp culture is intensive or super-intensive, productivity usually being five to ten times higher per unit area than that for freshwater prawn culture.

These higher risk systems consequently require dependable, complete diets. Making such feeds requires suitable equipment, high quality ingredients, efficient formulation and an experienced plant engineer. Thus high investment and qualified staff are essential.

Farm-made crustacean aquafeeds in Thailand include larval feeds, *Artemia* fattening feeds, and grow-out feeds for *Macrobrachium* and, to a lesser extent, *Penaeus* spp. Our paper describes the method of manufacture of some specific examples of these types of feeds but does not cover their use on farms in detail.

LARVAL FEEDS

Larval feeds are made for both freshwater prawns and marine shrimp and are confined to government hatcheries and small "backyard" hatcheries. Many of the more than 1,000 "backyard hatcheries" in Thailand are no longer small-scale. As they increase in size there is a tendency for them to convert to the use of commercial larval diets, because of the labour involved in making large quantities of larval feed on-farm. However, many hatcheries continue to make their own larval feed because the commercially available feeds are expensive, varying between US$ 24-80/kg, and their quality and texture is variable. In addition, they are still not complete feeds; in other words, live feeds remain an essential feature of hatcheries. For example, one commercial micro-encapsulated feed requires 15-30 *Artemia* nauplii/larva/day, in addition to large quantities of "artificial" feed (Tables 1 and 2).

Farm-made larval feeds in Thailand originate from the work of the Chacheongsao Fisheries Station, with UNDP/FAO assistance, in the late 70's and early 80's. Basically they consist of an egg custard, made from a mixture of whole eggs and the dried milk marketed for human babies (Figure 1). Sometimes a vitamin mix (three commercial shrimp vitamin premixes are marketed in Thailand) is added, though the other ingredients, such as mussel meat and fish flesh which used to be added a decade ago, have fallen out of fashion. Like the "artificial" feeds, farm-made larval feeds for crustacea are used in conjunction with *Artemia* nauplii, which are fed at 10-20 nauplii/ml/day. Feeding rates for egg custard are about 2-2.5 times as great as those recommended for the commercial encapsulated larval diets (Table 1) because they are moist. Since farm-made feeds require time to make, the first feed of the day consists of *Artemia*. Conversely the first feed of the day, when commercial larval feeds are used, is of dry feed (Table 3).

Table 1. Larval feeding schedule for black tiger shrimp (*Penaeus monodon*) according to water quantity

Stage	Artificial feed					Natural live feed**	
	Quantity (g/t of water/day)				Frequency (#/day)		
	microcapsule size			flake*		Diatoms; *Chaetoceros*, *Skeletonema* (cells/ml)	*Artemia* (nauplii per ml)
	No. 1	No. 2	No. 3				
Nauplius 6	-	-	-	-	1	40,000	-
Zoea 1	4	-	-	-	4-5	40,000	-
Zoea 2	4	-	-	-	4-5	40,000	-
Zoea 3	5	1	-	-	4-5	40,000	-
Mysis 1	-	5	-	-	4-5	20,000	0.5
Mysis 2	-	7	-	-	4-5		0.5
Mysis 3	-	7	3	-	4-5		0.5
P 1	-	-	12	-	4-5		0.5
P 2	-	-	12	-	4-5		1.0
P 3	-	-	12	2	4-5		1.0
P 4	-	-	12	2	4-5		1.0
P 5	-	-	12	4	4-5		1.0
P 6	-	-	2	6	4-5		1.5
P 7	-	-	1	8	4-5		1.5
P 8	-	-	1	10	4-5		1.5
P 9	-	-	0.5	12	4-5		1.5
P 10	-	-	0.5	14	4-5		1.5
P 11	-	-	0.5	16	4-5		1.5
P 12	-	-	0.5	18	4-5		2.0
↓				↓	↓	↓	↓
P 20	-	-	-	20	4-5	20,000	2.0

* after P10 stage, give flake feed at 15% of body weight/day

** constant or average level

Table 2. Feeding schedule for black tiger shrimp *(Penàeus monodon)* according to number of fry

Stage	Size of microcapsules	Feeding quantity	Daily feeding frequency
Zoea*	No. 1	40-50 g/1 million zoea/day	4-5
Mysis**	No. 2 No. 3	0.12 mg/1 mysis/day or 120 g/1 million mysis/day	4-5
Postlarva	Flake	15% of body weight per day and increase at 12% per day	4-5

* should also feed with diatoms

** should also feed with 15-30 *Artemia*/larva/day

Table 3. Feeding regime for commercial encapsulated and farm-made larval crustacean feeds

	Type of feed	Feeding times						
Farm-made	*Artemia* nauplii	06.00		11.30		16.00		22.00
	egg custard		09.00		13.30		18.00	
Commercial	*Artemia* nauplii		08.00					22.00
	dry feed	06.00		10.00	14.00		18.00	24.00

Farm-made crustacean larval feeds are manufactured in Thailand as follows. Sixty eggs are cracked into a bucket (both egg-whites and egg-yolks are used), and powdered full-fat baby milk added at a rate of 2 teaspoonfuls per egg. The mixture is then whisked until it thickens and rises (Figure 2). It is then poured into trays and steamed for 30 minutes. Usually the trays are layered above the boiling water. The resultant steamed egg custard is then forced through a series of sieves using high pressure water and retained in a plankton net; sieve sizes vary according to the size of larvae to be fed (Table 4). The "egg custard" feed is then added to the larval tank (Figure 3).

Figure 1. Basic ingredients for farm-made crustacean larval feed

Figure 2. Farm-made egg custard feed for larval crustacea, after cooking and before sieving

Table 4. Size of egg custard particles fed to marine shrimp larvae and postlarvae

Stage	Particle size (microns)	Mesh size	*Artemia* nauplii (per ml)
Mysis 3 - P4	177	80	20
P5 - P13	350	42	↓
P13 - P20	500	35	10

Figure 3. Farm-made crustacean larval feed at feeding time

ARTEMIA CULTURING FEED

Another type of farm-made larval feed used in Thailand is for culturing *Artemia* nauplii to an adult size. These *Artemia* are used as a supplement to the normal feeding schedule for *Penaeus monodon*, from post-larval stage 12 (P12) onwards, and also for 1 cm (one month-old) sea bass and grouper larvae. The major ingredient used is rice bran; 4 kg is added to 10 litres of sea water with 2 kg of sea salt and blended for 10 minutes. The blended mixture is then poured into buckets so as to allow the larger particles to settle. After being filtered through a nylon net it is squeezed by hand, washed and squeezed again to obtain as much filtrate as possible. The solid material is discarded, while the filtrate is filtered again through a finer (60 micron) nylon mesh. The concentrated *Artemia* feed is then ready for use or can be stored under refrigeration (0.5°C) for later use. Before use, 20 litres of the concentrated feed is diluted to 500 litres with water. An enrichment mixture is then prepared, which consists of 1 egg yolk, and 50 ml of "EPA oil" (eicosapentaenoic acid; EPA oil is available in the local markets), diluted to 1 litre with water. One litre of the enrichment mixture is then added to each 500 litres of *Artemia* fattening feed. The complete feed is added to *Artemia* rearing tanks by hand, or automatically through pipes equipped with a submersible pump and a timer; the feed is applied every 2-3 hours.

GROW-OUT FEEDS FOR MARINE SHRIMP

As noted earlier, few Thai marine shrimp farms use farm-made aquafeeds now because they are characteristically intensive or super-intensive, requiring reliable, nutritionally complete, and water stable feeds. However, two farms which have made their own feeds are described here.

The first farm, which is situated in Rayong, previously used mincing machinery but does not do so now because the quantity of feed was thought to be too large (however, very large through-put mincers are available in China and Japan). The 100 kg/hr mincer used was not the limiting factor; the main problems encountered were with drying, and with the use and regular supply of trash fish. A typical formula used when the extrusion process was employed, which used trash fish as a binder and protein source and had a high moisture content (35-40 %), is given in Table 5.

The first version of their new feed plant was relatively simple and included a hammer mill, a pelleter and a drier. Basically, ingredients were mixed, elevated to a triple conditioner heated by heat lamps rather than steam, pelleted and dried in a rotary drier also heated by lamps. The plant, capable of producing 300 kg/hr, cost approximately US$ 15,000. A typical formulation for this first pelleting plant is presented in Table 6. No trash fish was used in the formulation but cooked rice was employed, which resulted in a moisture content of 23-25 %.

Table 5. Formula for extruded* shrimp feed made in Rayong

Ingredient	Inclusion rate (kg)	(%)
Good quality trash fish	20.0	23.8
Fish meal	20.0	23.8
Shrimp shell meal	10.0	11.9
Squid liver meal	4.0	4.8
Groundnut meal	6.0	7.1
Soybean meal	10.0	11.9
Wheat flour	5.0	5.9
Rice bran	8.0	9.5
Vitamin and mineral mix	0.5	0.6
Fish oil	0.5	0.6
Vitamin C	0.1	0.1
Total	84.1	100.0

* extrusion through mincer

Table 6. Shrimp feed formula used in the first pelleting plant at Rayong

Ingredient	Inclusion rate (%)
Fish meal	33.0
Shrimp shell meal	10.0
Squid meal	4.0
Soybean meal	16.0
Wheat gluten	5.0
Broken rice	6.0*
Brown sugar	0.5
Rice bran	21.9
Mineral mix	1.0
Vitamin mix	0.5
Vitamin C	0.1
Fish oil	0.5
Eggs**	1.5
Total	100.0

* approximately 20 kg after boiling
** 30 whole eggs, without shell

Later, a new 500 kg/hr production plant was installed which was modified from a chicken feed plant, adding an electrically heated rotary drier. Dry ingredients were ground, transferred to a horizontal mixer and then by screw elevator to a vertical mixer. The mixture then passed through a conditioner/feeder to a pelleter. Following pelleting the feed passed through two driers, one electrically heated and the second a vertical cooler/drier, before packaging. A typical example of the formulation used in the second pelleting plant is given in Table 7. However, there were further problems. The lack of trash fish or boiled rice in the formulation, caused by the necessity to reduce moisture to 14-16%, and the inadequate conditioner (which was the weak point in the whole system), meant that a mixture of three binders had to be used to obtain the necessary water stability; the binders used were guar gum (1%), aqua-gel (0.5%) and wheat gluten (7%). However, poor conditioning resulted in problems with the pelleter and this plant is now shut down.

Table 7. Shrimp feed formula used in the second (final) pelleting plant at Rayong

Ingredient	Inclusion rate (kg)
Fish meal	33.0
Shrimp head meal	10.0
Squid meal	4.0
Soybean meal	16.0
Wheat gluten	7.0
Rice bran	23.4
Mineral mix	2.5
Vitamin mix	0.5
Vitamin C	0.1
Fish oil	0.5
Eggs*	1.5
Guar gum	1.0
Aqua-gel	0.5
Total	100.0**

* 30 whole eggs, without shell

** 3 litres of water were also added to this formula during manufacturing

The experience of the Rayong plant neatly demonstrates the problems that can occur when a small production unit is scaled up to make complete feeds for intensive culture. Design faults quickly caused plant breakdown and the company could not afford to hire a good nutritionist or a good plant engineer. Keeping the plant simple would have been more satisfactory. In retrospect, it would have been perfectly feasible to continue to extrude the feed through a mincer rather than a pelleter. Often it is thought that this technique, which enables the use of moist ingredients such as trash fish, would not be feasible because of the large area required for sun-drying. However, as the next example shows, mincing can be followed by steam drying to solve this problem.

Our second example of farm-made feed manufacture for marine shrimp is from a small cooperative of poor farmers in Surat Thani. These farmers, unlike those which were supplied by the Rayong plant, do not rear their shrimp very intensively. Their feed unit was started in 1986 and employs ten workers producing 1-5 mt of shrimp feed per day. Three types of shrimp feeds are produced with 40-42% protein for use in the first month (from P15 onwards), 38% for the second and third months and 30% for use after that. No formulae were provided but the ingredients used include fish meal, soybean meal, broken rice, rice bran, shrimp head and shell meal, squid meal, yeast, fish liver oil, squid liver oil, a premix, dicalcium phosphate, and a binder.

The small size of the mixer used by these farmers necessitates ten separate mixing cycles of 100 kg to produce each one-ton batch. Thus the workers work in two shifts, one in the morning and one in the afternoon. The feed manufacturing equipment employed consists of an ingredient grinder, a mixer (100 kg capacity), a meat grinder (used for extrusion), a small boiler (used for boiling the broken rice included in the formulation to increase its digestibility), and a drier. Wood is used to fuel the main boiler, which generates steam for the steam jacketed three-phase drier. Ingredients are weighed on a simple scale (shrimp meal and soybean meal being first re-ground) and then added to the mixer. Major ingredients are inserted first, except broken rice, which is pre-boiled. Minor ingredients, such as squid meal, yeast, vitamin premix and binder are premixed by hand before being added to the mixer. Liquid ingredients (squid or fish liver oil) are then mixed in, and finally the boiled broken rice is added. After the last ingredient is added, mixing is continued for 15 minutes. The material is then discharged from the mixer before transfer to the grinder where the mixture is extruded in spaghetti or noodle-like form. The moist pellets are then elevated to the first two stages of the steam-jacketed drier and the semi-dried material then elevated again to the third-stage drier. Finally the dried material is discharged and the pellets left on the floor overnight to cool before being bagged for transport to the members of the cooperative. Though effective, this production unit could be improved by adding a cutter after the extruder and a cooler after the drier.

GROW-OUT FEEDS FOR FRESHWATER PRAWNS

Farm-made aquafeeds for *Macrobrachium* are common in Thailand. Compared to marine shrimp feeds, freshwater prawn feed formulations are simple because the culture system is comparatively semi-intensive. Where marine shrimp culture is still semi-intensive, as for example in some other Asian countries, similar farm-made aquafeeds are feasible.

Freshwater prawn farms which make their own feeds are common in the province of Suphanburi, north of Bangkok. Many farms also have their own inland hatcheries, thus presenting simple examples of vertical integration. At the first farm we are describing, ingredients are weighed and mixed together on the concrete floor. A "wok" is used to boil the broken rice. Two examples of the feed formulations used on this farm are given in Table 8. Each ingredient is carefully added to the apex of a heap of ingredients on the floor to ensure adequate initial mixing. The mixture is then shovelled into bowls for transfer to the mincer for extrusion (and further mixing) through a coarse die plate. The material is collected in deep baskets where it remains while a finer die plate is fitted. The result of the second extrusion is collected in shallow baskets (Figure 4), which are easier to carry to the sun-drying area. The moist pellets are then spread thinly on a concrete floor and turned regularly with a simple rake.

Other *Macrobrachium* farms in Suphanburi use a similar process. Our second example has been manufacturing feed for the past six years. The formulation is shown in Table 9. Feed sufficient for two or three days is made at one time, depending on the availability of trash fish. The mixed ingredients are extruded twice (coarse, followed by fine die plate) through a mincer (powered by a converted car engine) and then sun-dried. The feed is extruded in the morning (07.00-08.00) and, if sunshine is bright, drying is completed within one day on a concrete surface (Figure 5). If necessary, feed is collected and put out to dry again on a second day. The water stability of the pellet is reported to be up to two days. Feeding rate depends on visual observations; grow-out ponds being fed once per day (at 15.00-17.00) and nursery ponds twice per day. Feed is broadcast by hand from the pond bank and no fertilizers are applied. The average farm gate value of the crop in June 1992 was reported to be US$ 4,600/t.

Table 8. Two formulae for farm-made freshwater prawn feeds

Ingredient	Formula #1		Formula #2	
	(kg)	(%)	(kg)	(%)
Trash fish	100.0	29.61	100.0	28.35
Soybean meal	40.0	11.84	40.0	11.34
Fish meal	20.0	5.92	10.0	2.84
Corn meal	80.0	23.70	80.0	22.68
Di-calcium phosphate	2.0	0.59	2.0	0.57
Antibiotic (oxytetracycline)	0.2	0.06	-	-
Vitamin and mineral mix	0.5	0.15	0.5	0.14
Vitamin C	-	-	0.2	0.06
Broken rice (boiled)	30.0*	8.88	30.0*	8.51
Chicken layers feed	50.0	14.81	60.0	17.01
Piglet feed	15.0	4.44	15.0	4.25
Shrimp shell meal	-	-	15.0	4.25
Total (approximate)	337.7	100.00	352.7	100.00

* weight before boiling

Figure 4. Extrusion of farm-made freshwater prawn feed

Table 9. Approximate formulation for farm-made *Macrobrachium* feed* in Mr. Prasit Kothon's farm, Suphanburi, Thailand

Ingredient	Cost (US$/t)	Inclusion rate (kg)	(%)
Trash fish	220	100	44.2
Chicken feed	240	60	26.6
Broken rice	168	30	13.3
Fish meal	720	20	8.8
Piglet concentrate**	427	15	6.6
Premix***	4,400	1	0.5
Total****	480	226	100.0

* lipid not less than 4; protein not less than 25; fibre not more than 8%

** composition not known

*** contains vitamins A, D, C, E and an unspecified antibiotic (oxytetracycline ?). Available locally and made specifically for *Macrobrachium*. Manufacturer suggests inclusion in the diet at 0.5-1.0%

**** it was said that no water was added; trash fish provides sufficient moisture

Figure 5. Sun-drying a 450 kg batch of freshwater prawn feed

CONCLUSIONS

Examples of a farm-made crustacean larval feed, an *Artemia* fattening feed, and a number of grow-out feeds for marine shrimp and freshwater prawns have been provided and the manufacturing process described*. Experience in Thailand indicates that, as the hatchery grows large or the (marine shrimp) farming system becomes intensive, operators move towards the purchase of commercial feeds, mainly because the large-scale production of farm-made crustacean feeds is time-consuming and labour intensive.

In the case of small hatcheries, simple egg custards are still popular, both for marine shrimp and freshwater prawns, and this is likely to remain so until commercial larval feeds no longer require supplementation with *Artemia* nauplii or other live feeds. These egg custard feeds are readily applicable outside Thailand. One important fact is that farm-made feeds for crustacean larvae in Thailand include <u>both</u> the yolks and the whites of the chicken eggs incorporated. Many fish larval diets exclude the egg white, which is a mistake in our view. Egg white is a valuable, high-protein ingredient, which will not foul the water in larval tanks if the diet is prepared (homogenized) correctly. In fact the egg white helps to encapsulate the egg-yolk to make a water stable larval feed.

Farm-made feeds for marine shrimp culture have not really proved feasible when the culture system is intensive or super-intensive. Feeds for this type of culture, which must supply all the nutritional requirements of the shrimp, must be of consistently high quality to produce optimum growth rates, food conversion efficiency, and survival. Intensive shrimp culture is a high-risk business and poor or inconsistent quality feeds are anathema. Good quality shrimp feeds require high quality fish meal. Feed plants only making shrimp feeds cannot utilize poor quality batches of fish meal in feeds for other farmed animals. The production of reliable feeds for intensive systems of shrimp culture requires the installation of good production equipment in a well-designed plant, a reliable supply of high quality ingredients, careful least-cost formulation to ensure the economic provision of all known nutritional requirements, a qualified plant engineer and well trained production staff.

However, the use of farm-made feeds in less intensive crustacean farms remains a viable option. Even large farms or cooperatives of a number of smaller farms could overcome the space requirements for sun-drying extruded pellets by installing steam-fired (or electric) driers. For the small farm sun-drying on a

---

* Editors' note: when this paper was presented, over 70 slides were shown, which illustrated the manufacturing procedures in detail.

clean concrete surface is feasible, even in the rainy season. The common and long-term use of farm-made feeds by *Macrobrachium* farmers in Thailand demonstrates that they are an acceptable, practical and profitable option. This applies not only to freshwater prawn culture but also to marine shrimp farms where supplemental, rather than complete, feeds are appropriate to the culture system used. The systems which we have described can, with minor adaptation, also be used for producing farm-made fish feeds.

SELECTING EQUIPMENT FOR PRODUCING FARM-MADE AQUAFEEDS

John Wood

Grain and Food Processing Section
Natural Resources Institute
Central Ave., Chatham Maritime, Kent ME4 4TB, U.K.

WOOD, J. 1993. Selecting equipment for producing farm-made aquafeeds, p. 135-147. *In* M.B. New, A.G.J. Tacon and I. Csavas (eds.) Farm-made aquafeeds. Proceedings of the FAO/AADCP Regional Expert Consultation on Farm-Made Aquafeeds, 14-18 December 1992, Bangkok, Thailand. FAO-RAPA/AADCP, Bangkok, Thailand, 434 p.

INTRODUCTION

When addressing the subject of preparing feeds for fish it is important to recognise that we are in fact seeking to prepare a foodstuff with specific nutritional and physical properties to meet the differing feeding habits of a range of aquatic consumers. The purpose of the feed manufacturing process is therefore to prepare a food which, as far as possible, meets the gastronomic habits of the target consumer. The need to prepare feeds for slow and fast feeders, water surface, mid water or bottom feeding species has been well recognised.

If a farmer is to be successful in feed preparation it is important that he examines the options available to him. In practice, most farmers have come to believe that pelleted feed is the most desirable since this is the feed form which has resulted in high yields on what are perceived to be financially successful large commercial fish farms. This is an understandable conclusion to make but its validity on a nutritional and practical basis still requires solid confirmation for semi-intensive culture. Nevertheless, in this paper we will consider the factors affecting the selection of equipment, principally for making feeds of high water stability in pellet or noodle form, since these types of feed potentially enable the maximum number of consumers to be fed at any one time. This paper will outline the principles behind equipment selection, and in particular comment on the properties of feed raw materials which will influence such decisions.

WHAT DO FARMERS WANT?

In seeking to meet the requirements of the fish we must also examine the requirements of the farmer. What are his objectives and are they achievable? The two may not necessarily be compatible in all cases. Annex 1 summarises the primary objectives of the farmer concerning feed preparation, but the degree to which they are achievable will vary from farm to farm, region to region and country to country. There is no single solution that will be practical and cost effective for all farm-made aquafeeds. We are assuming that the farmer already has a feeding strategy but wishes to improve upon it.

FACTORS INFLUENCING THE SELECTION AND INSTALLATION OF FEED PROCESSING EQUIPMENT

The supplementary feeding of cultured fish has become an established practice in many parts of the world and numerous designs and sizes of machines have been used for farm and commercial feed manufacture. Some farmers have made wise choices while others have regretted expenditure on equipment which has failed to perform adequately with the raw materials available, or was of an inappropriate capacity.

If a farmer is starting a new venture there are important factors to be evaluated before any selection of specific equipment should be considered. The response to their evaluation will then form the framework for determining the machinery requirements to meet the feed production objectives within the financial resources available (see Annex 2).

This evaluation of interacting factors must be conducted with an understanding of the physical and functional properties of the raw materials available. These aspects will be discussed in more detail later in this paper.

PROCESSING EQUIPMENT OPTIONS

There are many and various options for processing equipment for aquafeed processing, as can be seen from the practices of aquaculturists in any locality. Processing equipment options include electric or petrol/diesel engine driven machinery, and also the use of hands and feet. The most common processing operations can be summarised as:

- raw material size reduction;
- raw material blending;
- feed forming;
- feed drying.

The range of equipment options within each processing operation are summarised in Table 1. It is not appropriate to describe these machinery options in detail since almost all are well known and specific details can be obtained from manufacturers.

Table 1. Options for feed processing equipment

Process operation	Equipment		Raw material/product
Size reduction	Mortar and pestle		dry or moist grinding or blending
	Mincer		wet materials *e.g.* trash fish/offals
	Hammer mill		coarse-fine dry materials
	Plate mill		coarse-fine dry materials
Blending	Physical	Hand	for small quantities variable efficiency
		Feet	
	Mechanical mixers	Bowl	moist doughs
		Horizontal	dry powders or moist crumbs
		Vertical	dry powders
Forming	Hand		dough ball
	Mincer		moist noodles
	Pelleter		dry pellets
	Cooker extruder		semi-moist/dry pellets or noodles
Drying	Solar		variable efficiency
	Mechanical		controlled drying

The choice of equipment will be limited firstly by financial resources, secondly by the desired water stability of the feed (and thus raw materials availability), and thirdly by the required scale of feed manufacture. However, on the assumption that finance is not limiting and all forms of machinery are potentially available, then the factor which will govern machinery choice is the spectrum of raw materials on offer for feed formulation and manufacture.

FUNCTIONAL PROPERTIES OF FEED RAW MATERIALS

Commercial fish feed manufacture is predominantly associated with the processing of dry ingredients and the manufacture of a dry product. This is not necessarily the case for farm feed manufacture. Commercial processors require dry products for long term storage and transport. Farm feeds can utilise local raw materials which may be:

- high in moisture and of short shelf life;
- of insufficient quantity to become commercially viable;
- grown on the farm or be a byproduct of local agriculture;
- available during certain seasons only;
- having functional properties which have not been damaged through industrial pre-treatment.

The question which must then be asked is: "In what way can the nutritional and physical characteristics of the raw materials available meet the desired nutritional requirements of the fish to be farmed, and be in a form which will stimulate feed intake and be water stable throughout the feeding period?"

To enable this question to be addressed we should consider the general functional properties.of feed nutrients and the effect which moist heat is likely to have on them. The functional properties of raw materials, or more specifically of feed nutrients, are those which affect the ability of feed materials to:

- cross bond with each other;
- alter their viscosity characteristics;
- change from granular to plastic consistency;
- release bound moisture;
- induce water stability.

The effect of moist heat on the functional properties of feed raw materials is important since this is the kind of treatment which many are subjected to during processing before they, or their byproducts, are actually considered as raw materials for fish feed (or other animal feed).

The important functional properties of the major nutrients are summarised in Table 2. These properties are, of necessity, expressed in general terms since there are differences in specific properties depending upon the source of the nutrient. From Table 2 it is evident that the nutrients with the most important functional properties are the proteins and the starches. In most circumstances, both nutrients will also be present in high proportion in the mixed diet and therefore have the potential to significantly alter its properties. Let us examine the functional properties of these two nutrients in a little more detail.

Table 2. Functional properties of feed nutrients

Nutrient	Normal state	Effect of moist heat	Change in functional property
Water	liquid	vapourization	energy transfer
Proteins	colloidal fibrous globular viscous	denaturing	soluble/hydratable to insoluble
Starches	inert granules	gelatinisation	insoluble granule to soluble gel
Fats	liquid or solid	some cross bonding with starch amylose	difficult to extract with organic solvents
Fibre	fibrous	minimal	minimal
Sugars	soluble solid or liquid	reaction with lysine caramelization	minimal
Minerals	soluble or insoluble solid	minimal	minimal
Vitamins	soluble or insoluble solids	some heat labile	minimal

Starches

In their native state starches, whether from cereal or root crops are found in the form of starch granules which are essentially inert when placed in cold water. They resist water absorption, and there is minimal adhesion between granules. In this form starches are also of low digestibility to aquaspecies, and have minimal properties for binding other feed components. It is only when starches are heated in water, causing the granules to rupture and gelatinise forming viscous pastes and gels, that starch has desirable properties for binding feeds.

However, since the starch has also become more soluble during the gelatinisation process (a property which is retained after rapid starch drying), feeds bound with gelatinised starch alone will disintegrate as the feed hydrates and continues to absorb water, although the degree of binding is considerably better than when feeds contain no gelatinised starch.

Proteins

In their native or natural form, proteins associated with plants tend to be "globular" in shape whereas animal proteins are more characteristically "fibrous" in form. Apart from those proteins which are totally insoluble, such as those comprising wool, silk and hair, proteins are soluble in water, salt solutions or mild alkaline or acid solutions. When subjected to heating, proteins tend to coagulate or denature. This is a process which is well recognised during the cooking of egg albumin, when the raw soluble protein is converted to an irreversibly insoluble protein gel.

PROCESSING OPTIONS

To illustrate the potential interactions of raw materials let us examine what processing options could be available to a farmer for fish feed production. For example, let us consider the use of three raw materials potentially available in many developing countries *i.e.* trash fish, sun dried cassava and full fat soya beans. These raw materials may be described in the following way:

- trash fish: hydrated, fibrous protein, with some fat associated, but no starch;
- cassava: non-hydrated starch granules with little protein or fat;
- full fat soya beans: non-hydrated globular protein with high association with fat and low in starch.

We will assume that each raw material will be used at the same dry matter level in each feed, and that the blend of raw materials will meet the desired nutritional requirements of the culture system being operated. The oil present in the soya beans is a required nutrient for the diet.

Option 1

Let us assume a process whereby all three raw materials are treated individually as industrial agro-products for processing and storage. They will then be blended to form a feed of the desired nutrient specification.

RAW MATERIAL	PROCESS OPTION	EFFECT OF PROCESS ON FEED BINDING
Trash fish (soluble protein)	Hot air dried to fish meal (coagulated protein) (denatured protein)	Loss of ability to cross link with other proteins
Cassava (inert granules)	No further processing (inert granules)	Inert granules with no gelling characteristics
Soya beans (hydratable protein)	Oil expelling; toasting of soya meal (coagulated protein)	Loss of ability to cross link with other proteins

Result: Three separate raw materials with no inherent ability to bond with each other. These materials will require a binder and careful steam conditioning to partially gelatinise the cassava before any degree of binding between particles can be obtained.

Option 2

RAW MATERIAL	PROCESS OPTION	EFFECT OF PROCESS ON FEED BINDING
Trash fish (soluble protein)	Sun dried to produce fishmeal (coagulated protein) (partially denatured)	Partial loss of ability to cross link with other proteins
Cassava (inert granules)	Cook in water (gelatinised starch)	Viscous base for aiding feed binding
Soya beans (hydratable protein)	Oil expelling; toasting of soya meal (coagulated protein)	Loss of ability to cross link with other proteins

Result: Improved stability of the feed due to partial hydration of some fish proteins and viscosity of gelatinised starch. Heat denatured soya proteins are inert and add little to feed stability. *(Note: Even slow sun drying results in partial denaturation of fish proteins such that rehydration does not result in the reformation of colloidal protein material).*

Option 3

RAW MATERIAL	PROCESS OPTION	EFFECT OF PROCESS ON FEED BINDING
Trash fish (soluble protein) Cassava (inert granules) Soya beans (hydratable protein)	Blend the materials together and co-extrude while heating	Formation of intermeshing matrix of proteins and starch which are respectively denatured and gelatinised simultaneously

Result: This process will give the most effective bonding between the feed components, and thus the maximum feed stability. The process is in effect that which occurs during the process of cooker extrusion, and is the reason why this process has become so popular amongst commercial fish feed manufacturers. Furthermore, when using full fat soya as a raw material, the process enables the destruction of the trypsin inhibitors which would otherwise significantly depress protein digestion and assimilation by the fish.

A further advantage of this form of processing is that it can incorporate wet fish as an ingredient into the feed mixture without the need for pre-drying, or the need to extract the oil from soya beans prior to their use.

ARE THESE METHODS APPROPRIATE FOR FARM-SCALE AQUAFEED MANUFACTURE?

The foregoing options illustrate the desirability for the controlled processing of protein- and starch-containing raw materials by methods which will maximise the effects of their functional properties. Feed materials must be examined in terms of their physical or functional properties and not solely as sources of nutrients.

On many fish farms, feeds are being prepared which take into account the factors discussed above. The benefits of cooking starchy bases such as cassava or rice have been recognised, both for developing the viscous gelatinisation of starch and the resulting improvement in digestibility.

However, the stability of formulated feeds which include cooked starches may also be due to a degree of starch retrogradation during cooling and sun drying. During retrogradation, starch gels tend to lose their consistency as starch molecules and form into crystalline structures (for example, starch retrogradation is often associated with the staling of bread). Once a starch has retrograded it is often more difficult to solubilise, and is less readily hydrolysed by enzymes. From a fish feed perspective therefore, starch retrogradation in association with a protein matrix may enhance feed resistance to disintegration in water. The negative effect will be reduced starch digestibility.

Where farmers use trash fish as their base, they are introducing a valuable source of both nutrients and functional characteristics to the feed. For example: a Thai farmer produced his own feed for freshwater prawns from the following ingredients: wet trash fish (44.4%), fish meal (13.3%), rice bran (6.8%), soya meal (8.9%), poultry feed (13.3%), and broken rice (13.3%). The rice was cooked to a thick paste and then mixed with the ground dry ingredients and the minced trash fish. The moist dough was then formed into strands through a mincer die plate, and the product sun dried on a concrete pad. The dried feed.

which had a water stability of more than 24 hours in static water before disintegration took place, was bound with gelatinised and partially retrograded starch and by the matrix of partially denatured fish proteins produced during sun drying. Of particular interest is the addition of poultry feed. The farmer commented that it had been added to the feed mix to overcome the tendency of the fibrous rice bran to break the structure of the pellet during drying. Practically and nutritionally, the farmer may have had the same success by removing the rice bran and poultry feed from the diet and replacing them by soya meal and broken rice.

The process used in the preparation of this feed was quite labour intensive but, since the farmer was able to obtain most of his dry materials in a pre-ground form, his machinery requirements were reduced to a mincer and a simple heated bowl for cooking the rice. The mincer also served as the device for shaping the feed dough into long strands for sun drying prior to crumbling to feeding size pellets, and for storage.

This situation is, however, not typical of fishfarms which are situated away from coastal regions, where supplies of trash fish are non-existent. The only ingredients available may be the byproducts of other agro-industries which have heat treated the materials and denatured the proteins during processing, *e.g.* in oilseed cakes. Under these circumstances farmers may be able to obtain sufficient raw materials to meet the nutrient requirement of the culture system being operated, but have few materials which will help in developing any degree of water stability. The most sophisticated machinery will be of little use to the farmer unless he is able to obtain additional raw materials to induce binding, while maintaining the nutritional integrity of the desired feed.

ENVIRONMENTAL ASPECTS

As aquaculture expands throughout the world there is a requirement that farmers use their water resources wisely. This applies not only for the protection of their own stock, but for the protection of those who will abstract pond discharge waters for other agricultural uses, and also for the protection of the consumers of cultured species.

One of the first environmental requirements of an aquaculture farm is to maintain the quality of the aquatic environment by attempting to ensure that feed added to the pond reaches the target animal. This is the purpose of preparing water stable feeds.

A further requirement is to ensure the microbiological safety of aquatic feeding systems. Since many animal proteins are associated with potential pathogens such as salmonella, it is desirable that animal (including fish) proteins should be pasteurised before they are ingested by cultured fish. This is particularly important should the trash fish have been grown in septic ponds or other water courses potentially contaminated with faecal material. Similarly animal viscera, such as poultry offal, should also be treated before it is consumed by fish or other aquatic species. At the moment, few farm feed processors have the facilities to pasteurise protein material of this kind while including it as valuable nutrients and functional protein within feeds.

There is therefore a need to develop low cost farm scale feed processing equipment which will enable raw materials to be processed to produce feeds with acceptable microbiological and water stability characteristics. Of particular benefit would be a heated die plate for attachment to a mincer or similar feed forming machine which would act as a feed pasteurisation unit while also developing a degree of starch gelatinisation and protein denaturation. However, such a device could not replace the functions of a cooker extruder.

CONCLUSIONS

In presenting the above information an attempt has been made to demonstrate that the selection of machinery for the manufacture of farm made aquafeeds is not a simple exercise of selecting items from a catalogue in accordance with the quantity of feed desired. It is important that machinery is selected in relation to the properties of the raw materials available for formulating and processing. We are aware of what is desirable, but the task is to select the most appropriate equipment for use with non-ideal raw materials. This challenge will be much greater for some farmers than others.

Annex 1. Primary objectives of a farmer concerning feed preparation

Objectives	Potential for achievement
Prepare feeds of ideal nutritional and physical quality which are cheaper than existing feeds	Maybe
Use raw materials which are locally and constantly available	Doubtful
Produce feeds that are attractive and non-polluting	Possibly
Utilize processing equipment which has low capital and running costs and is available locally	Perhaps
Take minimal time and effort for processing	Uncertain
Have technology which is known only to him, is specific to his needs, and gives him an advantage over competitors	Unlikely
Have a technology which is operable by family members	Hopefully

Annex 2. Preliminary factors for quantifying desired aquafeed production capability

---

Identify and evaluate the following:

- target animals and their feeding behaviour;
- raw material availability and continuity of supply;
- formulation options;
- feasibility of obtaining the desired feed conversion and crop yield;
- output of feed required in relation to fish growth phase;
- desired frequency of manufacture in relation to fish growth phase;
- characteristics of the proposed site: access, power supply and its reliability, building design, storage facilities for raw materials and finished feeds, security;
- desired lifespan of equipment;
- possible future developments/expansion on site;
- locally made or imported equipment;
- process equipment options, flexibility, power requirements;
- compatibility of options for formulations and equipment;
- access to finance.

ECONOMICS OF ON-FARM AQUAFEED PREPARATION AND USE

Kee-Chai Chong

INDECO/Winrock International
Agency for Agricultural Research and Development
Jalan Salak 22, Bogor 16151, Indonesia

CHONG, K.C. 1993. Economics of on-farm aquafeed preparation and use, p. 148-160. *In* M.B. New, A.G.J. Tacon and I. Csavas (eds.) Farm-made aquafeeds. Proceedings of the FAO/AADCP Regional Expert Consultation on Farm-Made Aquafeeds, 14-18 December 1992, Bangkok, Thailand. FAO-RAPA/AADCP, Bangkok,Thailand,434 p.

INTRODUCTION

Background

Fish (used generically here to refer to finfish and shellfish), when fed a balanced diet, clearly respond well to the feed provided. The artificial (compound) aquaculture feed industry came into existence to supply the feed needed by fish farmers. These feeds are invariably based on 25-65% fish meal. Apart from micronutrient premixes fish meal is usually the most expensive of all commonly used feed ingredients.

Because they are commercially manufactured, feed quality can be controlled in terms of uniformity, nutrient composition and use convenience. Because of these advantages, the manufacturing process is costly and becoming increasingly so as the supply of fish meal declines and demand for fish meal increases, together with fish meal-based feeds. Feeds are often expensive because they tend to be over-formulated. This tendency arises due to the lack of information on the nutrient requirement of the fish under culture conditions.

At the outset it should be clearly stated that, contrary to the popular and widespread myth that fish meal is not (or should not be) used in farm-made aquaculture feed, fish meal can be included as one of the ingredients in on-farm feed formulation. However, the economic gains or returns from including or excluding it should be closely examined. The term fish meal is taken here to include both fish meal itself and trash fish.

I term aquaculture feeds as both single ingredient and compound feeds. The former consists mainly of natural feeds and is thus of variable quality while the latter is usually artificially manufactured and thus of more uniform and high quality. A distinction is also made between factory-manufactured aquaculture feed and on-farm prepared aquaculture feed. Since it first became commercially available, factory-manufactured aquaculture feed has been accepted and used as a necessary input in fish farming, in spite of its apparent high cost. Such aquaculture feed is generally used in the production of high-value (and normally low-volume) species. Because of this, its use is economical and profitable.

On the other hand, it is claimed that it is relatively cheaper to prepare your own feed than to buy commercially available feed, even when the capital cost of an on-farm feedmill is taken into account and even if the feedmill is underutilised! Such is the economic impetus and rationale for on-farm feed preparation. There is a 250-ha shrimp farm in Indonesia with its own shrimp feedmill which shows this to be the case. The implication is that the profit margin of commercial aquaculture feed is still large.

The apparently high cost of high-protein fish meal-based commercial aquafeed is still generally recognised as the main constraint to greater fish output from culture. Because of growing scarcity of fish meal supply due to overfishing and (traditional) competitive demand, for livestock feed and other uses, alternative aquaculture feeds and/or sources of feed ingredients are being actively explored. One of the options is on-farm prepared feed, using locally available feed ingredients, preferably with little or no fish as a protein and energy source.

The economic importance of feed on final farm profit is too critical to be ignored or taken lightly. At present, commercial aquafeeds cost an average of US$ 0.50/kg (from US$ 0.30-1.00/kg). Feed cost accounts for an average of 30-60% of production cost, sometimes even higher. As such, profit is highly sensitive to feed cost. Every improvement in feed conversion efficiency means a bigger profit margin for the fish producer.

Supplier induced demand

The purchase and use of commercial aquafeeds remains very much a supplier induced demand. By this is meant that the feed suppliers induce fish farmers to buy the feed. Fish farmers do so because they need to feed their fish. They (actually) do not have much choice as there seem few viable alternatives. This is especially so if they want higher yields using intensive systems. One alternative, which is to rely on natural feed, does not give as good a result as feeding commercial feed.

Another alternative, which is to fertilize the fish ponds with inorganic and organic manures to stimulate natural productivity in the water column, also does not give as good a result as commercial aquafeed. Given such production circumstances, fish farmers having access to capital rely on the feed supplied by manufacturers. Feed manufacturers are able to induce fish farmers to buy their feed through persuasive sales advice and recommendations because they are not as informed on fish nutrition.

Because of this supplier-induced demand, there is a real tendency for fish farmers to over-feed, either knowingly or unknowingly. This arises because farmers tend to follow manufacturers' feeding schedules, causing unnecessary waste. In addition, such feeds tend to be over-formulated and become uneconomical. More and more fish farmers and manufacturers are recognising the economic significance of feed.

It can be observed that the overall economic status of feed manufacturers is generally better than that of fish farmers. This is because of the supplier-induced demand for feed phenomenon. Fish farmers, being price-takers, have little or no influence on feed price determination. As a result, manufacturers tend to become prosperous more readily. In addition, farmers rarely keep records of feed use; if they do, the data is seldom rigorously analysed to improve farm performance.

Ratafia and Purinton (1989) estimated that there were over 118 feed companies worldwide producing aquafeeds. This is a low estimate as many small feed companies, especially in developing countries, have not been accounted for. The number of aquafeeds companies is projected to grow as the aquaculture industry itself expands. As aquafeed demand is at present largely induced by suppliers, producers need to closely scrutinize their feed costs as a cost centre in their farm management economic analysis. Economic farm performance indicators can be developed to guide them in the use of feed.

Purpose of study

Where on-farm aquafeeds are concerned, formulation, preparation and use are still largely governed by technical and technological considerations rather than economic criteria. There remains a general lack of information on the nutritional requirements of fish under culture conditions. By incorporating economic performance indicators, greater profits can be obtained.

The economics of on-farm aquafeeds is examined in this paper as a viable alternative to commercial feed and as one which is environmentally sustainable and inexpensive.

ECONOMIC BASIS OF ON-FARM FEED PREPARATION AND USE

Supplementary feed-based aquaculture

According to Akiyama (1991), most of the 15 million t global aquaculture output is produced without the benefit of artificial feeds. Most is produced by using fertilizers, both inorganic and organic. Reliance on organic manures to grow fish is quite prevalent in Asia, especially in China, and is usually identified with integrated crop-livestock-fish culture. Since Asia accounted for 85% of the world's aquaculture production in 1988 (Csavas, 1991), it can be inferred that approximately 80% of global aquaculture production is produced without using commercial aquafeeds. Thus, great potential exists to increase the use of feed to produce more fish. Feed can significantly increase yield. As pointed out above, fish meal is used in feed formulation for both commercial and farm-made aquafeeds. Commercial aquafeeds are heavily dependent on fish meal as a key feed ingredients.

However, on-farm aquafeed holds greater promise in a world increasingly concerned with environmental quality and natural resources system productivity, stability, sustainability and equitability. Even though farm-made feed can be environmentally more damaging than commercial feed, because of its higher biological oxygen demand (BOD), the former should be actively promoted as a viable alternative because of its use of byproducts and waste recycling basis. Farm-made aquafeed uses locally available raw materials as ingredients which are found mainly on or in the vicinity of the farm.

Since such wet ingredients are usually incorporated without further processing and the resulting ground feed is directly fed to the fish, its moisture content is very high. The use of such feeds, combined with improper feeding practices, not only results in unnecessary waste but also leads to more rapid (bottom) pond soil deterioration due to leftover uneaten feed. Constituting organic matter of relatively high protein content, waste feeds become rapidly fouled, giving rise to poor water quality, in turn creating stressful growing conditions. Loss of appetite, poor or slow/stunted growth and disease outbreaks causing high mortality can occur, all translating into economic loss.

Although effluents from intensive fish farms have been alarmingly viewed as an environmental problem, they may not necessarily be so. Ways and means can be found to strip the nutrient-rich pond effluents for recycling, including for fish farming (Chong, 1992a).

At present, many types of aquaculture feeds with many brand names are available in the market. For now, the market appears to be a captive one, as fish producers are more or less dependent on them. This is especially true in countries where feed technology is not well developed yet.

Economic value of feed in aquaculture

This paper is not concerned with pond fertilization but with the use of feed, especially the prospects and contribution of farm-made aquafeeds to grow fish. It will try to answer the million dollar question: is farm-made feed use economical and can such feed replace commercial feed in aquaculture?

The use of feed in salmon, trout, catfish, shrimp and eel culture, among others has been shown to be profitable. However, these commercial feeds rely heavily on fish meal as a major ingredient. As a result of the acceptance of feed as a key input in aquaculture, a sizeable aquafeed industry has developed and prospered. From less than 1 million t output in the early 1970's the industry is reputed to have grown by more than fourfold to 4 million t* in 1988 (Megisson, 1990). At an average price of US$ 500/t, this would be worth about US$ 2 billion. Because of the profitable use of feed for high value species the aquafeed industry, and fish farmers persuaded by feed salesmen, have recently turned their attention to the use of feed for the production of other species of lower market value.

Aquafeed is now available for carp, tilapia, milkfish and other similar species but these feeds still rely on fish meal for protein and energy although at lower levels. Fish meal is becoming scarce and expensive. Can non-fish meal ingredients be found as possible replacements? Is fish meal indispensible? Is the use of non-fish meal-based feed formulated and prepared on the farm economical and profitable?

On-farm feed economics

The preparation and use of farm-made aquafeeds can benefit from close observations of the feeding behaviour of the fish under natural conditions. The types of natural fish foods available in the wild and fish preference, feeding frequency and quantity consumed, are excellent nutrient/dietary indicators and an economic basis for the systematic evaluation, preparation and use of on-farm aquafeeds using locally available non-fish meal raw materials.

In the wild, fish do not completely depend on other fish for nutrition; other animal and plant organisms also form part of the diet. The aquatic food web is complex and interdependent. A balanced nutritional regime is more appropriate than one based on a single food item. Also information available from both private and publicly supported research and development should be used in

---

* Editors' note: however, we believe this to have been an over-estimate (see New and Csavas 1993).

formulating on-farm aquafeed. Such information can be applied to either grow the necessary feed ingredients on the fish farm itself or to gather them from the wild, or to purchase them from suppliers.

Once the available ingredients and their nutrient profile are known, including their pro- and anti-nutritive factors, the economics of on-farm feed production must be calculated and its use compared to commercial aquafeed. Such economic comparison must be made against the background of the farm gate value of the fish produced, which may differ according to the quality of the final product and be affected by the type of feed used. Different feeds may affect the flavour and texture of the fish flesh produced.

The unit cost of on-farm feed production, unit price of commercial feed and the cost of fish production using each type as well as the unit profitability of each system of fish production must be compared before one type of feed is selected. Growth and production performance of the fish cultured using the two different feeds have to be taken into account.

The environmental cost reflected in the social cost of using each feed must also be closely examined before one is selected. There are potential trade-offs between the two types of feeds in terms of the use of raw materials and their affect on the environment, for example, the fisheries on which fish meal reduction is dependent. Shrimp feed cost is projected to increase due to the growing competing demand for the largely stagnating supply of fish meal caused by the overfishing of the fisheries resources. Thus greater effort should be exerted to find or develop substitute ingredients. Mussels and soybean meal, among others, have been evaluated highly promising fish meal substitutes for aquafeed in Thailand. In Bangladesh, fish meal is unavailable and many other human foodstuffs such as slaughterhouse waste, including fresh coagulated blood, have successfully been used as fish feed, especially for fry and fingerlings.

Feed ingredient interchangeability

Feed ingredient interchangeability refers to the ability of feed ingredients to become partial or complete substitutes for each other. Of the 40 or so essential nutrients required by fish (including crustaceans), precise knowledge on quantitative (and possibly qualitative) requirements is available for only a few tropical species (World Bank 1991).

Methodologies are now available not only to assess the nutrient requirements and nutritional status of fish but also to assess the nutrient profile and bio-availability of feed ingredients such as digestibility, biological value of proteins, serum or tissue nutrient or enzyme concentrations and activities (World Bank 1991) and potential pro- and anti-nutritive factors. Finding inexpensive substitutes for fish meal and other expensive or increasingly scarce ingredients for fish feed remains an important task for fish nutritionists and aquaculturists. Fortunately, data compiled on raw materials suitable for livestock feed manufacture are also useful for evaluating their use in fish feed. This is especially so for on-farm feed preparation. Much of the scientific literature on livestock feed ingredients is in the public domain.

There is in fact voluminous information on ingredients for the formulation and preparation of on-farm fish feed. The main problem lies in how to get it to the farmers. There is also no lack of information on overcoming anti-nutritive factors of candidate ingredients through processing, either by physical, chemical or mechanical means. Numerous studies have been done on the replacement of fish meal by plant proteins in aquaculture feed. Also, as mentioned above, slaughterhouse waste is an excellent source of animal protein.

Training fish farmers on how to utilise such information is also of equal, if not more importance. This aspect is the least developed area of aquaculture extension in many developing countries. It has been overlooked until recently partly due to the availability of commercial aquafeed. It thus calls for greater government attention to upgrade fish farmers' skill in being self-sufficient in the different aspects of fish farming.

Basic considerations

On-farm aquafeed suffers from certain limitations not true of commercial feed. For example, it seldom, if ever, benefits from the normally well-supported feed and nutrition research and development programmes of the commercial feed manufacturers.

On-farm feed producers usually rely on whatever published information is available, which is usually scanty being highly proprietary in nature. This is in direct contrast to the information available on livestock nutrition and animal husbandry. Public domain information is produced by publicly-supported research and development in government fish feed laboratories or universities. Although not proprietary, the usefulness and degree of detail of such information is frequently limited. It is also well-known that feed companies going into joint-venture in other countries do not reveal their entire formulation. They retain certain ingredient "mixes" which have to be imported from their country of origin. It has frequently been alleged that, even when they seem to share their formulation, it is usually their old formula.

Farm-made feed is variable in quality and generally found to be more costly per unit of nutrient based on growth and final fish production, that is profit performance. It seldom benefits from economies of scale which can reduce production cost. Also, for those feed ingredients which have to be bought, the quantity discounts available for bulk purchases cannot be obtained. Such disadvantages, however, are counter-balanced by the relatively low cost of grinding and milling (inclusive of capital cost).

The pros and cons of farm-made aquafeed *vis-a-vis* commercial feed discussed so far do not necessarily imply that the former is unattractive. The problems farm-feed faces can be overcome by making more information available to fish farmers.

It is reported that the feed processing cost of commercial feed is very high in relation to other cost components like feed ingredient cost. For farm-made feeds, ingredient quality and cost is the most important consideration. Very little information is available on the proximate chemical analysis of potential less-conventional ingredients, which are not traded in the market place but are available in the vicinity of the farm.

Many commercial aquafeeds are over-formulated and are thus expensive, beyond the capital or financial capability of small fish farmers. Many farmers do not apply any production inputs, either those which increase or maintain yield. In many developing countries, most fish farms are owned by small farmers and are grossly underutilised. By encouraging them to use more inputs such as fertilizers and feeds (*e.g.* farm-made aquafeed), much greater output can be obtained thus increasing the supply of fish.

According to Akiyama (1991), currently available commercial feeds are believed to be using excessive amounts of protein for energy. He also stated that total phosphorus levels are excessive and not highly available to aquatic animals.

Having considered all the above factors, the remaining consideration is the nutritional and economic performance of the two kinds of feeds: farm-made *versus* commercial.

ON-FARM AQUAFEED FORMULATION

Ingredients and formulation

To encourage fish farmers, especially small fish farmers, to produce their own aquafeed, the cost must be low and its quality comparable to commercial feed. Thus, the promotion of farm-made feeds concerns sourcing locally available ingredients to provide cheap protein, carbohydrate and fat without excessive sacrifice of the required amino acid requirements of the fish. The protein content of such feeds should be utilised for growth and not to provide energy, which should preferably come from the fat and carbohydrate component. The aquafeed formulated should fulfil the nutritional requirements of the fish being grown.

Non-fish meal feed ingredients

Major candidate feed ingredients consist of macrophytes either of aquatic or terrestrial origin, agricultural, industrial and household/restaurant wastes and byproducts as well as animal organisms either of aquatic or terrestrial origin.

Kitchen ponds and live feed

During the author's survey of milkfish farmers in Taiwan, Province of China, the Philippines and Indonesia in the late 1970's and early 1980's fish farmers purchased "lumut" or filamentous algae and aquatic plants from other fish farmers who specialised in their cultivation. This is especially widespread in the Philippines where a "jeepney" load of "lumut" can often be seen. "Lumut" is also purchased to feed to siganid fish in the Philippines. The little snails, aquatic organisms, insects and plankton trapped in the "lumut" also provide valuable nutrients to the fish.

The concept of the kitchen pond where live food is produced to feed the fish constitutes a farm-made aquafeed option which merits further research and development. In Thailand, *Artemia* and other microscopic aquatic organisms are cultured in special ponds and the enriched pond water is then pumped into grow-out fish ponds as feed. This is actually an extension of pond fertilization during pond preparation. Further, there are also commercial formulations of such organisms available for the inoculation of pond water to boost planktonic food.

Plant origin feed ingredients

Feed ingredients of plant origin which have been traditionally used in farm-made aquafeed include the following:

- grains and bran (*e.g.* broken rice and rice bran);
- oil cakes and meals (*e.g.* copra cake and soybean meal);
- grasses;
- vegetables;
- tapioca (*Cassava manihot*) leaves;
- Ipil ipil (*Leucaena leucocephala*) leaves;
- *Azolla* sp;
- *Lemna* sp;
- *Hydrilla verticillata*;
- *Ceratophyllum demersum*;
- *Spirogyra maxima*;
- brewery wastes and byproducts.

Only certain species of fish, with specialized anatomical and physiological mouth parts and digestive apparatus for food ingestion, mastication, digestion and absorption can make full use of such feed ingredients.

In his review of the use of non-conventional plant feedstuffs in fish feed, Wee (1991) found that it is not possible to utilise them at high levels without comprising growth and production. In other words, such feedstuffs, which are low in protein and fat, cannot completely replace such conventional feed ingredients as fish meal. However, there are methods to overcome such deficiencies or to enhance their nutritive values, by proper processing to increase the bioavailability of nutrients and reduce anti-nutritional factors and by the inclusion of appropriate additives. Soybean meal is a good example of a byproduct (of soybean oil extraction) which has been successfully processed into a feed ingredient which is increasingly becoming a partial substitute for fish meal in aquafeeds.

According to Akiyama (1991), aquatic animals are more sensitive to feed quality than terrestrial animals. This implies that closer attention must be given to aquafeed formulation, especially when fish farmers are preparing them. At the same time, the ingredients must also be cost-effective.

CONCLUSIONS

As noted above, aquafeeds are costly or expensive inputs for fish culture. The demand for aquafeed at present is still largely induced by feed suppliers. As such, fish farmers have to guard against overfeeding caused by too closely following the feeding recommendations of feed suppliers. Fish farmers are well advised to closely scrutinize their feed costs to improve their profits.

In this paper, a distinction has been made between commercial aquafeed and farm-made aquafeed. Commercial aquafeed has long been accepted as a necessary production factor in fish farming, in spite of its high cost. Such aquafeed is generally used in the production of high-value species for which it is economical and profitable.

The high cost of high-protein fish meal-based commercial aquafeed is still generally recognised as the main constraint to increased fish output from culture. Because of growing scarcity of fish meal due to overfishing and (traditional) competitive demand for fish meal for livestock feed and other uses, alternative aquafeeds and/or sources of non-fish meal ingredients are being actively explored. One of these is farm-made feed using locally available ingredients.

The economic importance of feed on final farm profit is too critical to be taken lightly. At present, commercial aquafeeds cost an average of US$ 0.50/kg and feeding accounts for an average of 30-60% of production costs. As such, profit is highly sensitive to feed cost. Every improvement in feed conversion efficiency means a bigger profit margin.

The purchase and use of commercial aquafeed remains very much a supplier-induced demand. As such feed cost, as a cost centre in the overall fish production costs, remains outside the control of fish farmers. Feed prices are determined by the feed suppliers; fish farmers, especially small-scale operators, have little or no influence on the final determination of feed prices. Furthermore, feed suppliers encourage fish farmers to use their feeds by providing samples for feeding trials, extending credit and easy terms of payment and other such product promotional schemes.

Because of this supplier-induced demand, there is a real tendency for fish farmers to over-feed, either knowingly or unknowingly. This arises because farmers tend to follow feed suppliers feeding schedules and their guidelines on quantity and frequency. In addition, few farmers keep records as a means to evaluate their farm performance and even if they do so, seldom critically analyse the data.

The economics of farm-made aquafeed has been discussed and found to be not only a viable alternative to commercial aquafeed but one which is environmentally sustainable and inexpensive.

LIST OF REFERENCES

Akiyama, D.M. 1991. Future considerations for the aquaculture feed industry, p. 5-9. *In* D.M. Akiyama and R.K.H. Tan (eds.) Proceedings of the Aquaculture Feed Processing and Nutrition Workshop held in Thailand and Indonesia, 19-25 September 1991. American Soybean Association, Singapore.

Akiyama, D.M. and R.K.H. Tan (eds.) 1991. Proceedings of the Aquaculture Feed Processing and Nutrition Workshop held in Thailand and Indonesia, 19-25 September 1991. American Soybean Association, Singapore. 241 p.

Anon. 1988. Fish feeds. Infofish International 3/88:18-20.

Bostock, T. 1991. Better feeds for small-scale shrimp farmers. Bay of Bengal News 42:22-26.

Chong, K.C. 1992a. The DGF fisheries planning model, FAO/UNDP Project on Assistance in Agricultural Planning (INS/88/001), Food and Agriculture Organization of the United Nations and Directorate-General of Fisheries (DGF) and Bureau of Planning, Ministry of Agriculture, Jakarta, Indonesia.

Chong, K.C. 1992b. Improving profitability of shrimp aquaculture. Paper presented at the Asian Productivity Organization (APO) Seminar on Aquaculture, 25 August - 4 September, 1992, Tokyo, Japan.

Crampton, V., H. Kreiberg and J. Powell. 1990. Feed cost control in salmonids, p. 99-108. *In* Proceedings of the International Congress on Aquaculture, 4-7 September 1990, Vancouver, British Columbia, Canada.

Cruz, E.M. 1992. Preliminary assessment of the production and economics of culturing common carp (*Cyprinus carpio*) in running water system in North Sumatra, Indonesia. J. Aqua. Trop. 7:7-14.

Csavas, I. 1991. Physical and nutritional properties of aquafeeds. Paper presented at VICTAM-Asia '91 Symposium, Feed Production Tomorrow, 24-26 January 1991, Bangkok, Thailand.

De Silva, S. S. 1989. Reducing feed costs in semi-intensive aquaculture systems in the tropics. Naga, The ICLARM Quarterly 12(4):6-7.

De Silva, S.S. (ed.). 1991. Fish nutrition research in Asia. Proceedings of the Fourth Asian Fish Nutrition Workshop, Asian Fisheries Society, Philippines and International Development Research Centre, Canada. 205 p.

FAO. 1979. Training in fish feed technology. A report of the FAO/UNDP Training Courses in Fish Feed Technology held in the University of Washington, Seattle, U.S.A., 9 October-15 December 1978. ADCP/REP/79/8. FAO, Rome, Italy. 16 p.

Megisson, P.A. 1990. Future world of aquaculture. Paper presented at Rovithai International Shrimp Seminar, July 1990, Bangkok, Thailand.

New, M.B. and I. Csavas. 1993. Aquafeeds in Asia - a regional overview, p. 1-23. *In* M.B. New, A.G.J. Tacon and I. Csavas (eds.) Farm-made aquafeeds. Proceedings of the FAO/AADCP Regional Expert Consultation on Farm-Made Aquafeeds, 14-18 December 1992, Bangkok, Thailand. FAO-RAPA/AADCP, Bangkok, Thailand.

Ratafia, M. and T. Purinton. 1989. Emerging aquaculture markets. Aquaculture Magazine 15(4):32-46.

Tacon, A.G.J. 1992. The food and feeding of farmed fish and shrimp: an annotated selection of FAO Field Documents, 1973-1991, FAO Fisheries Circular No. 849, FIRI/C849. FAO, Rome, Italy. 92 p.

Tacon, A.G.J. and C.B. Cowey. 1985. Protein and amino acid requirements, p. 155-183. *In* P. Tytler and P. Calow, (eds.) Fish Energetics - New Perspectives: Croom Helm Press Ltd., London.

Wee, K.L. 1991. Use of non-conventional feedstuff of plant origin as fish feeds - is it practical and economically feasible? p. 13-32. *In* S.S. De Silva (ed.) Fish Nutrition Research in Asia. Proceedings of the Fourth Asian Fish Nutrition Workshop, Asian Fish. Soc. Spec. Publ. 5. Asian Fisheries Society, Manila, Philippines.

World Bank. 1991. Tropical aquaculture development: research needs, World Bank Tech. Pap. No. 151 (Fisheries Series), The World Bank/United Nations Development Programme/ Commission of the European Communities/Food and Agriculture Organization, Washington DC., USA. 52 p.

AQUAFEEDS AND FEEDING STRATEGIES IN BANGLADESH

M. Zaher and M. A. Mazid

Freshwater Station, Fisheries Research Institute
Mymensingh, Bangladesh

ZAHER, M. and M.A. MAZID. 1993. Aquafeeds and feeding strategies in Bangladesh, p. 161-180. *In* M.B. New, A.G.J. Tacon and I. Csavas (eds.) Farm-made aquafeeds. Proceedings of the FAO/AADCP Regional Expert Consultation on Farm-Made Aquafeeds, 14-18 December 1992, Bangkok, Thailand. FAO-RAPA/AADCP, Bangkok, Thailand, 434 p.

INTRODUCTION

Fish and fisheries play a significant role in the economy of Bangladesh in terms of nutrition, income, employment and foreign exchange earnings. The people of Bangladesh depend on fish as the principal source of animal protein. Fish contribute about 80% of the nation's animal protein intake, nearly 6% to GDP and more than 12% to the export earnings of the country (G.O.B. 1990). Approximately 1.4 million people depend on fisheries as their primary source of income and another 11 million are involved in seasonal or part-time fishing and other ancillary fishery activities. About 73% of the rural households are involved in subsistence fishing. The sector provides employment to about 8% of the population.

The country has abundant water resources with 1.5 million ha of perennial and 2.8 million ha of seasonal inland waters, in addition to a coastline of 480 km. There are approximately 1.3 million freshwater ponds covering a total area of about 150,000 ha, out of which nearly 46% are under culture, 30% are culturable and 25% are derelict (F.R.S.S. 1987). At present fish production from these ponds is low, at 600-800 kg/ha/yr. The average fish production was 1,175 kg/ha for cultured ponds, 300 kg/ha for culturable and 175 kg/ha for derelict ponds. The country has 5,000 ha of oxbow lakes (baors) which also contribute to aquaculture production.

There are about 2.5 million ha of coastal lands, large parts of which are suitable for shrimp farming. Coastal aquaculture of black tiger shrimp (*Penaeus monodon*) has been identified by the government as a priority area. Shrimp farming in "beries", "ghers" or "ghonas" (*i.e.* areas impounded by dikes) of the coastal area is an old traditional practice. However, its rapid expansion in the past few years is the most remarkable development in the fishery sector of Bangladesh, which is linked with increasing demand for shrimp in the world market. The production in 1987-88 (18,000 t) increased by a factor of nine over 1982-83 (2,000 t). This increase was largely due to an expansion in the area of coastal shrimp farming, from only about 20,000 ha of brackishwater ponds in 1980 (Mahmood and Chowdhury 1989), to an area of about 150,000 ha in 1989. Unfortunately, the shrimp yields of the coastal farms are very low (120-200 kg/ha/yr), mainly due to lack of feed and proper management.

Fish production

Bangladesh ranks third in inland fish production among the countries of the world (Islam 1989, Ahsanullah 1989). The current level of production (1990) of fish and shellfish is estimated to be about 850,000 t, out of which 610,000 t (72%) comes from inland waters and 240,000 t (28%) from marine fisheries. Of the inland production, 420,000 t (69%) is from capture fisheries and 190,000 t (31%) from culture (G.O.B. 1991). Capture fisheries show a decreasing trend, while output of culture fisheries increases.

Aquaculture production

Aquaculture production systems in Bangladesh are diversified in type, character and source, constituted by ponds, tanks, oxbow lakes (baors) and coastal shrimp farms. Details of aquaculture production from 1985-86 to 1989-90 are shown in Table 1 and Figure 1.

Table 1. Aquaculture production in Bangladesh

Water bodies	Water area (1,000 ha)	Production (1,000 t) 1985-86	1986-87	1987-88	1988-89	1989-90
Ponds	147	124	143	147	152	161
Oxbow lakes (baors)	5	1	1	1	1	1
Coastal shrimp farms	150	20	22	23	24	27
Total	302	145	166	171	177	189

Source: G.O.B. (1990)

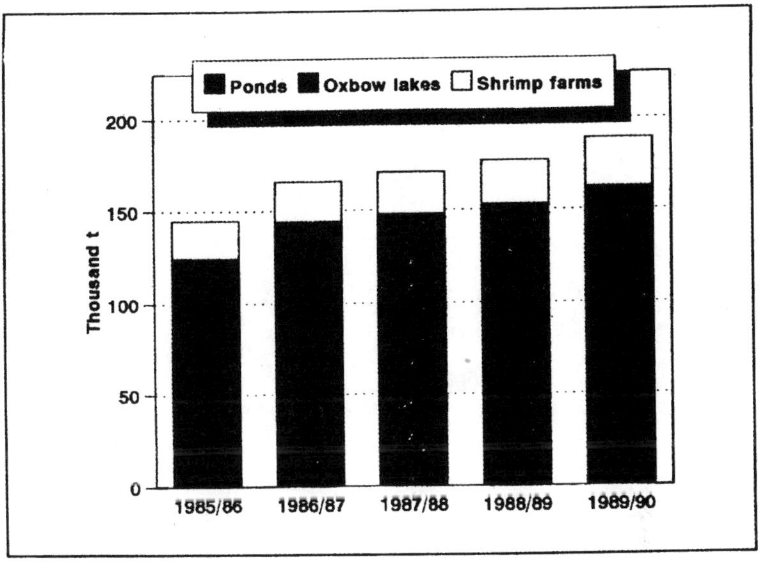

Figure 1. Aquaculture production in Bangladesh

Major cultured species

Pond aquaculture in Bangladesh comprises mainly Indian major carps (*Labeo rohita, Catla catla, Cirrhinus mrigala, Labeo calbasu, etc.*), catfishes (*Clarias batrachus, Heteropneustes fossilis, Mystus aor, M. tengra, Ompok pabda*), snakeheads (*Channa punctatus, C. striatus, C. marulius*), climbing perch (*Anabas testudineus*), freshwater prawns (*Macrobrachium rosenbergii*) and many small indigenous species. Pond culture of *Macrobrachium* is a new practice in Bangladesh. Some exotic fishes, like tilapia (*Oreochromis niloticus*), silver barb (*Puntius gonionotus*), common carp (*Cyprinus carpio*), silver carp (*Hypopthalmichthys molitrix*), grass carp (*Ctenopharyngodon idella*), African catfish (*Clarias gariepinus*) *etc.*, are being also cultured in ponds. Tilapia and silver barb are cultured in small water bodies, such as ditches and shallow ponds. Indian major carps, Chinese carps, catfishes, snakeheads and small indigenous fishes are used in oxbow lake (baor) aquaculture. Coastal aquaculture is based mainly on black tiger shrimp (*Penaeus monodon*). *Macrobrachium* culture is also practiced in low salinity areas.

Indian major carps and Chinese carps contribute about 47% while shrimp, the main source of export earnings from aquaculture, contribute about 14% to the total aquaculture production (Figure 2). Individual production data of different species are not available. Details of aquaculture production by major species groups are given in Table 2.

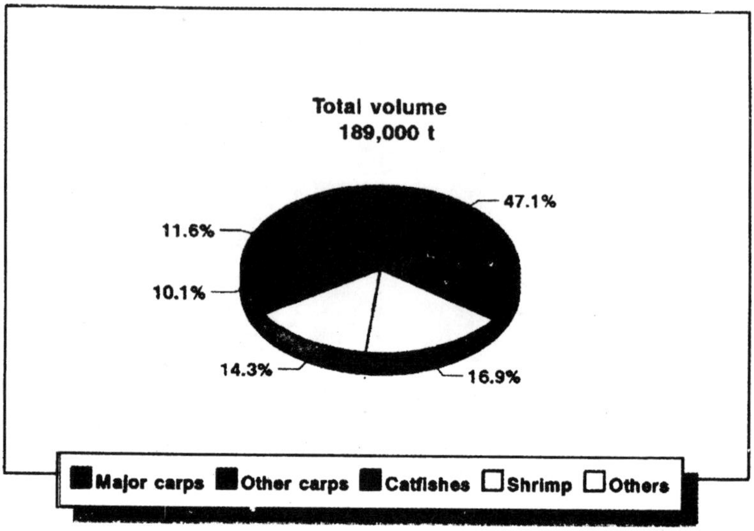

Figure 2. Aquaculture production by major species groups in Bangladesh in 1989/90

Table 2. Aquaculture production by major species groups in Bangladesh in 1989/90

Species groups	Production (1,000 t)	Percentage of total (%)
Major carps	89	47.1
Other carps	22	11.6
Catfishes	19	10.1
Shrimp	27	14.3
Others*	32	16.9
Total	189	100.0

* "others" include tilapia, silver barb, snakehead, *etc*.

Source: Department of Fisheries, Dhaka

Importance of aquafeeds

Against a backdrop of rapid population increase without concomitant increase in fish production, the per capita consumption of fish has dropped from 33 kg in 1963-64 to 20.5 kg in 1989-90, compared to the recommended level of 38 kg (G.O.B., 1985). As a principal source of animal protein, fish production has to be increased manifold to meet the requirement in Bangladesh. In order to achieve the targeted production of 1.2 million t by the terminal year of the Fourth Five Year Plan (1995) production has to be increased by nearly 25% over 1990. Scientific opinion is that fish production as high as 3,700-4,500 kg/ha could be obtained by using semi-intensive polyculture in ponds with supplementary feeding. This demonstrates a real possibility of increasing production and reveals the potential importance of aquafeeds in Bangladesh. Recently some shrimp farmers began to use supplementary and/or complete feeds for shrimp culture, which increased production from 200 kg/ha to 1,200-1,600 kg/ha.

Contribution of feed to aquaculture

The use of feed is increasing with the expansion of aquaculture in Bangladesh. Nursery farmers use rice bran and mustard oil cake for carps. The unit feed price is US$ 0.10-0.15/kg, and the total costs of feeding are US$ 325-672/ha, 10-20% of the total production cost of US$ 3,416-3,524/ha. In polyculture, farmers use a mixture of rice bran and mustard oil cake, which costs US$ 0.06-0.10/kg. The total production cost is US$ 1,833/ha and the total cost of feeding is US$ 510/ha (28%). Semi-intensive culture of tilapia and silver barbs with supplementary feed, composed of rice bran and mustard oil cake (US$ 0.08/kg of feed), has a total production cost of US$ 3,115/ha, while cost of feeding is US$ 1,524 (49%) resulting in an increase in production from 560 kg/ha/yr to 1,300 kg/ha/yr. Some shrimp farmers use complete pelleted shrimp feeds priced at US$ 1.00/kg. Their total cost of feeding is US$ 610/ha, 39% of the total production cost of US$ 1,574/ha.Feeding increases the yield from 120-200 kg/ha to 1,200-1,600 kg/ha. Farmers shift gradually from no feed, through the use of farm-made feeds, to factory-made feeds.

AVAILABILITY OF FEED INGREDIENTS

Since the demand for suitable fish feeds is increasing day by day with the gradual intensification of aquaculture practices, a nationwide survey was conducted by the Fisheries Research Institute to identify potential fish feed ingredients based on their availability, price and primary nutritional value. Eighty-three different types of ingredients, both of plant and animal origin, were studied. The survey covered both traditional ingredients and non-conventional items, such as kitchen wastes, slaughterhouse wastes, processed wastes from the food industry, aquatic weeds, *etc.*, which are not being properly utilized. Most of these conventional and non-conventional ingredients are abundantly available in Bangladesh at an affordable price for making good quality fish feeds.

As Bangladesh is mainly agro-based, a large variety of agricultural wastes and by-products, such as pulse brans, oil seeds and oil cakes, and molasses are being used as fish feeds. In addition, the country has a vast water area supporting huge quantities of aquatic plants and weeds. Aquatic plants, like water hyacinth, duckweed, water velvet, etc., have been tried as fish feed ingredients. Feed ingredients of animal origin, like fish meal, are used as sources of animal protein in aquafeeds. Others, including poultry by-products, slaughterhouse wastes, blood meal, bone meal, *etc.*, have been evaluated for their possible use. Even in small amounts, they may greatly improve the nutritional value of the entire diet.

Information on price and availability of major fish feed ingredients are shown in Tables 3 and 4. It was observed that the availability and price of agricultural by-products suitable for fish feeds is largely seasonal. Although most are available throughout the year and all over the country, some are very localized. On the other hand, most have peak seasons, when their prices are lower than in the off-season.

Rice bran is extensively used for feeding fish, being the cheapest and most commonly available feed ingredient. Wheat bran is also used in feed formulation. Mustard oil cake and sesame oil cake are sufficiently available year-round, throughout the country, to be used in fish, poultry and livestock feeds, either individually and in combinations with other ingredients. Molasses and flour are used as binders.

Proximate composition of potential fish feed ingredients was determined by A.O.A.C. methods (A.O.A.C. 1985). Although the information generated does not provide a complete picture of their nutritional value, it is adequate for their selection for feed formulation by farmers and for further research on the formulation and manufacture of good quality fish feeds by scientists. Proximate analyses of the major fish feed ingredients available in Bangladesh are presented in Table 5.

The import of major feed ingredients is still not allowed by the government. However, the matter is under active review.

Table 3. **Availability and price of major potential plant ingredients for aquafeeds in 1988 in Bangladesh**

Ingredients	Quantity (1,000 t/year)	Price Retail (US$/kg)	Price Wholesale (US$/kg)	Availability (peak season)
Water hyacinth	n.a.	*	-	Nationwide, all year (Jun-Aug)
Duckweed	n.a.	*	-	Nationwide, all year (Jun-Aug)
Water velvet	n.a.	*	-	Nationwide, seasonal (Jun-Aug)
Grass pea/ch. vetch bran	14	0.10	0.09	Faridpur, Rajshahi, Jessore, all year (Nov-Jan)
Split pea/lentil bran	9	0.10	0.09	Faridpur, Rajshahi, Jessore, all year (Jan-Mar)
Black gram bean bran	3	0.13	0.10	Faridpur, Rajshahi, Jessore, all year (Nov-Jan)
Mustard oil cake	80	0.12	0.10	Nationwide, all year (Jan-Mar)
Coconut oil cake	2-3	0.15	0.10	Khulna, Barisal, Chittagong, all year (Jan-Mar)
Soybean grits	0.3	0.25	0.23	Faridpur, Noakhali, seasonal (Nov-Jan)
Linseed oil cake	5	0.15	0.14	Southern region, all year (Mar-May)
Sesame oil cake	12	0.14	0.13	Southern region, all year (Jul-Dec)
Molasses	90	0.11	0.10	Around sugar mills of BSFIC, all year
Rice and rice bran	1,500	0.08	0.07	Nationwide, all year (harvesting)
Barley	6.0	0.38	0.35	Nationwide, all year (harvesting)
Wheat flour	1,900	0.20	0.18	Nationwide, all year (harvesting)
Wheat bran	n.a.	0.10	0.09	Nationwide, all year (harvesting)
Flour	n.a.	0.25	0.23	Nationwide, all year (harvesting)

* obtainable free of charge; however, commercially collected volumes are priced

Source: F.R.I. (1989)

Table 4. Availability and price of major potential animal ingredients for aquafeeds in 1988 in Bangladesh

Ingredients	Quantity (1,000 t/year)	Price Retail (US$/kg)	Price Wholesale (US$/kg)	Availability (peak season)
Slaughterhouse waste	(4,627,000 head)	0.32	n.a.	Nationwide, all year (festivals)
Kitchen wastes	n.a.	*	n.a.	Hotels, restaurants, all year (festivals)
Blood meal	0.4	*	n.a.	Slaughterhouses, all year (festivals)
Bone meal	4.8	0.13	0.10	Chittagong, Khulna, Rangpur all year (festivals)
Fish meal (A1 grade)	0.3	0.50	0.50	BDFC plant, Chittagong, all year
Fish meal (A2 grade)	0.2	0.41	0.41	BDFC plant, Cox's Bazar, all year
Fish meal (B grade)	0.2	0.36	0.36	BDFC plants, Khulna, Dhaka, all year
Fish silage	n.a.	*	-	Coastal districts, all year (Nov-Feb)
Shrimp waste (raw)	4-5	0.10	0.09	Chittagong, Khulna, Barisal all year
Crab meal	n.a.	*	-	Chittagong, Khulna, Barisal seasonal (Jun-Aug)
Silkworm pupae	n.a.	0.25	0.25	Silk industry, Rajshahi, all year (Jun-Aug)

* obtainable free of charge; however, commercially collected volumes are priced

Source: F.R.I. (1989)

Table 5. Proximate composition of major feed ingredients in Bangladesh

Ingredient	Moisture (%)	Protein (% on D.M.)	Lipid (% on D.M.)	Fibre (% on D.M.)	Ash (% on D.M.)	NFE (% on D.M.)	Gross energy (kcal/kg)
Water hyacinth leaves	85.73	10.73	2.20	18.20	10.80	58.07	4,031
Duckweed	92.00	14.02	1.92	11.08	12.10	60.88	3,969
Water velvet	94.95	19.27	3.49	9.50	17.55	50.19	3,912
Grass pea/ch.vetch bran	12.17	12.07	1.15	37.77	7.80	41.21	4,270
Lentil bran	15.67	19.45	0.47	25.86	7.33	46.89	4,286
Black gram bean bran	12.93	18.80	0.55	22.83	7.47	50.35	4,252
Mustard oil cake	14.46	30.33	13.44	12.12	9.73	34.38	4,978
Coconut oil cake	0.53	18.19	10.20	11.65	6.47	53.49	4,723
Soybean grits	7.53	23.82	14.83	18.49	4.87	37.99	5,188
Linseed oil cake	4.20	26.45	7.78	14.33	15.51	35.93	4,386
Sesame seed oil cake	13.13	27.20	13.18	11.18	13.47	34.97	4,753
Molasses	31.67	4.45	0.00	0.00	11.93	83.62	3,624
Slaughterhouse waste	81.47	51.60	21.14	1.69	12.53	13.04	5,566
Kitchen waste	74.17	27.87	7.83	8.69	25.95	29.66	3,950
Blood meal	8.47	63.15	0.56	0.00	20.70	15.59	4,250.
Bone meal	7.49	17.50	5.19	3.53	65.73	8.05	1,988
Fish meal (A1 grade)	13.75	59.61	11.22	2.17	22.76	4.24	4,754
Fish meal (A2 grade)	17.63	50.81	7.62	1.54	25.89	14.14	4,274
Fish meal (B grade)	18.64	44.74	7.87	1.73	28.84	16.82	4,068
Fish silage	77.96	54.95	20.37	0.97	16.06	7.65	5,432
Crab meal	70.68	31.97	7.78	11.13	39.94	9.18	3,471
Silkworm pupae	68.47	52.46	23.23	6.17	7.33	10.81	5,939
Frog wastes	7.80	62.12	14.69	1.61	11.21	10.37	5,444
Rice bran	11.67	10.26	10.45	20.85	16.40	42.04	4,235
Rice starch	96.06	3.51	0.63	0.00	3.38	92.48	3,988
Atta (wheat flour)	9.93	17.78	3.90	1.12	1.60	75.60	4,488
Wheat bran (fine)	10.67	14.57	4.43	9.71	4.93	66.36	4,394
Wheat bran (coarse)	12.68	12.92	4.51	10.98	5.20	66.75	4,366
Flour	10.27	19.73	0.38	0.00	0.40	79.49	4,368

Source: F.R.I. (1989)

AQUAFEED MANUFACTURING INDUSTRY

The success of intensive and semi-intensive fish culture depends to a large extent on the application of suitable feeds, although other considerations, such as stocking density, the ecology of the water body, the physico-chemical conditions of the water, etc., also have a bearing. Fish feeds provide nutrients for optimum fish growth and bring higher economic return to farmers.

Although Bangladesh is rich in water resources suitable for aquaculture, intensive aquaculture is almost absent and semi-intensive aquaculture is only a recent trend. Thus the demand for compounded complete and supplementary feeds is very low, being only 3,000-4,000 t/yr. Moreover, farmers make their own aquafeeds on-farm. Because of low demand, the fish feed industry has not yet developed. Recently (1990) an aquafeed mill named Saudi-Bangla Fish Feed Mill was established by the Saudi-Bangla Industrial and Agricultural Investment Company, as a joint venture between Saudi Arabia and Bangladesh. Its equipment was supplied by the Nanlien Corporation, Taiwan, and the mill was installed under the consultancy and engineering services of the supplier. Originally the mill was designed to produce shrimp feed, with a capacity of 6,000 t/yr but fish feed manufacture is also feasible and said to be in effect.

The major ingredients for manufacturing the compound shrimp and fish feeds of this mill are local. Some minor ingredients, like growth promoting substances (di-calcium phosphate, di-sodium phosphate, *etc.*), fungicides, and binders, are imported from Singapore, Taiwan and Thailand. The ingredients used for (complete) shrimp feeds include fish meal, shrimp shell meal, soybean meal, wheat flour, di-calcium phosphate, and vitamin/mineral premix. In (supplementary) fish feed formulae, fish meal, shrimp shell meal, sesame oil cake, coconut oil cake, rice bran, wheat bran, wheat flour, di-sodium phosphate and vitamin/mineral premix are used. Both shrimp and fish feeds are produced by the company in dry, pelleted form. Details of the feeds made are given in Tables 6 and 7. The address of this sole aquafeed manufacturer is: Saudi-Bangla Fish Feed Mill, Bhaluka, Mymensingh, Bangladesh. Its head office address is: Saudi-Bangladesh Industrial and Agricultural Investment Co., Steel House (5th Floor), Kawran Bazar, Dhaka, Bangladesh.

ON-FARM FEED FORMULATION AND MANUFACTURING

In traditional fish culture, farmers stock their ponds but usually do not use any supplemental feed. With increasing demand and market value, farmers are now beginning to use supplementary feeding to increase the production of cultured fish.

Table 6. Prices and output of aquafeeds from the Saudi-Bangla Fish Feed Factory (Bangladesh)

	Production 1991 (actual) (t)	1992 (10 month) (t)	1992 (target) (t)	Price (1992) (US$/kg)
Fish Feed (protein)*				
Nursery (30%)	100	300	n.a.	0.64
Grower (26%)	100	300	n.a.	0.38
Sub-total	200	600	900	
Shrimp Feed (protein)				
Nursery (40%)	50	100	n.a.	1.14
Starter-1 (39%)	100	200	n.a.	1.02
Starter-2 (38%)	100	200	n.a.	1.02
Grower (36%)	150	400	n.a.	1.02
Finisher (35%)	100	300	n.a.	1.02
Sub-total	500	1,200	2,000	
Total production (t)	700	1,800	2,900	

* sold for both carp and catfish

Table 7. Proximate analyses of aquafeeds manufactured by the Saudi-Bangla Fish Feed Mill (Bangladesh)

	Composition (as received) Moisture (%)	Lipid (%)	Protein (%)	Ash (%)	Fibre (%)	NFE (%)	Energy kcal/kg
Fish Feed							
Nursery	6	10	30	11	8	35	3,500
Grow-out	7	6	26	12	10	39	3,000
Shrimp Feed	6-8	3-9	35-40	12-16	n.a.	30-38	3,000-3,500

The formulae of farm-made feeds, used by farmers or researchers for various cultivated species, are diverse. Some farmers use only rice bran for both carp nurseries and grow-out ponds. Some also use rice bran for tilapia and silver barb culture. Sometimes substitution of wheat bran for rice bran is tried by scientists and farmers for tilapia and silver barb culture and also for carp nursery operations. Some carp nursery operators use only mustard oil cake or sesame oil cake, soaking it with water for several hours before spreading it over the pond surface.

Other carp nursery operators formulate feeds with various proportions of mustard oil cake and rice bran, ranging from 40:60 to 60:40, in a dry, non-pelleted form. Sometimes sesame oil cake and wheat bran are substituted. Scientists using a carp nursery feed composed of rice bran (50%), mustard oil cake (25%) and fish meal (25%) got higher survival and better growth than with farmers' feeds (Haque and Hossain 1991). Another carp nursery feed formulated of silk-worm pupae (25%), sesame oil cake (50%), wheat bran (21%), vitamin/ mineral premix (1%) and molasses (3%) when fed to rohu (*Labeo rohita*) fry was found to promote higher growth and survival (Zaher *et al.* 1992). An alternative improved nursery feed is shown on Figure 3.

In carp polyculture farmers use a mixture of rice bran (50-80%) with mustard oil cake or sesame oil cake (20-50%) (Figure 4). The hand-made feed is usually formed into wet, non-pelleted balls (Figure 5) but sometimes it is prepared as a dry, non-pelleted mixture. Most fish farmers feed their broodstock with a supplementary feed comprising of 50% rice bran or wheat bran, 45% mustard oil cake and 5% fish meal (Islam 1989).

In catfish (*Clarias batrachus, Clarias gariepinus, Ompok pabda, Mystus tengra*) nurseries, live feeds, such as *Tubifex* and sometimes zooplankton are used. For catfish grow-out some farmers use a mixture of 40% fish meal, 40% mustard oil cake and 20% rice bran in non-pelleted moist form.

Farm-made shrimp feeds with 60% animal viscera, 20% rice bran and 20% wheat flour, or with 40% fish meal, 35% rice bran and 25% wheat flour are fed moist, or sun-dried, stored and used later (Karim 1986).

Farmers do not use any equipment to prepare feed, they are made by hand. Locally made sieves are used to sift the ingredients. On the other hand scientists use a mixer and mincer (Figure 6). A simple pellet mill is also used in research work, supplied by David Brown, Radicon, England.

Figure 3. Improved formulated carp nursery feed with 30% protein content (50.0% mustard oil cake, 24.9% rice bran, 21.1% fish meal, 3.0% molasses and 1.0% vitamin premix)

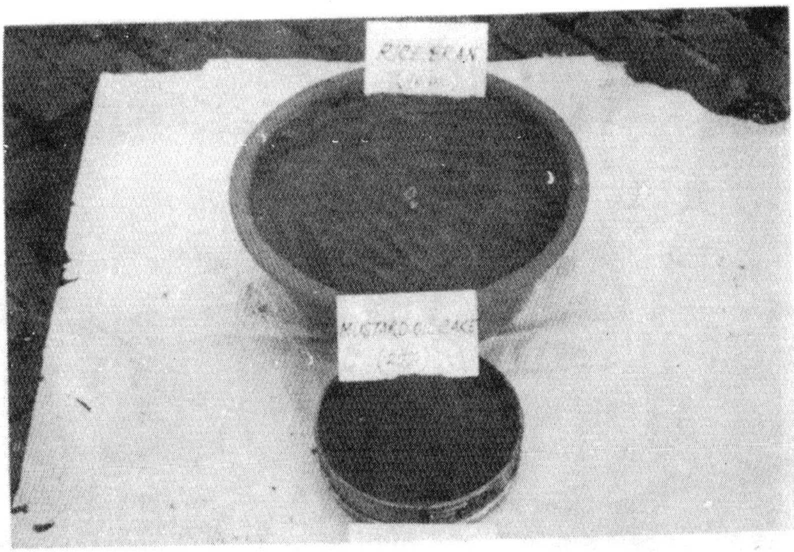

Figure 4. Traditional farm-made feed mixture (75% rice bran and 25% mustard oil cake) for carp polyculture

Figure 5. Manual preparation of farm-made moist feed balls (75% rice bran, 25% mustard oil cake)

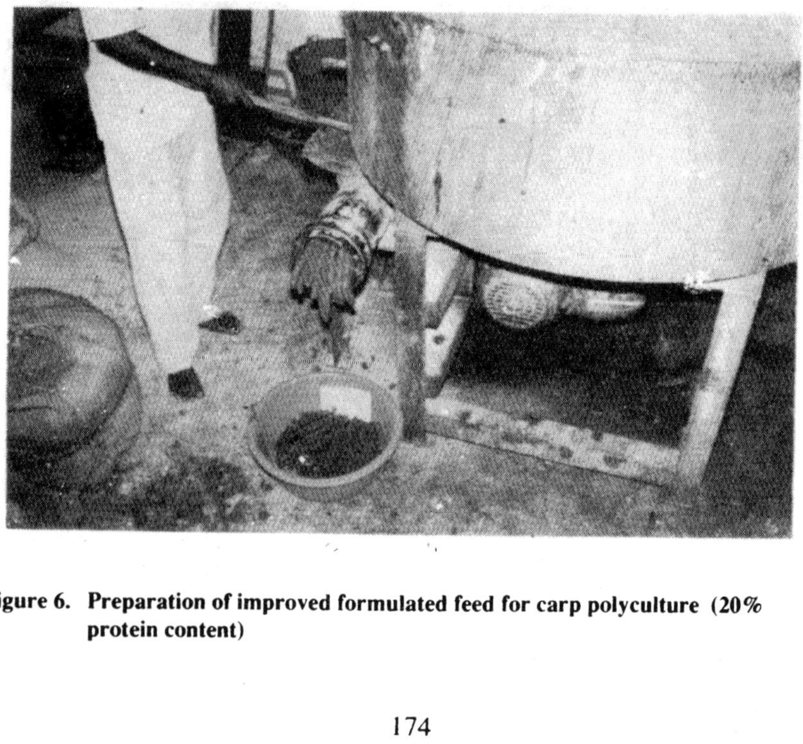

Figure 6. Preparation of improved formulated feed for carp polyculture (20% protein content)

ON-FARM FEEDING STRATEGIES

Proper utilization of feed largely depends on feeding rates and frequencies. For nursery management of carps, farmers apply feed at 3 times the total stocking biomass, and increase this volume by 2 kg/ha every 10 days until the end of the 60 day nursing period, with a stocking density of 2-3 kg/ha and a fertilization rate of 1,500-2,500 kg/ha of cow dung, or 25 kg/ha of urea and 19 kg/ha of TSP, every 10 days. Feeding is usually twice a day, morning and afternoon. Feeding rates and frequencies vary from farmer to farmer. Some farmers apply feed only once a day, while others apply feed every alternate day or irregularly.

For the grow-out of carps and catfish farmers apply feed at the rate of 2-8% of the body weight, once a day, usually in the morning. Sometimes feed is given every alternate day or without any schedule. Water exchange is not practiced, nor is there provision for it.

In shrimp farming farmers apply feed at 7-15% of body weight per day for nursery management and 5-10% per day for starter, both twice a day (usually early morning and evening). Grower and finisher feeds are fed at 3-5%, once a day, either in the early morning or evening.

MAJOR INSTITUTIONS INVOLVED IN AQUAFEED RESEARCH AND DEVELOPMENT

The Fish Feed and Nutrition Division of the Fisheries Research Institute has been involved in aquafeed research and development since its inception in 1986. The Fisheries Faculty of the Bangladesh Agricultural University has also been involved in basic aquafeed research. Names and addresses of these two institutions and the principal investigators, with their area of study, are given in Annex 1.

CURRENT NATIONAL REGULATIONS

As the manufacture of aquafeeds is new to Bangladesh, no specific regulations are existent to control their processing or sale. The government has allowed the aquafeed industry to develop under industrial regulations. Existing import policies do not permit the import of aquafeeds or feed ingredients, other than chemical-like ingredients. Each manufactured product must be tested and certified by the Bangladesh Standard and Testing Institution before marketing under the regulations in practice.

CURRENT PROBLEMS AND CONSTRAINTS

Though Bangladesh has vast water bodies suitable for aquaculture, fish and shrimp culture is still traditional in nature. Intensive aquaculture is not yet practiced and semi-intensive aquaculture is a recent thrust. Current problems and constraints in the manufacturing and use of aquafeeds in the country are:

- demand for and use of manufactured aquafeeds is very low and not yet conducive to the development of commercial production;
- information on the nutritional requirements of Indian major carps for grow-out fish are available but not for nursery and broodfish. For catfish and other target species, nutritional studies are incomplete and knowledge is insufficient for the formulation and manufacture of feeds;
- information on the digestibility of locally available ingredients is not available;
- some important ingredients, such as good quality fish meal, soybean meal, growth promoting substances, fungicides and preservatives are not available;
- some potential fish feed ingredients, such as blood meal, viscera meal, slaughterhouse wastes, poultry processing wastes, silkworm pupae, *etc.*, are not available in processed and refined form;
- research on the development of suitable aquafeeds is incomplete and hence no technology is available to produce good quality aquafeeds on a commercial basis;
- skilled manpower in fish feed manufacturing and fish nutrition is insufficient;
- funding for research and development is inadequate. Lack of proper equipment and machinery seriously hinders aquafeed research and development;
- current transportation and marketing systems for commercial aquafeeds are not well organized;
- extension work to motivate farmers to use aquafeeds in order to increase production and economic benefits is inadequate;
- above all, the poor socio-economic conditions of farmers does not permit the use of feed for fish and shrimp culture. Bangladesh is a poor developing country where the *per caput* income is very low; farmers cannot afford considerable expenditure on aquafeeds. This is the greatest constraint on the manufacturing and use of aquafeeds.

TRENDS IN AQUAFEED MANUFACTURING AND USE

The Fisheries Research Institute, the Directorate of Fisheries, the Bangladesh Agricultural University and various non-governmental organizations (NGO's) have been working to increase aquaculture production in Bangladesh. Farmers, hatchery and farm owners are now more aware of the need to use aquafeeds to increase production. An increasing trend in the use of shrimp feed in coastal shrimp farming is being observed. About 10-15% of the shrimp farmers now use supplementary or complete feeds and achieve production levels of 1,000-2,000 kg/ha compared with 120-200 kg/ha without feed. This encouraging result inspires others to use shrimp feeds also.

Recently, pond aquaculture has increased tremendously. Many ponds are now under improved aquaculture practices, with the supervision of the above-mentioned organizations. Production has increased from 600-800 kg/ha without feeding to 2,000-3,000 kg/ha when farmers use aquafeeds. Farmers are now becoming more interested in carp polyculture, fish seed production and catfish culture. Trends in applying aquafeeds in pond culture, especially in carp polyculture and fish seed production, are encouraging. The use of fish feeds in pond aquaculture has increased from 5-10% to 15-20%.

Due to increasing demand for aquafeeds, investors have now taken keen interest in the establishment of new mills. Two private enterprises, Rakamary Matsha Khamar and Bengal Fisheries International, have already planned to establish new fish/shrimp feed mills. In this connection, the proprietor of Rakamary Matsha Khamar has already communicated with FRI and the Saudi-Bangla Fish Feed Mill and has visited Singapore. There is hope that the new mills will be established in the near future.

LIST OF REFERENCES

Ahsanullah, M. 1989. Fisheries extension and credit in Bangladesh, p 84-92. *In* Proceedings of the SAARC Workshop on Fish Seed Production, 11-12 June 1989. Fisheries Research Institute, Mymensingh, Bangladesh.

Anon. 1986. p. 21-49. *In* Nutritive values of local feedstuffs. Institute of Nutrition and Food Science, Dhaka University, Dhaka, Bangladesh.

A.O.A.C. 1965. Official methods of analysis. Association of Official Agricultural Chemists, 10th ed., Washington D.C., USA. 957 p.

F.F.R.S. 1987. Fish catch statistics of Bangladesh (1984-85). Fisheries Resources Survey System, Department of Fisheries, Dhaka, Bangladesh. 13 p.

F.R.I. 1989. Survey of potential fish feed ingredients of Bangladesh on the basis of their availability and biochemical composition. Research Project Report No. 1; Fisheries Research Institute, Mymensingh, Bangladesh. 70 p.

G.O.B. 1986. The Third Five Year Plan, Government of Bangladesh. 1986-1990. Chapter V.C. Fisheries. Dhaka, Bangladesh. 11 p.

G.O.B. 1991. The Fourth Five Year Plan, Government of Bangladesh, 1991-95. Chapter V.C. Fisheries. Dhaka, Bangladesh. 13 p.

Haque, M.Z. and M.M. Hossain. 1991. Progress Report on the development of nursery pond management methods for optimizing production of fry/fingerlings of carps. Manuscript. Fisheries Research Institute, Mymensingh, Bangladesh. 15 p.

Islam, M.A. 1989. Fish seed production in Bangladesh, p 1-12. *In* Proceedings of the SAARC Workshop on Fish Seed Production, Fisheries Research Institute, Mymensingh, 11- 12 June.

Karim, M. 1986. Brackishwater shrimp culture demonstration in Bangladesh, BOBP/REP/5, FAO/SIDA, Madras, India. 44 p.

Mahmood, N. 1990. Potential and strategies for brackishwater aquaculture development in Bangladesh. Paper presented at the National Workshop on Aquaculture Strategies in Bangladesh. Bangladesh Agricultural Research Council, Dhaka, Bangladesh.

Mahmood, N. and M.R. Chowdhury. 1989. An overview of shrimp fishery of Bangladesh. Biological aspects. Paper presented at the Workshop on the Environmental Aspects of Agricultural Development and Surface Water Systems. Organized by the Bangladesh Centre for Advanced Studies at the BARC Auditorium, Dhaka, 8-9 November 1989. Manuscript.

Mazid, M.A. 1992. Aquaculture development in Bangladesh. The Bangladesh Observer, Dhaka. 25 September, 1992.

New, M.B. 1987. Feed and feeding of fish and shrimp. ADCP/REP/87/28. FAO, Rome, Italy. 275 p.

Statistical Yearbook of Bangladesh. 1989. Bangladesh Bureau of Statistics, Statistic Division, Ministry of Planning, Dhaka, Bangladesh. 365 p.

Zaher, M., M.A. Mazid and S. Gheyasuddin. 1992. Progress Report on development of cost-effective feeds from indigenous raw materials for fingerlings and adult carps. Manuscript. Fisheries Research Institute, Mymensingh, Bangladesh. 21 p.

Annex 1. Institutions, principal investigators involved, and areas of aquafeed research and development in Bangladesh

Fisheries Research Institute (FRI) P.O. Keyotkhali Mymensingh-2201	
Dr. M.A. Mazid Director FRI	Fish nutrition (nutritional requirements of Indian major carps and tilapia)
Dr. A.K.M. Aminullah Bhuiyan Consultant Department of Fisheries 2nd Aquaculture Project	Fish processing (survey of potential fish feed ingredients of Bangladesh on the basis of their availability and biochemical composition)
Mr. Muhammad Zaher Senior Scientific Officer FRI Freshwater Station	Fish feed technology (development of feeds from indigenous materials for commercially important freshwater fish)
Bangladesh Agricultural University (BAU) Faculty of Fisheries Mymensingh-2202	
Dr. Syed Gheyasuddin Professor BAU Dept. of Biochemistry	Biochemistry (development of cost effective feeds for carp fingerlings and adults from indigenous raw materials)
Dr. Mohammad Rezaul Hassan Associate Professor BAU Dept. of Fisheries Biology and Limnology	Fish biology (diet formulation for *Labeo rohita* using various commercially available, indigenous ingredients - IFS project)

Dr. Mohammad Arshad Hossain Associate Professor BAU Dept. of Fisheries Technology	Fish nutrition (diet formulation for intensive *Heteropneustes fossilis* culture using indigenous feed ingredients - IFS project)
Mr. Md. Ahsan Bin Habib Associate Professor BAU Dept. of Fisheries Biology and Limnology	Fish biology (culture of *Clarias batrachus* using formulated feeds - IFS project)
Dr. Muhammad Asadur Rahman Professor BAU Dept. of Fisheries Technology	Fish processing (diet formulation for intensive culture of *Clarias batrachus* and *Heteropneustes fossilis* using indigenous ingredients)
Dr. Fazlul Awal Mollah Associate Professor BAU Dept. of Fisheries Biology and Limnology	Fish breeding (formulation of nursery feed for *Clarias batrachus*)

AQUAFEEDS AND FEEDING STRATEGIES IN CAMBODIA

S. Nuov[1] and M.C.Nandeesha[2]

[1]Department of Fisheries, Ministry of Agriculture
Phnom Penh, Cambodia

[2]College of Fisheries, University of Agricultural Sciences
Mangalore - 575 002, Karnataka, India

NUOV, S. and M.C. NANDEESHA. 1993. Aquafeeds and feeding strategies in Cambodia, p. 181-200. *In* M.B. New, A.G.J. Tacon and I. Csavas (eds.) Farm-made aquafeeds. Proceedings of the FAO/AADCP Regional Expert Consultation on Farm-Made Aquafeeds, 14 - 18 December 1992, Bangkok, Thailand. FAO - RAPA/AADCP, Bangkok, Thailand, 434 p.

INTRODUCTION

Cambodia has acquired a special place in history for the sufferings it has undergone in this modern age not only due to the internal conflicts of its own people, but also to its isolation by the international community. These factors have impeded development in almost all spheres, the worst affected being the food sector. Planning for the rehabilitation and reconstruction of Cambodia has been initiated by the international community and it is anticipated that, like other sectors, fisheries will also receive support.

In order to provide a better understanding of the existing situation in Cambodia, some information on its general and agricultural status has been included in this section of the paper.

Agriculture

Cambodia, with an area of 17.64 million ha and a population of 9.01 million, is primarily an agrarian state with 80% of the country's labour force being employed in agricultural sector (Table 1). Nearly 70% of the land area is still reported to be covered by forests and only about 3.63 million ha are under agriculture. Rice is the most important crop cultivated with an area of about 1.9 million ha. The climate and topography have not provided much opportunity for crop diversification from rice to other food or cash crops in most parts of the cultivable land (Table 2). The average yield of paddy is only 1,200 kg/ha.

Although, before the war, rice was an important export item, presently there is a shortage even for internal consumption. In terms of production, agriculture contributes nearly 50% of the GDP (Table 1). Per caput income is estimated to be about US$ 180. Nearly 30% of the families are headed by women because so many men lost their lives during the war. While there is a large trade imbalance, inflation is now running at 190%. These unusual circumstances pose several problems to developmental activities.

Table 1. General geographic, demographic and economic indicators of Cambodia in 1992

Total land area (million ha)	17.64
Agricultural land area (million ha)	3.63
Forest and woodland (million ha)	13.37
Population (million)	9.01
Female population (%)	51.70
Density of population (person/km)	49.00
Rural population (%)	91.68
Urban population (%)	8.32
Population growth rate (%)	2.80
Total gross domestic product (million US$)	1,964.50
Agriculture GDP (%)	49.86
Livestock GDP (%)	6.59
Fisheries GDP (%)	2.04
Agricultural production (%)	45.00
Per caput GDP (US$)	180
Foreign trade	
- convertible area (1991)	
- exports (million US$)	25.50
- imports (million US$)	52.50
- non-convertible area (1990)	
- exports (million Roubles)	20.10
- imports (million Roubles)	103.30

Exchange rate (October, 1992) : US$ 1.00 = Riel 2,050

Sources: various reports and government departments

Table 2. Production of major crops in Cambodia in 1991

Crop	(1,000 t)
Rice	2,400
Maize	56
Cassava	75
Sweet potato	54
Vegetables	249
Mungbean	27
Groundnut	33
Soybean	6
Sesame	304
Sugarcane	9
Tobacco	6
Jute	1
Rubber	24

Source: Department of Agriculture, Phnom Penh

Animal husbandry

Besides agriculture, animal husbandry forms an important component in the Cambodian economy and contributes about 6.6% to the GDP (Table 1). The numbers of livestock, which had declined drastically during the 1970's is now reported to have reached pre-war levels (Table 3). They form an important source of animal protein for the more affluent section of the population, aside from their contribution to animal energy used in agriculture and through the generation of manure. Most of the livestock is of traditional breeds and attempts are now being made to introduce improved varieties and production techniques.

Table 3. Total livestock in Cambodia in 1991

Livestock	(Million)
Cows	2,234
Buffaloes	737
Pigs	1,516
Poultry	8,164
Draught oxen	1,017
Draught buffaloes	483

Source: Livestock Department, Phnom Penh

Fisheries

Fish and rice are important staple diets of Cambodians. This food habit has been largely due to the influence of nature as the country is bestowed with a large number of rivers, the Great Lake and a vast amount of flooded areas, which still provide relatively good sources of fish. Although Cambodia has a coastline of 425 km, marine fish has been of less importance in the Cambodian diet as most people prefer freshwater fish. Fish consumption was reported to be as high as 20-25 kg/*caput* in 1960's, but this declined to about 13 kg/*caput* /yr in 1991. Total fish production in 1991 was estimated to be about 100,071 t out of which 74,700 t came from the freshwater sector. Figure 1 illustrates the annual variations in freshwater fish production between 1940 and 1991. There is no data available for some years, particularly for the period 1970-1979 as these were years of intensive unrest and organized fishing was probably almost non-existent. It is widely argued that there is little or no scope to return to the 1957 capture fisheries production level of 130,000 t. At best, with appropriate scientific measures, it may be possible to maintain the current production. Degradation of natural spawning grounds through extensive destruction of flooded forests and a large number of "Pol Pot canals" constructed during the Pol Pot regime have been mostly blamed for the decline in natural catches.

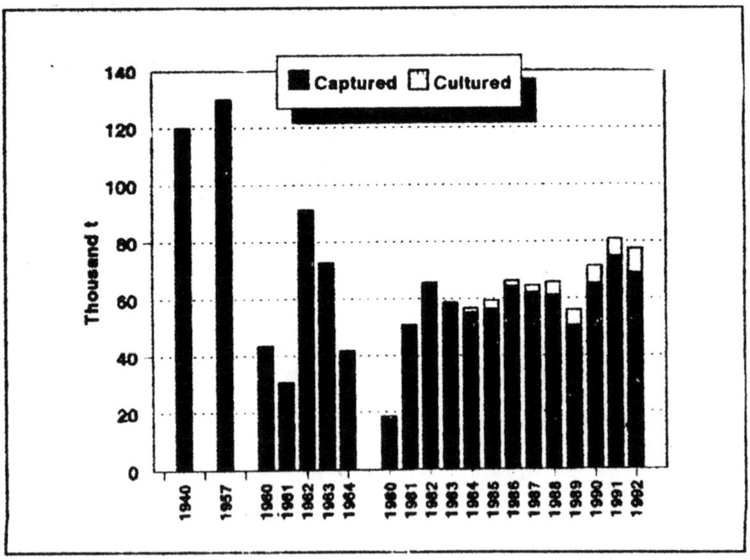

Figure 1. Inland fish production in Cambodia from 1940 to 1992

Aquaculture

Aquaculture in Cambodia is relatively young as compared to other Asian countries. Although attempts were made during the 1950's to introduce pond fish culture, mainly through introductions of tilapia and Chinese carps, it did not widely catch on, probably due to the abundant availability of captured fish. Bardach (1959) reported that a few farmers in some rubber estates were successful in raising Java tilapia (*Oreochromis* spp.). It was also reported that the culture of Chinese carps in ponds around Phnom Penh was successful as these fish were popular with the Chinese community. In contrast to pond culture, the culture of fish in cages has been in practice for nearly a century and it is believed that cage culture originated in Cambodia (Chua and Tech 1990). However, this activity almost disappeared during the 1970's and it only started again in the 1980's. Pond culture, mainly of pangasid catfishes, was also started by a few enterprising farmers around Phnom Penh in the 1980's and this activity is slowly expanding to other areas of the country. Freshwater aquaculture production has increased by a factor of four over the years 1984-1991 (Table 4). Feed is the most important cost component (more than 70%) of existing aquaculture systems. Coastal aquaculture activity is yet to be initiated but the recent liberalized policy of the government is expected to attract large investors to this sub-sector.

Table 4. Aquaculture production in Cambodia

Year	(t)
1984	1,610
1985	3,000
1986	2,200
1987	2,500
1988	4,600
1989	5,538
1990	6,400
1991	6,700

Source: Department of Fisheries, Phnom Penh

Species cultured

Both indigenous and exotic species are cultured, the former being dominant (Table 5). Among the exotics, tilapia culture is widely attempted by small-scale farmers. Culture of other exotic species is restricted by the availability of fish seed. Among the indigenous groups, pond culture of *Pangasius* spp. is more prevalent around Phnom Penh city. Cage culture of *Pangasius* spp. and *Channa micropeltes* is widely practiced in several parts of the country. Cyprinids are the next important species group cultured in cages. Culture of other species is also seen in some areas, in both cages and ponds, but their contribution to aquaculture production is still very small or, in some cases, is on an experimental basis only by some farmers.

Table 5. Indigenous and exotic fish species cultured in Cambodia

Scientific name	Common name	Cultured in ponds	Cultured in cages
Indigenous species			
Pangasius sutchi	Striped catfish	+	+
Pangasius larnaudii	Black ear catfish	-	+
Leptobarbus hoevenii	Hoeven's slender carp	+	+
Cirrhinus microlepis	Mud carp	+	+
Puntius gonionotus	Silver barb	+	+
Puntius altus		+	+
Clarias batrachus	Walking catfish	+	+
Trichogaster pectoralis	Snakeskin gourami	+	+
Channa micropeltes	Giant snakehead	+	+
Anabas testudineus	Climbing perch	+	-
Notopterus chitala	Featherback	-	+
Exotic species			
Aristichthys nobilis	Bighead carp	+	-
Hypophthalmichthys molitrix	Silver carp	+	-
Ctenopharyngodon idella	Grass carp	+	-
Cyprinus carpio	Common carp	+	-
Oreochromis niloticus	Nile tilapia	+	-
Oreochromis mossambicus	Java tilapia	+	-

AVAILABILITY OF FEED INGREDIENTS

No published data is available on the accessibility of various types of ingredients. The following information is based on the observations of the authors as well as information provided by various NGO's working in the country.

Plant resources

Since paddy is the most widely cultured crop (Table 2), rice bran of different qualities is an important feed ingredient available throughout the country. Its quantity is estimated to vary between 0.2 and 0.3 million t. Considerable amounts of broken rice are also produced during milling and these are used by some farmers, particularly for the early age groups of fish. Normally two grades of rice bran are obtained during the milling process: first grade and second grade, the former with a considerable percentage of broken rice. Most of the maize produced is exported; little is consumed domestically. Some farmers also use waste corn flour as fish feed when it is cheaply available.

Soybean is used only in the nurseries of fish seed production farms. Although some oilseed crops are also grown, as yet there is no organized oil extraction in the country. Coconut is one of the important plantation crops, but there is no data available on the extent of production.

Farmers engaged in small-scale aquaculture use a variety of wastes available from agricultural crops. Soft aquatic vegetation, like *Lemna, Azolla, Hydrilla* and morning glory are also utilized as feed for herbivorous and omnivorous fish. A large portion of the water bodies is infested with a variety of aquatic vegetation, the most prevalent being water hyacinth. These aquatic weeds can form a potential source of feed ingredients if suitable techniques for their use could be demonstrated.

Animal resources

Among the animal resources of feed ingredients, fresh fish and/or dried fish are the most commonly used. While there is no organized production of fish meal, most farmers buy the required quantity of fresh fish during the peak fishing season and use it as feed after drying. In the cage culture of both pangasids and snakeheads, the fish are fed with fresh fish throughout the culture period. Fish caught from the wild are classified into three categories based on size: 1st grade, 2nd grade and 3rd grade. While 1st and 2nd grade fish are generally used for human consumption, 3rd grade fish are utilized for making fish sauce and fish paste. Low quality fish from this group is generally utilized for feeding pangasid catfishes and snakeheads in fresh form or is converted into dried fish for feeding *Pangasius* in ponds.

The quantity of the different grades of fish harvested during 1989-1991 is presented in Table 6. It appears that, based on a food conversion ratio of 5:1, 70-80% of the 3rd grade fish produced in the country is presently utilized for feeding pangasid catfishes and snakeheads. In addition to fish, waste fish heads, available when dressing fish for fish paste (prahok), are also utilized by the farmers, mainly for feeding pangasids raised in ponds. Other potential sources of animal feeds include poultry offal, blood meal, meat and bone offal, rumen contents, pig offal, the manure of these animals, etc. However, none of these wastes are presently utilized in fish culture, either due to unorganized slaughtering or to difficulties encountered in the collection and transportation of these wastes to the culture sites.

Table 6. Quantity and quality of fish produced in Cambodia

Year	1st grade	2nd grade	3rd grade
1989	10,200	13,200	27,100
1990	9,660	15,643	39,778
1991	13,465	17,622	43,585

Source: Department of Fisheries, Phnom Penh

It should be noted that most parts of the slaughtered animals still have a greater value as human food. Only pig and cattle manure is utilized by a few farmers in rural areas for pond fertilization. Available information on the proximate composition of a few feed ingredients is presented in Table 7.

Table 7. Proximate composition of some feed ingredients available in Cambodia (%, on "as received" basis)*

Ingredient	Moisture	Protein	Fat	Ash
1st grade rice bran	8.8	10.3	11.3	12.6
2nd grade rice bran	7.9	6.0	4.4	11.9
	(7.3-8.4)	(5.2-6.6)	(1.6-9.0)	(11.1-12.7)
Dried fish	5.3	45.3	27.0	17.1
	(3.0-7.2)	(44.4-46.0)	(24.6-29.1)	(16.9-17.4)
Fish powder	7.6	29.0	10.3	21.7
Cut fish heads	4.2	31.8	24.1	30.7
	(4.0-5.6)	(25.4-38.5)	(19.0-28.7)	(28.8-34.9)
Feed mixture**	7.6	10.9	5.2	13.3
Corn powder	8.8	8.4	4.5	11.1
Full fat soybean	8.1	43.6	16.2	6.7
	(5.6-10.7)	(42.6-44.9)	(14.0-18.6)	
Colocassia powder	10.6	21.5	6.3	n.a.
Water hyacinth powder	11.5	21.4	2.5	n.a.
Morning glory powder	12.2	19.5	2.7	n.a.
Green plant manure	11.9	24.6		n.a.

* figures represent average of at least two measurements; figures in parentheses indicate range of values observed

** feed mixture sampled at Chak Angre Station is a mix of dried fish, rice bran and corn meal

Source: Nandeesha (1991)

ON-FARM FEED MANUFACTURING AND FEEDING STRATEGIES

There is no commercial feed production in Cambodia and most of the feed is farm-made on a daily basis and fed moist. Broadly, the following five types of fish culture systems can be distinguished in the country:

- pond culture of pangasid catfish;
- pond culture of tilapia and silver barb;
- cage culture of pangasid catfish;
- cage culture of snakeheads;
- cage culture of other finfish species.

Some information on the culture practices is available in recent publications (Csavas 1990; Edwards 1991; Nandeesha 1991; Anon. 1992; Nandeesha *et al.* 1992). In the current paper, the above systems are examined further, from the view point of the feeding strategies adopted by the farmers. In all of the above systems feed constitutes an important component and production is largely governed by its quality and quantity.

Feeding strategies in pond culture of pangasids

The culture of pangasid catfishes, mostly consisting of *Pangasius sutchi*, is prevalent in ponds located around the capital of the country, Phnom Penh. Culture of pangasids is gaining momentum even in other parts of the country. The ponds generally vary from 600-2,000 m^2. Fingerlings, weighing more than 50 g, are stocked during the monsoon months (from June to October) at a density of more than 3,000 kg/ha. The stocking density and the weight of fingerlings varies considerably from farmer to farmer. Most ponds have a stagnant water system and possess no drainage facility. They are either filled with rainwater or water is pumped from the nearby lake. Most farmers do not use fertilizers. However, some apply urea at the rate of 5-6 kg/pond of about 800 m^2 area once in two months to keep the water green. Some farmers engaged in *Pangasius* culture have latrines overhanging their ponds.

The feeding strategy adopted is largely based on the availability of feed ingredients rather than on any scientific consideration. Fish are fed during the first two months with first quality rice bran. Some farmers also use a 9:1 ratio of broken rice mixed with rice bran during this period. From the third month onwards fish are mostly fed second quality rice bran up to November. Rice bran is always fed in cooked form. The bran, mixed with a small quantity of water, is cooked in an open vessel until a proper dough is obtained. Firewood is generally used for cooking. The cooked dough is allowed to cool and then rolled into small balls and fed to the fish in wet form. Feeding is done during the cool hours of the day, generally in the evening.

With the onset of the fishing season, feeding shifts from rice bran to small fish, as the latter become abundantly and cheaply available. Farmers either buy fresh fish and sun-dry it or buy dried fish and store them for daily feeding. In addition to fish, cut fish heads obtained as a by-product in fish paste manufacturing, are also used by some farmers. Fresh or dried fish is fed generally without any processing but some farmers mix the dried fish with 10-20% rice bran and feed in an uncooked form. The feeding of dried fish continues up to March-April of the following year. Feeding dry fish is drastically reduced or discontinued from April onwards, since farmers believe that the higher water temperatures prevailing later are not conducive to this practice. Rice bran alone, or with a small added quantity of dried fish, continues to be fed until the ponds are harvested.

Feed quantity is not accurately calculated, based on body weight. During the first two months, fish are given feed at approximately 10% of body weight and the quantity is gradually reduced as the fish increase in weight. Towards the end of the culture period the feeding level is about 1.5-2.0%. Fish mortality is a common problem encountered from April onwards due to higher biomass, increased water temperature, the accumulation of organic wastes and lack of water exchange. Unhygienic drying and storage of the fish used for feeding often leads to contamination and spoilage. As can be seen in Table 7, the fat content of both dried fish and fish heads is high.

Fish are harvested after they have attained a weight of more than 1 kg. Significant variability in the weight of fish is common due to variations in the size of fish at stocking. The harvesting period normally coincides with the monsoon, as there is a high market demand for food fish in this period due to the closed fishing season in the natural waters. Production varies widely depending on the intensity of the management practices adopted. Some of the production data provided by farmers indicate that a net production of above 35 t/ha is quite common in well managed ponds. The food conversion ratio varies from 4.5-6:1.

In earlier years farmers tended to stock tilapia together with *Pangasius*. Presently, the majority of farmers avoid this practice because of the uncontrolled breeding of tilapia and the poor demand for small-sized fish in the market. However, when tilapia are stocked, they are fed with uncooked rice bran broadcast on the pond surface. Some farmers also stock carps with *Pangasius*, but there is no special feeding practice adopted for cyprinids. These fish appear to derive most of their food through the natural food biomass produced in the pond by the unconsumed food and by faecal fertilization. Non-availability of carp seed and lack of information on appropriate species combination has been the major constraint in the spread of composite culture techniques in Cambodia.

Feeding strategies in cage culture of pangasids

Cage culture of *Pangasius* spp., consisting largely of *P. sutchi* and, to a lesser extent, *P. larnaudii* is common in several parts of Cambodia. A large part of the cage culture activity is concentrated in and around the Great Lake where cage size is generally large (26-900 m^3). Some farmers also specialize in the nursery rearing of *Pangasius* seed collected from rivers. Cages are generally stocked with fish of more than 100 g at a stocking rate of 4-12 kg/m^3. Fish are usually fed trash fish when it is available. A large quantity of small fish is spread on the cage surface and allowed to rot because farmers believe that *Pangasius* feed voraciously on spoiled fish. Under such circumstances, fish are normally fed only once every two or three days.

When trash fish is not available, cooked rice bran is used (Figures 2 and 3).. A pelletizing machine is installed on some of the larger cage sites.

Figure 2. Cooking feed for cage cultured pangasid catfish

Figure 3. Hand-feeding pangasid catfish in a boat-shaped floating cage

Cooked rice bran is pressed through a mincer to roll it into small balls which are allowed to fall directly into the cage (Figure 4). When even rice bran is not available, aquatic vegetation, consisting mainly of *Lemna, Azolla* and morning glory, is used as feed. Aquatic vegetation is only used as a last resort, since farmers have experienced poor growth rate in fish fed continuously with plant materials over a long period.

35-60 kg/m^3 of fish is produced in 8-12 months. The food conversion ratio is reported to be 4-6:1. Farmers' instinct tells them that fish need to be fed more than 4 kg feed for each kilogramme of fish produced and they tend to achieve this ratio by providing more feed during the glut season. Fish mortality during the summer months, particularly during April and May, is a common phenomenon due to high temperatures and reduced water exchange, coupled with the deterioration of water quality because of the discharge of wastes from the cages, which are often crowded in one locality.

Figure 4. Locally manufactured feeding machine for large catfish cages

Feeding strategies in snakehead culture

Two species of snakehead, namely *Channa micropeltes* and *C. striatus*, are cultured in cages. The former is the most popular due to its faster growth. Unlike the culture of pangasids, snakehead rearing is entirely dependent on an animal source of protein, namely trash fish. Cages used for snakehead culture are smaller (7-211 m^3). Fish are normally stocked during the monsoon months (June to August) with fish of more than 50 g at 20-30 kg/m^3. Fresh trash fish caught from the rivers is fed daily, normally *ad libitum*. When trash fish is unavailable, the snakehead are starved for periods of up to several weeks. Larger trash fish are chopped before feeding. Some farmers also use suspended feeding trays within the cage.

The feed conversion rate (FCR) is 4.5-6:1. Again farmers make the general assumption that a minimum of 4 kg of trash fish is needed to produce 1 kg of marketable fish. During periods of trash fish glut farmers feed their fish more than once a day to reach these levels. Occasionally, some farmers feed rats captured from paddy fields, particularly when trash fish becomes scarce. Recently, the culture of *Pangasius* along with snakehead but in a separate cage, is gaining popularity; this utilizes leftover trash fish from the snakehead cages.

Fish achieve more than one kilogramme in 8 to 12 months. Growth is dependent on the availability of feed and stocking density. Mortality during the summer months is the most common problem due to high water temperatures and reduced water exchange. Most cages are in crowded conditions resulting in severe water pollution due to the discharge of wastes.

Feeding strategies in pond culture of tilapia and silver barb

Fish culture has been initiated during the last 3-4 years in ponds constructed under the Food for Work programme of UNICEF and the World Food Programme. Tilapia and/or silver barb are generally stocked during the monsoon, when ponds can be filled. For various reasons, pond culture has not been uniformly successful. Long-term severe water turbidity which inhibits the level of natural productivity is experienced. Despite this, some farmers have been able to achieve production rates of 60-100 kg in ponds of less than 300 m^2.

The successful farmers, besides fertilizing the ponds regularly with organic manure, feed the fish with rice bran. The quantity increases gradually, as the fish grow, from 0.5 kg/day to 3.0 kg/day. No definite feeding ratios are adopted. The quantity provided is mainly governed by the availability and cost of rice bran. In addition to rice bran, fish are also sometimes fed corn meal, termites, red ants and aquatic vegetation. Almost all the fish pond owners also rear pigs; these are usually given a higher priority for rice bran than fish.

Feeding strategies in the culture of other species

Besides pangasid catfishes and snakehead, *Puntius gonionotus*, *Puntius altus* and *Leptobarbus hoeveni* are cultured in many cages. Small cages (50 m^3) are stocked with a total of 8-10 kg/m^3 (average weight 3-5 g) of cyprinids, and reared for 8-10 months. Rice bran is commonly used as feed, as well as aquatic vegetation and kitchen wastes. 40-50 kg/m^3 is generally obtained.

Clarias batrachus, *Cirrihinus microlepis* and *Notopterus chitala* are also cultured on a small scale, largely as scavenger species in *Pangasius* cages. Attempts to culture sand goby (*Oxyeleotris marmorata*), have not been successful. Currently, cages are used to hold larger sand gobies caught from rivers until they are exported. Even during this short holding period, heavy mortality is common.

Alternating pen/cage culture is another system, seen only in the two northeastern provinces in the Great Lake. As the Great Lake seasonally fluctuates farmers use pens during the dry season and transfer the fish to cages during the wet season when the water level is higher. Fish are heavily fed with trash fish in the pens, and with rice bran in the cages.

Attempts to rear climbing perch (*Anabas testudineus*) and a few indigenous cyprinids in ponds have also been made. Although exotics such as common carp (*Cyprinus carpio*), silver carp (*Hypopthalmichthys molitrix*), grass carp (*Ctenopharyngodon idella*) and bighead carp (*Aristichthys nobilis*) are in demand, due to the lack of seed of these species and a poor extension system to demonstrate their suitability in culture systems, their culture is limited. In those seed farms which exist fish are fed mainly with rice bran; in addition ponds are periodically fertilized. Rice-fish culture is also being tried on a small scale by some farmers, but is not yet established.

MAJOR INSTITUTIONS INVOLVED IN AQUAFEED RESEARCH AND DEVELOPMENT

The country lost the majority of its educated manpower during the Pol Pot regime. The Fisheries Institute, which used to offer a degree course in fisheries, was also destroyed. With the surviving trained personnel, fisheries development programmes restarted only in 1980. A large number of untrained people had to be recruited to manage the fisheries sector. The Fisheries Institute was restarted, with the assistance of the Soviet Union, in 1984 to provide trained manpower. Presently, a four and a half year degree programme in fisheries is offered by the Chamcar Daung Institute of Agriculture and a three year diploma course in fisheries is provided by the Prek Leap Agricultural School. At the latter some in-service training is also offered.

Although some aspects of fish nutrition are taught to students at these two schools, they have not yet started a research programme. A National Freshwater Fisheries Research Station has been established by the Department of Fisheries in Phnom Penh to conduct research on various aspects of fisheries, including nutrition. However, no major research activity has begun yet due to the lack of trained manpower and financial resources. More recently PADEK, a non-governmental organization, has initiated an aquaculture development programme, which includes a major component on fish nutrition and breeding.

CURRENT NATIONAL REGULATIONS

The Department of Fisheries has complete control over fishing in natural waters. The water bodies are divided into fishing lots which are auctioned annually for the fishing season, which begins in October and ends in May. Hence, trash fish is available only during this season. During the closed season, fishermen resort to illegal fishing to obtain feed for cultured snakeheads. Some attempts have been made by the Department of Fisheries to regulate the number of snakehead cages to avoid this practice. Besides this, there are no regulations with regard to the manufacture or quality control of fish feeds.

CURRENT PROBLEMS AND CONSTRAINTS

Existing feeding techniques result in a considerable wastage of feed, even though the supply of feed is often inadequate. The availability of trash fish is constantly declining due to the increasing demand for all varieties of fish for human food. Overcrowding of cages at some localities and consequent impairment of the water quality has been causing considerable production losses. In common with other parts of the region, the nutritional requirements of pond-grown fish in the presence of natural food are unknown.

The variety and quantity of raw materials available as aquafeed ingredients is limited due to competing demand. Most animal "waste" products are still used for human consumption; very little is left over for aquaculture after the animal husbandry sector has taken first pick. Furthermore, due to poor transportation facilities, ingredients are not always available at the places and times when they are required. Even simple machines for feed processing and manufacturing are not easily available.

There is no established credit system. As a result, many farmers borrow money at a high interest rate; most of their profits go to the moneylenders. Finally, due to the lack of trained manpower and resources, there has been no progress in research and development

TRENDS IN AQUAFEED MANUFACTURING AND USE

Trash fish availability will become a major constraint and it is likely that omnivorous and herbivorous species will become more popular than carnivorous species. Research will then be required, to evolve cost-effective diets and feeding strategies. Research must concentrate on deriving the maximum benefit from the natural food produced in the system, because of the national shortage of feed ingredients. Since Cambodia is heading towards a free market economy, it is likely that foreign feed companies will enter the market to develop the aquaculture of valuable fish species for export. Development of on-farm feed manufacturing and feeding strategies is likely to be priority areas of future research. The Department of Fisheries has no definite plans as yet; planning must also depend on external assistance, since expertise is not readily available within the country.

CONCLUSIONS

Although international isolation has contributed to the slow growth of aquaculture in Cambodia, enterprising farmers have made significant progress with some species. Non-availability or shortages of fish in markets, particularly in areas far from rivers, is quite common. Even those living close to rivers do not get abundant supplies of fish throughout the year. A recent study conducted by the Mekong Committee (Anon. 1992) indicated that even to meet the minimum human nutritional requirements in animal food, the present fish production of 100,000 t has to be doubled.

Since there is very little scope to increase fish production from capture fisheries, aquaculture must fill this gap. As fish is popular among Cambodians, the transfer of appropriate technologies would assist the expansion of aquaculture. Fish culture is already recognized by successful farmers as more profitable than agriculture, due to its higher rate of return. In view of the scope for expansion both in commercial aquaculture and family level feed production systems, there is an urgent need to address the following priority fish feeding problems.

Ingredient availability

There is a paucity of information on the type, quantity, quality, seasonality and cost of available feed ingredients. This information is essential to formulate a good aquafeed development programme. A detailed survey is therefore essential to prepare a national feed map using the guidelines suggested by Tacon *et al.* 1987.

Feed management

Existing cage culture practices require closer examination. With the increasing demand for fish, trash fish availability will be reduced. It is therefore necessary to find a substitute diet for snakeheads. Existing feeding practices for pangasids result in considerable wastage of food. The establishment of a factory to make fish meal available at competitive prices throughout the year would help to solve this problem. Since farmers are already familiar with on-farm feed manufacturing the development of cost-effective feed formulae, based on those ingredients which are locally available, would help the industry. The possibility of using some of the cheap marine fish for the production of fish meal should also be considered.

The feeding practices adopted in the pond culture of pangasid catfishes is also dependent on the seasonal availability of feed ingredients and results in considerable wastage. The provision of balanced diets throughout the year would enable farmers to increase production at reduced cost.

Natural food

There is also scope to increase the production of herbivorous and omnivorous fish for culture in cages through improvements in their diets and feeding strategies. Presently, farmers feed mainly rice bran, obtaining moderately good production levels. Natural food inputs are also significant. Quantifying the contribution of natural food organisms would aid the development of cost-effective diets.

The development of aquafeeds and feeding strategies based on ingredients which are easily accessible to farmers and the level to which the nutritional requirements of fish could be met from natural food would pave the way for further expansion of aquaculture (New 1989; De Silva and Davy 1992; Tacon 1992).

Unconventional ingredients

The pond culture of other species also faces shortages of suitable feed or ingredients. While rice bran continues to be an important feed ingredient in rural areas, competition from pig raising often results in its scarcity for aquaculture. Some farmers collect termites and red ants and feed them to fish and it could be worthwhile studying cost effective methods of culturing such promising live food organisms for use as fish feed. Aquatic and terrestrial vegetation may provide good sources of feed for rural aquaculture. The development and/or introduction of simple technologies to culture aquatic vegetation would also help the farmers.

Research and development

There appears to be a number of indigenous fish species suitable for culture in both cages and ponds. Screening such species and studying their nutritional requirements, both in the absence and presence of natural food, should be a long term programme for investigation.

Shrimp culture is reported to have commenced on a small scale with the participation of some foreign companies. Cambodia will thus gain from the experience of other countries in the region in both shrimp culture and the development of a shrimp feed industry.

Financial and human resources

Due to inadequate financial resources and trained manpower there has been no progress in research work. It is urgent to train people in all aspects of aquaculture, including fish nutrition and diet development. External support and expertise is necessary until Cambodia is able to support itself.

LIST OF REFERENCES

Anon. 1992. Fisheries in the lower Mekong basin (Review of the fishery sector in the lower Mekong basin) - Main report. Interim Committee for Coordination of Investigations of the lower Mekong basin, Mekong Secretariat, Bangkok. 92 p.

Bardach, J. 1959. Report on the fisheries in Cambodia. USOM/ Cambodia, Phnom Penh, Cambodia. 80 p.

De Silva, S.S. and F.B. Davy. 1992. Fish nutrition research for semi-intensive culture systems in Asia. Asian Fisheries Science 5:129-144.

Chua, T.E. and T. Tech. 1990. Aquaculture in Asia - quo vadis? p. 13-30. *In* M.M. Joseph (ed.) Aquaculture in Asia. Asian Fisheries Society, Indian Branch, Mangalore, India.

Csavas, I. 1990. Back to office report of the fisheries specialist of the Interim Mekong Committee's fact finding mission to Cambodia, 1-12 November, 1990. 14 p.

Edwards, P. 1991. Report on a trip to Cambodia. Asian Institute of Technology, Bangkok, Thailand. 63 p.

Nandeesha, M.C. 1991. A report on aquaculture in Cambodia - status, constraints and opportunities. CIDSE and PADEK, Phnom Penh, Cambodia. 85 p.

Nandeesha, M.C., H.P.C. Shetty and T.S. Tana. 1992. Freshwater aquaculture in Cambodia. Paper presented at the Third Asian Fisheries Forum, 26-30 October 1992, Singapore.

New, M.B. 1989. Formulated aquaculture feeds in Asia : some thoughts on comparative economics, industrial potential, problems and research needs in relation to the small-scale farmer, p. 19-30. *In* Report of the Workshop on Shrimp and Finfish Feed Development, 25-29 October, Johor Bahru, Malaysia. ASEAN/UNDP/FAO Regional Small-Scale Coastal Fisheries Development Project, Manila, Philippines.

Tacon, A.G.J. 1992. Aquaculture nutrition and feeding in developing countries: a practical approach to research and development. Paper presented at the IVth International Symposium on Nutrition and Feeding in Fish, 24-27 June 1992, Biarritz, France.

Tacon, A.G.J., G. Maciocci and J.E. Vinatea. 1987. National agricultural feed surveys for aquaculture planning and development in Latin America and the Caribbean. 1. Guidelines. FAO Project GCP/RLA/075/ITA, Field Document No. 1/E, Brasilia, Brazil. 11 p.

AQUAFEEDS AND FEEDING STRATEGIES IN CHINA

Wang Ping

Division of Science and Technology,
Bureau of Aquatic Products, Ministry of Agriculture
11 Nongzhanguan Nanli, Beijing 10026
People's Republic of China

WANG, P. 1993. Aquafeeds and feeding strategies in China, p. 201-212. *In* M.B. New, A.G.J. Tacon and I. Csavas (eds.) Farm-made aquafeeds. Proceedings of the FAO/AADCP Regional Expert Consultation on Farm-Made Aquafeeds, 14-18 December 1992, Bangkok, Thailand. FAO-RAPA/AADCP, Bangkok, Thailand, 434 p.

INTRODUCTION

China is located in the northwestern part of the Pacific, covering about 40 latitudes from the tropical through the temperate zone. It faces the Bohai, Yellow, East China, and South China Seas as well as the Pacific in the eastern coast of Taiwan. The coastline stretches from the mouth of the Yalu River, Liaoning Province, to the mouth of Beilun River of Guangsi Zhuang Autonomous Region, a distance of about 3,200 km. This long curved coastline forms many excellent bays and harbours which provide tremendous potential for fish culture. China also has 20 million ha of inland water resources. Rivers, lakes, reservoirs and ponds can be found all over the country; 5.8 million ha is suitable for aquaculture according to current technology. Moreover, there is plenty of low lying and waste land which can be excavated for fish farming.

China is abundant in aquatic organisms. There are more than 1,500 species of marine finfish and 800 from freshwater. Of these, roughly fifty are of high economic value. In addition to finfish there are thousands of valuable crustaceans, molluscs, seaweeds and other aquatic organisms.

China has a long aquaculture history and plays a very important role in the global fish farming industry. Some 2,400 years ago Fan Li, the premier of Yue State, wrote the earliest book on pisciculture. Now, China's aquaculture production is the greatest in the world, contributing 6.53 million tons, more than 50% of aquaculture production worldwide. It is well-known that China is the first country to have developed techniques for the artificial breeding of carps which made fish farming bloom not only in China but also in the whole world. China's freshwater fish fry production in 1991 was 157.8 billion, 142.1 billion of which came from artificial breeding. The area under aquaculture production was 4.3 million ha. Total fisheries production in 1991 was 13.5 million t, of which aquaculture contributed 6.5 million t (Table 1 and Figure 1). According to the State Plan, fisheries production will reach 18 million t in the year 2000; 70% of the output will have to come from fish farming due to the limitation of offshore fisheries resources.

Table 1. Annual production of aquatic products in China (1,000 t)

Year	Total production	Marine		Inland	
		capture fisheries	aqua-culture	capture fisheries	aqua-culture
1950	912	536	10	300	66
1955	2,518	1,549	107	543	319
1960	3,038	1,749	121	668	500
1965	2,984	1,910	104	456	514
1970	3,185	2,097	184	322	582
1975	4,412	3,068	279	312	573
1980	4,497	2,813	444	338	901
1985	7,052	3,485	712	475	2,379
1988	10,609	4,633	1,424	654	3,897
1989	11,517	5,036	1,576	734	4,170
1990	12,371	5,509	1,624	779	4,459
1991	13,540	6,096	1,905	913	4,626

Since 1979, Chinese government policy has been to emphasize and encourage the protection of natural fisheries resources and to develop aquaculture to meet the increasing demand, both domestic and international, for aquatic products. Aquaculture production and area has increased greatly. Now there are 30 different species cultured in China, the most important being freshwater fish, including silver, bighead, common, and crucian carps, Chinese freshwater bream (Wuchang fish), mud carp, tilapia and black carp. These species contribute more than 90% of the total aquaculture output (Table 2). Coastal aquaculture production in 1989 (Table 3) was 1.6 million tons, of which 12% were crustaceans and 67% molluscs (Figure 2).

Shrimp culture has bloomed since the 1980's and has become one of the success stories of aquaculture, not only in China but also in the whole globe. By 1988 the value of China's cultured shrimp exports had risen to US$ 595 million per year (Table 4). Now, due to the land limitation and environmental problems, the government has shifted attention, from expanding the area of arable land used for aquaculture ponds, to encouraging fish farmers to fully utilize the ponds they have. Increasing the yield per unit area by using high yielding or intensive culture techniques and formulated feeds, and developing low-lying and waste land for aquaculture are promoted.

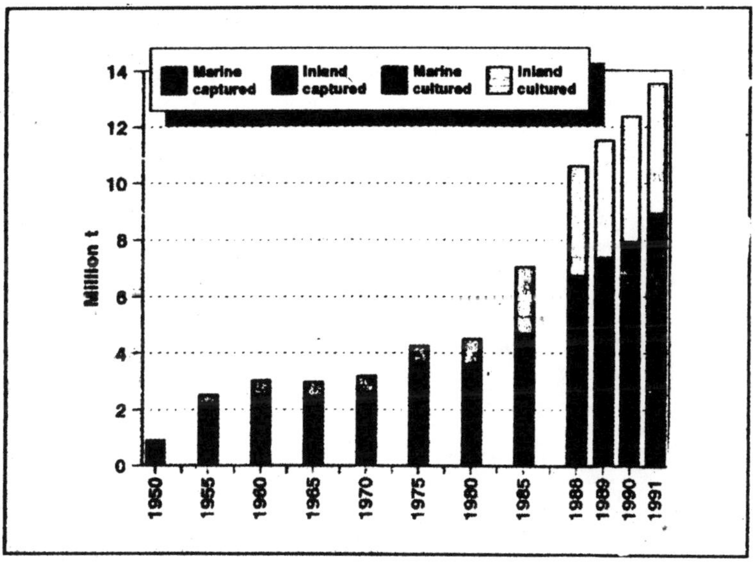

Figure 1. Aquatic production in China from 1950 to 1991

Feed manufacturing is one of the fastest developing agro-industries in China. According to 1991 national statistics, there are 9,154 feed mills in China which have a designed production capacity of 1 t/hr or more. Total feed production was 36 million t. Of this, formulated or compound feeds constituted 35 million t, 589,000 t were concentrates and 298,000 t feed premixes. Of the formulated feed, swine feeds contributed 44% and poultry feeds 48%, while aquafeeds only accounted for 3% (Table 5 and Figure 3). Guangdong Province produced the most formulated feed (4 million t), followed by Shandong Province with 3.21 million t and Jiangsu with 2.58 million t. Hunan Province is the primary producer of concentrates with 150,000 t, followed by Shaanxi Province. Sichuang Province produced 73,000 t of feed premixes, placing it first in the country, followed by Tianjin and Shanghai.

Table 2. Main species cultured in inland waters in China (1,000 t)

Species	1980	1985	1988	1990
Silver carp	405.7	999.3	1,481.0	2,056.9 *
Bighead carp	180.3	475.8	701.5	
Grass carp	135.2	356.9	584.6	1,023.2
Common carp	54.1	237.9	584.6	522.4
Crucian carp	27.0	71.4	116.9	211.6
Tilapia	9.0	23.8	39.0	106.1
Mud carp	18.0	47.6	77.9	-
Freshwater bream	45.1	119.0	194.9	161.6
Black carp	-	-	-	37.5
Others**	27.0	71.4	116.9	339.8
Total	901.4	2,403.1	3,897.3	4,459.1

* includes bighead carp, which were not reported seperately in 1990

** includes freshwater prawns, river crabs, freshwater pearl oysters and other cultured species

Table 3. Coastal aquaculture production in China

Species groups	1980 Area (ha)	1980 Production* (t)	1989 Area (ha)	1989 Production* (t)
Finfish	17,000	2,600	46,380	36,409
Crustaceans	10,000	2,600	155,390	190,223
Molluscs	86,000	177,000	204,140	1,055,327
Seaweeds	20,000	262,000	17,300	293,680
Total	133,000	444,200	423,210	1,575,639

* data are given according to Chinese standards, *e.g.* seaweeds are measured in dry weight and some of the molluscs with their shells off

Table 4. Shrimp exports of China

	1980	1981	1982	1983	1984	1985	1986	1987	1988
Value (US$ million)	179	184	122	89	113	111	258	380	595
Quantity (1,000 t)	22.6	23.5	14.8	10.8	17.0	19.4	41.0	65.0	108.5

The structure of China's feed industry is quite complicated (Figure 4). Feed mills belong to various Ministries or so called "systems". Two of the most important systems, which control feed ingredients and production, are the Ministry of Commerce and the Ministry of Agriculture. The Ministry of Commerce (and the companies and enterprises under its control) is the main channel for grain processing and marketing. Feed production from its feed mills is usually for commercial use. The Ministry of Agriculture controls grain production and feed consumers, so part of its production is for sale, and part is for its own farms' use. Feed mills under these two "systems" account for more than 93% of Chinese feed mills.

Table 5. Formulated feed production in China in 1991

Type of feed	Production (1,000 t)	(%)
Swine feed	15,373.6	44
Poultry feed	16,771.2	48
Aquafeed	1,048.2	3
Others	1,747.0	5
Total	34,940.0	100

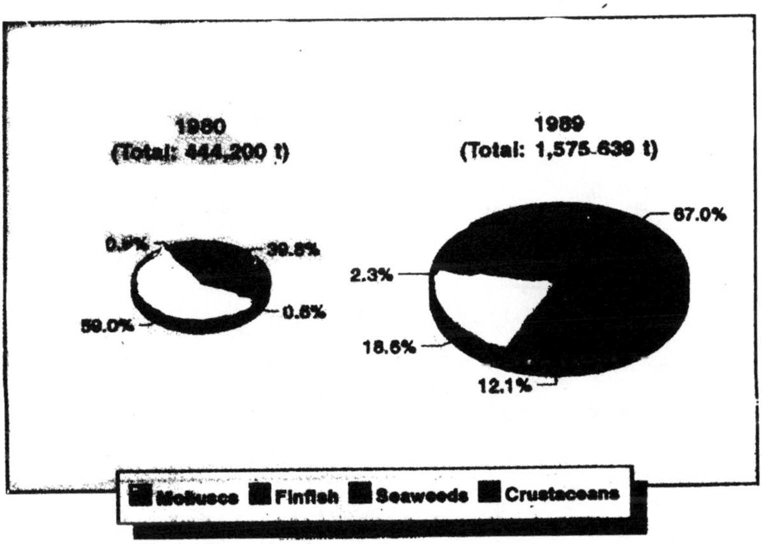

Figure 2. Coastal aquaculture production in China in 1980 and 1989

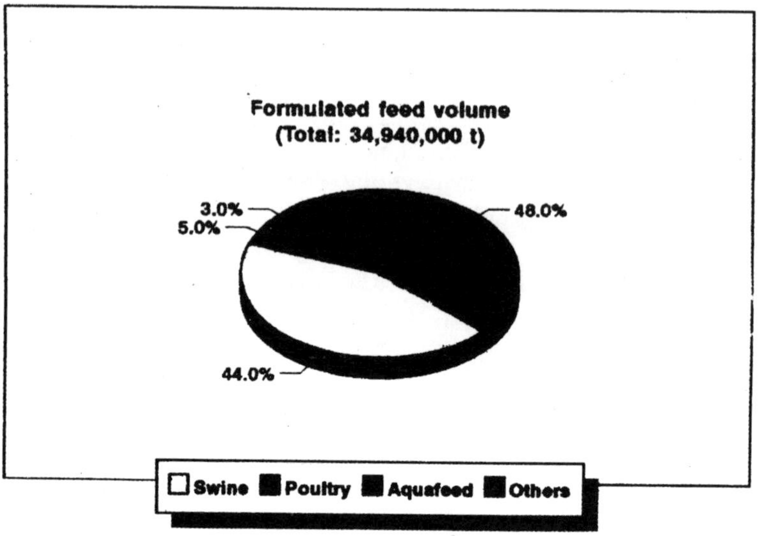

Figure 3. Formulated feed production in China in 1991

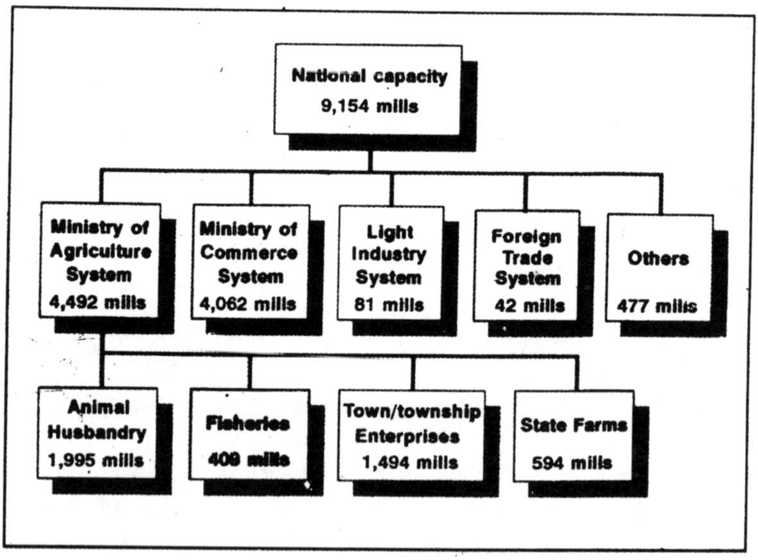

Figure 4. Structure of feed milling industry in China in 1991 (mills over 1 t/hour capacity)

AVAILABILITY OF FEED INGREDIENTS

Feed ingredients used in China vary according to the target animal species, to feed mill locations and to the availability of products and by-products. The main animal protein resources for shrimp and marine finfish are fish meal (either produced in China or imported from South America) and yeast. Bone meal and other animal processing wastes are mainly used for freshwater fish culture. Due to the limited availability of animal proteins, plant proteins also play a very important role in aquafeeds, especially for freshwater fish. The traditional plant proteins used in aquafeeds are soybean meal or cake, rapeseed cake and cottonseed cake. Scientists are working on the replacment of fish meal in shrimp feeds by plant proteins, in order to reduce feed costs, and they have got quite exciting results. With the rapid growth of aquaculture, feed protein resources, especially high quality animal proteins, are in great demand. Total fish meal utilization in 1991 was 706,542 t, of which domestic production was only 74,000 t, accounting for 10% of the total (Table 6). 633,000 t were imported from other countries, mainly from South America, at a cost of US$ 313 million.

Table 6. Imports and exports of feed ingredients in China

Ingredients	1990		1991	
	Imports	Exports	Imports	Exports
Volume (t)				
Fish meal	223,500	570	632,747	89
Soybean cake/meal	153	1,957,791	412	2,188,193
Rapeseed cake	-	333,089	-	540,435
Feed additives	15,358	9,718	34,999	10,662
Total	239,011	2,301,168	668,158	2,739,379
Value (US$1,000)				
Fish meal	103,045	390	312,756	45
Soybean cake/meal	55	391,632	118	399,465
Rape seed cake	-	36,615		60,771
Feed additives	34,331	9,299	67,067	11,932
Total	137,431	437,936	379,941	472,213

China's annual oil cake production is about 10 million t. Soybean cake contributes more than 60% of the total, cottonseed cake accounting for 30% and rapeseed cake 10%. For economic reasons, more oil cake has been used for shrimp and fish feeds, especially soybean cake. In the southern part of China, rapeseed cake and cottonseed cake are the main plant proteins used but in the northern part of the country it is soybean cake. Cottonseed cake and rapeseed cake are said to be toxic to livestock and other kinds of terrestrial animals but, according to research results, aquatic animals are less sensitive than other livestock. In the southern part of China fish farmers spread crushed rapeseed or cottonseed cake into the fish ponds as supplementary feed for fish in integrated farming systems. As commonly used compound feed ingredients in China, rapeseed or cottonseed cakes account for more than 60% of feed formulae and are thus the principal protein resource.

AQUAFEED MANUFACTURING INDUSTRY

Aquafeed manufacturing is a newly emerging branch of the feed industry. Feed has been used for aquaculture starting from the early 1980's but has been greatly promoted in the last ten years. Through the development of aquaculture, especially the extension of intensive fish and shrimp farming and the new fish culture techniques of cage/pen culture, aquafeeed production has become an independent arm of the feed industry. According to statistics which are incomplete, due to the complicated structure of the industry noted above, there were more than 1,600 aquafeed mills by 1991, with a total designed single-shift production capacity of nearly 2 million t, and an actual production of about 700,000 t of aquafeed (Table 7). Of these, two hundred were shrimp feed mills with 700,000 t of capacity, twenty produced eel feed and nearly 1,400 produced feeds for freshwater fish. In addition, there were eleven aquafeed additive mills and eight micro-pellet mills. Currently, it is estimated that 15% of the total feeds used in aquaculture are formulated commercial feeds. State planning aims to increase this to 40% by 1995; this means an increase to about 3.5 million t of formulated aquafeed per year.

The aquafeed industry has evolved from producing single feeds for a few cultured species to producing many types of feeds, of different sizes and nutrient composition, to meet the special requirements of the different developmental stages of grass carp, common carp, Chinese freshwater bream, tilapia, rainbow trout, eel, shrimp, abalone, etc. Shrimp feed production has increased very quickly, due to the high market value of shrimp. Of total formulated aquafeed production in 1991, shrimp feed constituted 300,000 t (43%). China's eel feed market used to be monopolized by foreign feed mills but, in 1991, China produced 20,000 t of eel feed itself, accounting for 3% of the total. Freshwater fish culture accounts for more than 70% of total aquaculture output but, at 380,000 t, only 54% of the total formulated aquafeed output.

Table 7. Aquafeed milling capacity and production in China

Year	Single-shift production capacity (1,000 t)	Actual production (1,000 t)
1983	43	n.a.
1988	1,540	770
1989	1,780	650
1991	1,870	700

Shrimp and eel feed together contribute over 46% of the total formulated feed production because of the high market value of these species groups. Traditional fish culture methods are generally employed in freshwater fish farming, even though the government has put great emphasis on the expanded use of formulated feeds. Due to the relatively low value of the aquaculture output, the use of aquafeeds for freshwater fish is mainly confined to farms employing intensive methods of fish production. Currently, feed costs usually contribute more than 60% of total Chinese production costs in shrimp or freshwater fish culture.

Formulated feeds are generally used on commercial fish and shrimp production sites where there are always some large-scale feed mills nearby. These feed mills are usually owned by state and joint ventures and have good facilities and quality control systems. They purchase feed ingredients, additives and feed premixes from both domestic and international markets. In general, their feed formulation is based upon research results and the quality is good. However, the cost of their feeds is relatively higher than that of farm-made feeds. In freshwater fish production complete formulated feeds are used only in rainbow trout and eel farms, which are mainly intensive monoculture systems with running water and very high stocking densities. In these cases the fish can only get their required nutrients from the feed presented.

Shrimp feed is a special case, because most Chinese shrimp farms practice semi-intensive culture methods. Formulated feeds and unprocessed ingredients, like molluscs and trash fish, are fed to the shrimp ponds. Until now, there are no shrimp farms that use formulated feeds alone. Compared to fish feed, shrimp feed is expensive, about US$ 0.78/kg with an average conversion rate of 3:1.

Feed additives and premixes for use in feeds produced for shrimp, marine finfish, eels and other cultured aquatic organisms are usually imported from other countries. According to customs statistics, China imported 981 t of additives in 1991, valued at US$2.25 million. There are no special additives for aquaculture; most of the large-scale aquafeed mills are using additives prepared for livestock. It is said that Roche has developed premixes and additives specially for aquafeed but these have not yet been used in China. The small-scale feed mills and on-farm feed manufacturing workshops generally do not use additives or vitamins for freshwater fish but some add some commercially available vitamins manufactured for humans.

ON-FARM FEED FORMULATION AND MANUFACTURING

Small-scale feed mills or feed manufacturing workshops on fish farms are very simple, typically having a single grinder and one pelleting machine. Such feed mills usually use the grains and grain processing wastes which are locally available, such as wheat or rice bran, mixed with small amounts of fish meal and oil cakes. Their formulation commonly follows the advice of local fisheries extension agents and is based on the availability of feed ingredients in the vicinity. The feeds coming from these feed mills are quite cheap, about US$ 0.29/kg. Although they do not supply all the nutrient requirements, they are suitable for extensive or semi-intensive fish culture to supplement from the natural food produced in the pond. Since polyculture systems are used, it is difficult to calculate feed conversion rates but, in general, they are about 2.5:1.

MAJOR INSTITUTIONS INVOLVED IN AQUAFEED RESEARCH AND DEVELOPMENT

The Chinese government has paid great attention to fish and shrimp nutrition and to feed manufacturing for a long time. According to estimates, there are more than 100 scientists working on fish and shrimp nutrition research, and on the development of feed formulation and manufacturing for aquafeeds alone. Most fisheries research institutes are actively involved in fish and shrimp nutrition research and development projects, depending on local requirements and environments.

The **Heilongjiang Fisheries Research Institute** is located in the north of China where water temperatures are quite low. Its research projects on fish nutrition concentrate on rainbow trout. The Institute has published many papers on the results of its research work and has also developed and produced rainbow trout feed for fish farmers. The **Yellow Sea Fisheries Research Institute** is located on the coast of Shandong Province. Attention is focused on the nutritional requirements of, and feeds for, shrimp. Its research results are now the foundation of shrimp feed formulation in China and have been widely adopted by the shrimp feed manufacturing industry.

The **East China Sea Fisheries Research Institute** is in Shanghai, in one of the most developed areas of China. Its research and development projects are aimed not only at shrimp feed but also at eel feeds. The pelleted eel feeds developed by this institute are very well received by eel farmers and have greatly promoted the development of a domestic eel feed industry. At the same time, the Institute also works on the use of plant proteins to substitute for fish meal in aquafeeds in order to decrease their cost. The **South China Sea Fisheries Research Institute** also works on shrimp feeds but concentrates mainly on tropical species and marine finfish. The **Yangtze River Fisheries Research Institute** has many qualified researchers who are involved in freshwater fish nutrition and feed formulation. Their results are commonly used in manufacturing compound feeds for grass carp. Researchers at the **Shanghai Fisheries University** are working on the nutrition of black carp, having studied the amino acid and fatty acid requirements of this species. Based on their research, feeds for black carp have been manufactured.

The **Beijing Fisheries Research Institute** is very successful in common carp feed formulation and manufacturing. With the assistance of the American Soybean Association, it has developed techniques to replace fish meal by using extruded soybean meal. Based on this technique, feed costs can be greatly reduced. Common carp is the favorite species of the North which is also the production base of soybean. Currently, in the northern part of China, many feed mills are using the formulation and feed additives developed in the Beijing Fisheries Research Institute to produce common carp feed for fish farmers.

Besides the above mentioned institutions, the **Shanghai Fisheries Machinery and Instrument Institute**, a specialized institute working on fisheries machinery development and manufacturing, includes work on aquaculture equipment. Monitoring the needs of developing aquaculture for feeds, it designs and produces various feed pelleting machines and extruders for fish farmers. The Institute has developed a new kind of shrimp feed by using a special machine; the resultant feed sinks to the pond bottom and floats to the water surface later. This new feed promotes good water management because, when uneaten feed floats to the water surface, farmers can reduce their feeding rate thus both improving water quality and decreasing feed wastage. This research result, however, is not yet utilized in commercial shrimp feed mills.

CURRENT PROBLEMS AND CONSTRAINTS

China's aquafeed industry has developed very fast, gained much experience and learnt many lessons in the past ten years. Compared to the Chinese livestock feed industry, the aquafeed industry has a long way to go to meet the demands of a rapidly expanding aquaculture production. The problems and constraints are two-fold, from both within and outside the industry.

First of all, feed processing capacity is limited and the utilization rate of existing facilities is low. In 1991, the single shift aquafeed production capacity was 2 million tons, which accounted for only 6% of the country's total formulated animal feed production capacity. Actual aquafeed production was less than 3% of the total. Only 15% of the total volume of aquafeeds used in fish culture were formulated feeds. Assuming double-shift operation, the utilization rate of the built-in capacity was less than 28%.

The aquafeed supply structure is not ideal. Until now, feeds are generally available for the grow-out phase of fish and shrimp culture, but the production of feeds for younger age groups, especially for larvae and fingerlings, has only just begun. The domestic production base for feed premixes, binders and other additives, as well as for concentrates, is quite weak. currently falling behind requirements.

Domestic animal protein supply is inadequate to meet the requirements of aquaculture development. The import of fish meal has been steadily increasing over several years, and has caused increasing feed costs and required large amounts of foreign currency exchange.

Due to the lack of scientific management, the quality of the feeds produced is often poor. Except for some large-scale feed mills and joint ventures, most feed mills do not have trained personnel to provide professional management.

AQUAFEEDS AND FEEDING STRATEGIES IN INDIA

M.C. Nandeesha

College of Fisheries, University of Agricultural Sciences
Mangalore-575 002,Karnataka, India

NANDEESHA, M.C. 1993. Aquafeeds and feeding strategies in India, p. 213-254. *In* M.B. New, A.G.J. Tacon and I. Csavas (eds.) Farm-made aquafeeds. Proceedings of the FAO/AADCP Regional Expert Consultation on Farm-Made Aquafeeds, 14-18 December 1992, Bangkok, Thailand. FAO-RAPA/AADCP, Bangkok, Thailand, 434 p.

INTRODUCTION

India, with the second largest population in the world, has recently opted to move towards a free market economy with the hope of increasing the living standards of the people. Although population pressures had, and continue to have, a negative impact on developmental programmes, considerable success has been achieved in agriculture and in creating food security for the population. A summary of general indicators of the economic situation in 1991 is provided in Table 1.

Table 1. Geographic, demographic, agricultural and economic indicators of India in 1991*

Area (million km^2)	3.29
Population (million)	865
Female population (million)	400
Population growth rate (%)	2
Rural population (%)	74
GNP *per caput* (US$)	350
Agricultural land area (million ha)	1.81
Share of agricultural GNP (%)	45
Import as proportion of GNP (%)	7.3
Export as proportion of GNP (%)	4.8

* Exchange rate (as at October 1992) : US$ 1 = Rupee 31

Agriculture is the most important occupation of the rural population. Fisheries is considered not only from the view point of employment, but also a a means of producing nutritious food for the undernourished population. *Per caput* availability of fish is still only 2.08 kg, while the minimum quantity recommended to meet the basic nutritional requirement from fish is 11 kg/yr. *caput*.

In view of the limitations of capture fisheries production, aquaculture has been given the highest priority in the fisheries development plans of both central and state governments. Besides the open ocean along the 7,100 km coastline, a large area is available for different types and scales of aquaculture production, both in freshwater (Table 2) and brackishwater (Table 3).

Table 2. Freshwater resources of India

Type	Area
Coldwater streams/rivulets (km)	8,310
Coldwater lakes (ha)	24,450
Rivers (km)	29,000
Canals (million ha)	0.10
Lakes and reservoirs (million ha)	1.97
Derelict waters/oxbow lakes (million ha)	1.89
Paddy fields (million ha)	30.00
Flood prone/waterlogged areas (million ha)	1.00
Ponds and tanks (million ha)	2.21

Source: Sinha *et al.* (1991)

Table 3. Brackishwater area available for aquaculture, area under brackishwater culture and total production of shrimp in India

State	Estimated area (ha)	Area under culture (ha)	Production (t)
West Bengal	405,000	33,815	12,500
Orissa	80,000	7,075	4,100
Andhra Pradesh	150,000	6,000	7,350
Tamil Nadu	56,000	250	450
Pondicherry	800	10	5
Kerala	242,000	13,000	8,925
Karnataka	8,000	2,500	1,000
Goa	18,500	525	245
Maharashtra	80,000	1,800	800
Gujarat	376,000	125	125
Total	1,416,300	65,100	35,500

Source: MPEDA, Cochin

Until recently there has been no information on the actual volume of aquaculture production, particularly from the freshwater sector. Although the information available now still requires considerable improvement, at least a start has been made. Since fisheries is within the domain of state governments, the authenticity of data provided largely depends on the methodology adopted by each state. In the freshwater sector, total production is reported to be 0.98 million t (Table 4), consisting mainly of carps. Details by species are not available, except for common carp (*Cyprinus carpio*) which contributed 8.9% to the total aquaculture production. Other important cultured carps include catla (*Catla catla*), rohu (*Labeo rohita*), mrigal (*Cirrhinus mrigala*), silver carp (*Hypophthalmichthys molitrix*) and grass carp (*Ctenopharyngodon idella*).

Table 4. General situation of fisheries and aquaculture in India (1991-92)

Parameter (unit)	
Total fish production (million t)	4.14
Marine fish production (million t)	2.44
Inland fish production (million t)	1.70
Penaeid shrimp production (million t)	0.16
Non-penaeid shrimp production (million t)	0.15
Aquaculture production (million t)	0.98
Area under freshwater aquaculture (million ha)	0.59
Area under brackishwater aquaculture (million ha)	0.06
Average yield in freshwater aquaculture (kg/ha/yr)	1,656
Average yield in brackishwater aquaculture (kg/ha/yr)	545
Contribution of carps in freshwater aquaculture (%)	67.70
Total quantity of fish exported (million t)	0.17
Total export value (million US$)	455
Carp hatcheries (#)	257
Shrimp hatcheries (#)	21
Freshwater prawn hatcheries (#)	3

Source: Dehadrai (1992)

Carp culture is gaining popularity in several places with the practical demonstration by the farmers of Andhra Pradesh of its potential. Supplementary feed forms an important component in the culture of carps. Depending on the intensity of the culture operation, the cost of the feed has been reported to be between 57-87%, taking the material input costs only into consideration (Table 5 and Figure 1). However, when labour, electricity and fuel charges are also included, it would vary between 50% and 70% and constitute the single largest input cost (Anon. 1983; Veerina *et al.* 1992).

Table 5. **Percentage of major cost components in carp culture in different regions of India and in Andhra Pradesh (A.P.)**

Input	East	South	West	Central	A.P.1	A.P.2
Seed	5.80	5.80	5.28	3.80	7.13	2.64
Manure	22.51	9.05	7.83	12.33	33.07	24.39
Medication	-	-	-	-	2.86	2.49
Feed	71.69	85.15	86.89	83.87	56.94	70.48
Total	100.00	100.00	100.00	100.00	100.00	100.00

Sources: Anon. (1983), Veerina *et al.* (1993)

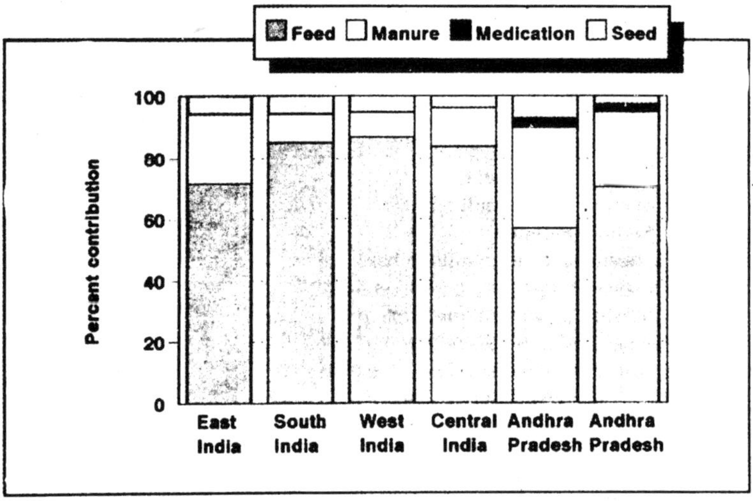

Figure 1. Percentage of major cost components in carp culture in different regions of India (Anon. 1983; Veerina *et al.* 1993)

Although attempts have been made to promote the culture of air breathing fishes (*Clarias batrachus* and *Heteropneustes fossilis*), this has not met with much success. In some derelict water bodies of Assam and West Bengal the culture of *Clarias batrachus* has been attempted. Trash fish and rice bran constitute important dietary ingredients. Some efforts are being made to propagate the culture of *Macrobrachium* spp., but this is still in a very early stage and is mainly attempted for polyculture with carps. In the brackishwater sector the culture of penaeid shrimp has increased rapidly over the past five years and many programmes have been formulated for its promotion. Presently, cultured shrimp contribute only 35,000 t to the total output. With the introduction of improved management practices, the use of artificial feed is gaining priority almost everywhere. Feed cost is estimated to vary between 42-58%, depending on the type of feed used and the intensity of other management practices (Table 6 and Figure 2). The most important species of shrimp cultured are *Penaeus monodon* and *P. indicus*. A number of other species have also been identified as potential candidates for culture (Table 7).

Table 6. Percentage of major cost components in different shrimp production systems in India

Cost component	Extensive without feed	Extensive with feed	Improved extensive doughball feed	Improved extensive pellet feed	Semi-intensive
Pond preparation	19.61	6.41	7.02	3.88	1.57
Seed	19.61	16.03	14.04	12.62	15.66
Feed	-	42.31	42.10	58.25	52.68
Energy	19.61	12.82	14.04	9.71	3.74
Labour	31.36	16.03	15.79	5.83	9.08
Harvest/marketing	4.91	2.40	2.63	2.91	2.73
Miscellaneous	4.90	4.00	4.38	6.80	14.54

Source: Viswakumar (1992)

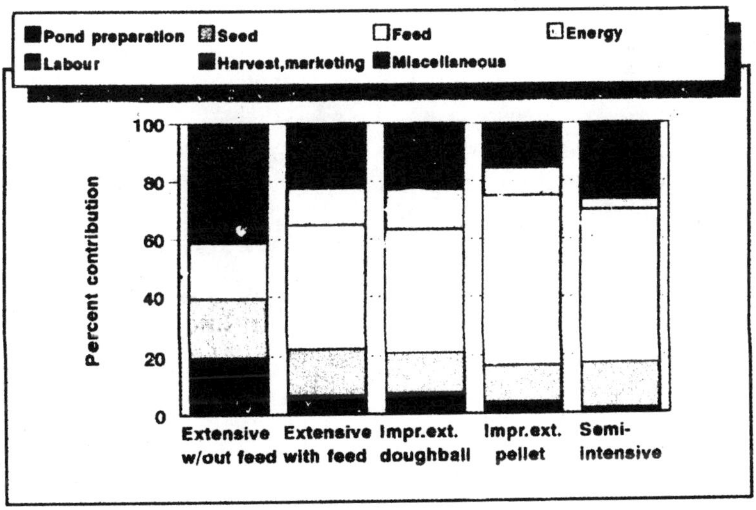

Figure 2. Percentage of major cost components in different shrimp production systems in India (Viswakumar 1992)

Table 7. Shrimp and prawn species with potential for culture in India

Shrimp	Prawns
Penaeus monodon	*Macrobrachium rosenbergii*
Penaeus indicus	*Macrobrachium malcolmsonii*
Penaeus semisulcatus	*Macrobrachium villosimanus*
Penaeus merguiensis	*Macrobrachium josephi*
Metapenaeus monoceros	*Macrobrachium choprai*
Metapenaeus affinis	*Macrobrachium idella*
Metapenaeus dobsoni	*Macrobrachium javanicum*
Metapenaeus brevicornis	
Parapenaeopsis stylifera	
Parapenaeopsis sulptilis	
Parapenaeopsis hardwickii	

Sources: Silas (1980); Tripathi (1992)

AVAILABILITY OF FEED INGREDIENTS

A wide variety of feed ingredients is available in India; their quantity and approximate cost is presented in Table 8. Paddy being an important crop cultivated in an area of 30 million ha, rice bran constitutes the largest quantity of feed ingredients produced in the country. Recently, rice oil solvent extraction units have been set up in many parts of the country. As a result, the cost of rice bran has increased almost everywhere. Not all the rice bran is being used for oil extraction and the actual quantity available as defatted rice bran is not known. In the northern part of India wheat bran is the most important feed ingredient available. Among the oil seed cakes, groundnut is available in the largest quantities. The area under soybean culture has increased over the years and is now estimated to produce nearly 840,000 t of soybean cake. However, it should be noted that soybean is also an important human food resource. Besides rice, other important cereals available include sorghum and maize. Though these ingredients are being used by some farmers in fish and shrimp culture, they also constitute important human food items.

Among the animal sources of feed ingredients, only fish meal is available in significant quantities. It is speculated that the availability of fish meal is likely to decrease further due to the increasing demand for fish for human consumption. Besides this, the quantity of fish landings from capture fisheries is unlikely to increase. There is an equally high demand for fish meal from the animal husbandry sector and it is estimated that the present fish meal production can satisfy only 50% of the needs of the poultry industry (Wood *et al.* 1989). Shrimp waste, mainly consisting of shrimp heads, is also available but difficulties in its collection from various processing centers reduces its practical availability to farmers. *Squilla*, an important by-catch on the west coast, mainly in Karnataka, is also in high demand from the poultry industry. Squid wastes, and clam and mussel meat are available only in small quantities; in addition, the latter two are important export items. Silkworm pupae are available in small quantities in a few southern states. The silkworm industry is rapidly expanding in several regions. As pupae are rich in oil, there are two factories established in Karnataka for its extraction for industrial use. The volume of fish oil produced in India is only 8,000 t; it is reported that the availability of oily fish, like sardines, for this purpose has drastically declined due to its increasing usage as human food.

Table 8. Major feed ingredients available in India in 1992

Ingredient	Quantity (t)	Cost (US$/kg)*
Rice bran	2,182,968	0.08
Defatted rice bran	n.a.	0.06
Wheat bran	1,200,000	0.09
Groundnut oil cake	1,479,700	0.13
Sunflower cake	178,215	0.11
Soybean meal	838,626	0.21
Rapeseed cake	767,661	0.12
Salseed cake	98,345	n.a.
Mustard cake	98,343	0.11
Cottonseed cake	n.a.	0.09
Maize	9,100,000	0.18
Sorghum	11,900,000	0.16
Fish meal	110,000	0.21
Shrimp waste	40,000	0.10
Squilla	40,000	0.10
Squid waste	n.a.	n.a.
Clam meat	300	n.a.
Mussel meat	200	n.a.
Acetes shrimp	20,000	n.a.
Poultry offal	60,000	0.06
Silkworm pupae	20,000	0.13
Defatted silkworm pupae	n.a.	0.19
Fish oil	8,000	0.65

* US$ 1 = Rupee 31

Source: Department of Economics and Statistics, Department of Agriculture and Cooperatives, New Delhi, and various other reports

Quality and proximate composition of ingredients

Several ingredients available in India are reported to be of too poor quality to produce high quality aquafeeds, especially for shrimp. The proximate composition of various plant (Table 9) and animal (Table 10) feed ingredients indicates the variability of these resources. This variability is due to a number of factors, including methods of handling and processing, and the nutritional status of the environment in which they were grown, in addition to variations in analytical methodologies.

Table 9. Proximate composition of selected feed ingredients of plant origin in India

Ingredient	Moisture (%)	Crude protein (%)	Crude fat (%)	Crude fibre (%)	Ash (%)	Nitrogen free extract (%)
Rice polish	12.6	14.5	17.3	7.5	n.a.	n.a.
Rice polish	10.0	12.2	16.0	9.0	6.0	46.8
Rice polish	8.4	11.4	15.3	11.0	12.9	41.0
Rice, broken	10.0	12.0	4.2	5.3	3.1	65.4
Rice bran	10.1	12.6	11.3	19.3	10.2	36.5
Rice bran	7.8	7.8	6.1	14.4	20.5	43.4
Rice bran	8.4	2.9	5.0	18.0	27.3	38.4
Rice bran	8.7	9.4	4.7	13.5	31.4	32.3
Defatted rice bran	7.2	12.1	1.3	15.2	23.8	40.4
Wheat bran	12.3	15.8	4.3	8.7	n.a.	n.a.
Wheat bran	10.0	13.5	2.6	12.2	3.0	58.7
Wheat bran	13.0	8.2	6.6	33.5	4.2	34.5
Wheat bran	9.3	12.6	7.5	11.9	4.2	54.5
Wheat, broken	9.0	11.5	1.9	4.0	0.2	73.4
Wheat flour	12.6	14.5	3.7	2.7	2.3	64.2
Groundnut cake	7.8	28.6	13.8	7.5	13.4	28.9
Groundnut cake	6.0	37.7	11.5	13.2	7.3	24.3
Groundnut cake	10.0	42.0	7.3	13.0	2.5	25.2
Groundnut cake	8.3	46.6	7.7	6.5	7.7	23.2
Groundnut cake	7.1	35.8	8.5	8.2	10.5	29.9
Groundnut extr.	7.0	48.0	2.0	11.2	2.7	29.1
Sunflower extr.	8.0	31.0	2.1	18.4	1.5	39.0
Sunflower extr.	10.2	30.1	2.9	24.7	6.5	25.6
Palm kernel cake	8.9	12.2	4.9	25.6	2.6	45.8
Soybean meal	11.8	46.3	1.3	5.0	n.a.	n.a.
Soybean meal	3.0	58.6	1.4	0.4	5.3	31.3
Soybean meal	10.0	46.0	0.9	7.3	0.6	35.2
Soy sauce waste	12.0	13.5	8.2	5.8	5.3	55.2

Table 9. (cont.)

Ingredient	Moisture (%)	Crude protein (%)	Crude fat (%)	Crude fibre (%)	Ash (%)	Nitrogen free extract (%)
Rapeseed cake	11.0	35.9	0.9	13.2	6.9	32.1
Salseed cake	8.6	8.2	2.9	1.7	10.2	68.4
Sesame cake	8.3	41.9	9.2	6.2	14.8	19.6
Sesame cake	10.0	29.0	12.9	18.3	10.0	19.8
Sesame cake	10.0	42.7	6.9	5.7	12.9	21.8
Mustard cake	8.5	30.8	9.3	6.2	10.3	34.9
Mustard cake	9.2	23.6	9.6	6.3	10.4	40.9
Cotton seed cake	7.0	37.0	6.7	13.0	1.0	35.3
Cotton seed cake	8.2	42.7	1.0	12.6	8.2	27.3
Gingely cake	9.0	34.0	7.8	7.9	3.1	38.2
Gingely extr.	7.0	40.0	2.0	9.7	2.9	38.4
Niger extr.	7.0	35.0	2.0	19.0	3.5	33.5
Copra cake	12.0	22.0	6.5	12.2	5.2	42.1
Copra cake	8.4	20.3	11.4	16.2	6.2	37.5
Copra cake	n.a.	22.0	6.0	12.0	2.1	n.a.
Tobacco seed extr.	7.7	30.6	0.3	-	13.7	47.7
Maize meal	13.5	9.5	4.0	4.0	1.5	67.5
Maize	10.4	4.6	7.8	3.5	1.0	72.7
Sorghum	10.0	9.0	2.8	3.0	0.1	75.1
Spirulina	8.7	50.5	1.0	2.1	11.0	26.7
Tapioca flour	11.5	3.1	2.3	2.0	2.3	78.8
Tapioca flour	8.0	1.8	1.3	1.8	0.2	86.9
Coffee pulp	12.3	14.0	1.2	20.8	8.2	43.5
Colocasia meal	5.8	24.6	4.5	8.2	9.9	47.0
Eichornia meal	3.3	19.5	2.3	18.3	9.3	47.3
Pistia meal	4.9	19.5	1.3	11.7	25.6	37.0
Leucaena meal	11.8	33.1	4.7	9.0	7.2	34.2
Mulberry leaf, dry	8.9	27.7	2.4	11.5	8.1	41.4
Salvinia meal	2.6	16.2	1.1	18.5	22.0	39.6

Sources: Chow (1982); Meyers (1988); Sukumaran *et al.* (1990); Singh (1991); information provided by Mysore Snack Foods, Bangalore and Nav Maharashtra Chakan Oil Mills Ltd., Pune; analyses carried out in author's laboratory

Table 10. Proximate composition of selected feed ingredients of animal origin in India

Ingredient	Moisture (%)	Crude protein (%)	Crude fat (%)	Crude fibre (%)	Ash (%)	Nitrogen free extract (%)
Fish meal	8.6	64.4	7.5	0.3	19.2	
Fish meal	9.5	53.6	5.4	3.1	20.9	7.5
Fish meal	9.2	56.1	2.5	17.8	2.5	11.9
Fish meal	14.6	37.2	2.7	22.7	6.7	16.1
Fish meal	10.6	14.4	8.7	30.0	16.5	19.8
Fish meal	14.0	47.8	10.3	2.6	18.3	7.0
Fish meal	10.0	72.0	10.0	0.5	n.a.	n.a.
Fish meal	9.0	50.0	7.0	1.0	4.0	29.0
Fish meal	10.0	45.0	8.0	1.2	6.0	29.8
Shrimp waste	10.0	28.0	2.7	12.5	n.a.	n.a.
Shrimp waste	9.0	22.5	3.6	35.3	18.6	11.0
Shrimp waste	3.6	34.2	6.7	12.2	27.9	15.4
Shrimp waste	15.6	28.3	1.1	7.1	31.6	16.3
Squilla meal	14.1	46.0	2.6	13.5	18.0	5.8
Squid meal	8.0	75.0	6.5	4.0	n.a.	n.a.
Clam meal	7.0	52.0	11.6	5.5	n.a.	n.a.
Clam meal	8.1	50.7	8.9	3.9	6.4	22.0
Silkworm pupae	7.1	43.9	25.7	4.2	15.8	3.3
Defatted sw. pupae	8.1	68.0	2.6	1.3	7.2	12.8
Blood meal	10.0	65.3	0.5	n.a.	n.a.	n.a.
Blood meal	12.9	76.6	1.1	1.0	3.8	4.6
Meat meal	8.0	50.0	4.4	6.8	5.0	25.8
Meat meal	10.0	71.2	13.3	0.7	n.a.	n.a.
Liver meal	7.0	65.0	3.4	1.2	2.4	21.0
Earthworm meal	5.0	51.7	3.4	12.8	12.5	14.6

Sources: Chow (1982); Meyers (1988); Sukumaran *et al.* (1990); Singh (1991); information provided by Mysore Snacks Ltd., Bangalore; analysis carried out in author's laboratory

Rice polishings, though rich both in protein and fat, are available only in small quantities, constituting less than 10% of the total quantity of rice bran. Most often the rice bran is adulterated with husk, mainly contributing fibre. As indicated earlier, rice bran is now extensively used for oil extraction and hence, in several regions, only the defatted rice bran is available for aquaculture use. Even in the case of oil seeds, solvent extraction processes are increasingly used, hence most (for which there is a heavy demand from the aquaculture sector) contain little fat. Among the various types, groundnut cake and mustard oil cake are the most widely used feed ingredients in aquaculture. However, the use of other oil cakes is also gaining popularity due to cost factors as well as the seasonal non-availability of the two types of oil cakes conventionally used.

Sorghum and maize are used to a smaller extent than rice bran, principally as carbohydrate sources. Tapioca flour, which is rich in carbohydrates, is mainly used as a binder. *Spirulina* is rich in protein as well as vitamins and minerals. Technology for the production of this single cell protein is available (Becker and Venkataraman 1984) and a private company in Karnataka has just started commercial production. However, its cost, and other nutritional considerations may inhibit its inclusion at high levels in feeds. The composition of a few non-conventional plant ingredients is also indicated in Table 9 since they are likely to become important sources of nutrients in the future in small-scale aquaculture operations. Most of the plant ingredients contain a high percentage of fibre which poses constraints on the production of quality feed.

Among the few animal feed ingredients available (Table 10), fish meal is considered the most important. Its composition varies widely, depending on the method of handling and processing. There is also the likelihood of adulteration. As a result the supply of standard quality fish meal is not assured. Only one factory in Karnataka is reported to produce fish meal with more than 60% protein. In all other places, besides having a low protein content, fish meals contain high amounts of fibre, mainly due to the inclusion of other products such as *Squilla* and shrimp wastes. The high ash content of fish meal is often due to contamination with sand since most is still produced by the traditional method of beach drying, followed by grinding.

Several of the companies producing good quality fish meal export it, making only low grade fish meal available in local markets. Similarly, the proximate composition of shrimp wastes also indicates differences in the proportions of shrimp head and shell. *Squilla* meal, produced largely on the west coast, is frequently contaminated with sand particles. However, these crustacean wastes are good sources of nutrients in shrimp diets (Ahamad Ali and Mohamed 1982).

Another upcoming industry in India is sericulture, producing a major by-product, silkworm pupae. Pupae contain significant protein and fat levels and have been found to be an excellent dietary ingredient for freshwater fishes (Jayaram and Shetty 1980). Silkworm pupae are rich in unsaturated fatty acids. Due to an increasing demand for oil from various industries, this ingredient is also extracted. Defatted silkworm pupae, with a high percentage of protein, is presently used in the poultry industry. Other useful animal feed ingredients, like clam meat, squid meat and meat meal, are available only in small quantities, and all are also important sources of human food.

Presently there is no organized collection of wastes from slaughterhouses. However, due to the good nutrient content of such wastes and the increasing demand for animal protein, it is anticipated that they will be used in aquaculture. Meat meal and liver meal are only used to a minimal level in shrimp culture due to their high cost and their demand for human food. Earthworm meal, though potentially a good source of animal protein, is not yet available in commercial quantities. Some efforts have been made to promote earthworm cultivation for the production of wormicompost, but these have not yet been successful.

Import of feed ingredients

While official information is not readily available on the import of feed ingredients for the aquafeed industry, at least two manufacturers are known to be importing fish meal for shrimp feed production. The industry argues that, because of the poor quality of domestic fish meal, its duty-free import should be permitted for use in shrimp feeds since most of the cultured shrimp is exported.

AQUAFEED MANUFACTURING INDUSTRY

Aquafeed manufacturing in India is less than a decade old. Previously few aquaculture operations depended heavily on commercial feeds. However, more than 20 companies became involved in the aquafeed business recently (Annex 1). Carp culture was once the only major type of aquaculture operation in India and it was practised as a family food source rather than as a commercial enterprise. Rice bran and oil cakes of various origins were used as feeds whenever the economy of production permitted or the market price of the cultured fish was high.

Large scale carp culture has only been practised for about ten years in Andhra Pradesh. Feed forms an important input and is required in large quantities. In order to satisfy the demand, a large number of small-scale and a few large-scale feed manufacturers have appeared. Most of the carp feed produced is of a supplementary nature, manufactured by the machinery available for cattle and poultry feed production. Though a few companies produce pelleted feeds, they have poor water stability. Mash feeds have been quite popular, both with the producers and the users, due to the simplicity of their production and their low cost. The percentage of carp farmers using factory-made feed is less than 8%, even in Andhra Pradesh (Veerina *et al.* 1993). Most farmers prefer to make their own, cheaper mix on the farm. No official estimate is available of the total volume of factory-made carp feed, but one commercially sponsored survey indicates a maximum of 35,000 t. However, the total feed usage in Andhra Pradesh carp farms alone is estimated to be over 200,000 t.

The proximate composition of factory-made carp feeds is reported to be 20-30% protein, 2-4% fat, 10-15% fibre, 30-40% carbohydrate and 8-10% ash. Often these feeds are claimed to have been enriched with lysine, methionine, vitamins and minerals. Depending on the protein content, cost varies between US$ 0.10-0.16/kg. No single feed has been able to win the confidence of farmers thus far. Farmers are not willing to pay more than US$ 0.13-0.16/kg for carp feed because of the low market price of fish (US$ 0.65/kg). Producing quality feed for that price is not economic. Compounders claim food conversion ratios of 2.5-3.0:1. Most of the farm-made feeds, whose composition is manipulated according to the growth of fish and the cost of raw materials, have almost as good, sometimes even better FCR's (Veerina *et al.* 1993).

Shrimp feed production has been tentatively expanding in the past five years. Currently, there are more than 20 companies at various stages of development but few are reported to be making good progress. The major constraint they face is the non-availability of quality feed ingredients. Many companies are working in partnership or close collaboration with foreign companies to capture the large anticipated Indian shrimp feed market. The availability and cost of some of the shrimp feeds available is provided in Table 11, while the ingredients used in some brands are listed in Table 12. Most of the feeds are marketed at less than US$ 0.32/kg, but some cost twice as much. Although some companies produce various grades of feed, most produce only one type, for use in semi-intensive systems.

Most existing feed mills use poultry feed manufacturing equipment to produce simple pelleted shrimp feeds and claim good water stability and feed conversion ratios. Farmers' observations do not always tally with these claims. Two of the newly established companies are reported to be using extrusion technology to produce good quality water-stable pellets. A commercial market survey indicates that about 20,000 t of shrimp feed were produced in 1991. Most was used in Andhra Pradesh, Tamil Nadu and West Bengal.

Table 11. Proximate composition of commercial shrimp feeds in India

Company	Crude protein (%)	Crude fat (%)	Crude fibre (%)	Ash (%)	Gross energy (kcal/kg)	Stability (hours)	FCR	Cost (US$/kg)*
Jalasampada feeds	38.0	3.0	3.0	15	3,200	8	2.0	0.29
Godrej Agrovet Pvt. Ltd.	38.0	3.7	7.7	n.a.	2,200	n.a.	n.a.	0.22
Provimi Products Pvt. Ltd.	37.0	6.0	3.5	15	n.a.	n.a.	n.a.	0.40
Laxmi Agro Products	25-28	n.a.	n.a.	n.a.	n.a.	n.a.	n.a.	0.24
Tayo Matsya	30-45	4.5-7.0	3.5-4.5	14-17	3,000-3,200	8-10	1.2-1.4	0.65
Pioneer Aqua Farms	36-39	11.5	4.0-5.4	16.2-17.6	2,950-3,050	3.0-4.5	2.5-3.0	0.26
Higashimaru Feeds (Ind.)	36-42	5-7	2.5-4.0	12.0-17.0	3,100	6-8	1.7-2.0	0.96
Him Feeds	30-43	4-9	3-6	n.a.	n.a.	2-3	1.3-2.9	0.45

* US$ 1 = Rupee 31

Sources: feed manufacturing companies

Table 12. Raw materials used by the Indian shrimp feed industry

Ingredients	Higashi-maru	Jalasampada Feeds	Pioneer Aqua Farms	Tayo Matsya	Him Feeds
Fish meal	X	X	X	X	X
Squid meal	X		X	X	X
Clam meal	X				
Clam meat			X		
Shrimp head meal				X	X
Squilla meal			X		
Meat meal		X			X
Wheat flour	X		X	X	X
Wheat bran	X			X	
Broken ricc				X	
Rice bran			X	X	X
Maize		X			
Maize flour				X	X
Maize gluten meal					X
Soybean flour			X	X	X
Soybean meal				X	
Tapioca flour	X	X			
Maida		X			
Avisa leaf meal		X			
Oil cakes		X	X	X	X
Fish oil	X		X		X
Soya oil					X
Shark liver oil					X
Soya lecithin	X				
Vitamins	X	X	X	X	X
Minerals	X	X	X	X	X
Additives		X		X	X

Sources: feed manufacturing companies

The quantity of feed imported annually is increasing rapidly (Table 13), reaching a record of 6,243 t in 1991. Presently most of these imported feeds are sold at about US$ 1.29/kg. There are about five brands of grow-out feed imported, namely Hanaqua, Trudina, Pingtai, VIP Feed and President Feed. Most is being used by large-scale farms for intensive culture.

ON-FARM FEED FORMULATION AND MANUFACTURING

A number of on-farm feed formulations have been used by farmers for the commercial culture of carps (Table 14) and shrimps (Tables 15-17). In the case of carps, rice bran and groundnut cake are the most important ingredients nationally. In the northeast, mustard oil cake is generally used instead of groundnut cake due to its better availability. These two ingredients are fed in an uncooked form. While some farmers only use rice bran, others use it in combination with 20-50% of oil cake, at 1-3% of fish body weight. In some cases farmers use fertilization alone to avoid high feed costs. Most do not use any vitamins and minerals. Although extension agents suggest the use of a 1:1 mixture of rice bran and oil cake and several other feed formulae have also been developed through research, farmers seem to prefer their own formulae due to cost constraints. The unavailability of some recommended ingredients in adequate quantities and at an affordable cost are also factors.

Table 13. Indian imports of shrimp feeds

Year	Quantity (t)
1988	121
1989	932
1990	1,213
1991	6,243

Source: MPEDA, Cochin

Table 14. Feed combinations used by the farmers of Andhra Pradesh in carp culture

Combination of ingredients	Percentage* of farmers using the combination
Defatted rice bran	8
Defatted rice bran and cotton seed cake	4
Defatted rice bran and extracted groundnut	3
Defatted rice bran and groundnut cake	75
Defatted rice bran and rice bran	9
Defatted rice bran and sunflower cake	1
No feed	1

* total not 100%, due to rounding up

Source: Veerina *et al*. (1993)

Table 15. Farm-made feed formulae adopted by shrimp farmers in India* (%)

Feed ingredients	A.P.#1	A.P.#2	A.P.#3	A.P.#4	A.P.#5
Fish meal	25	25	25	20	40
Squilla powder	5				
Shrimp waste	5				10
Blood meal	5				
Liver meal	5			10	
Fish oil	1.5				
Soya oil	1.5				
Soybean cake	15	25	20	10	30
Groundnut cake	10			10	
Defatted groundnut cake		25	25		
Sunflower cake					10
Corn meal			5		
Wheat flour	15	15	5	20	
Defatted rice bran			10		
Rice bran				15	
Rice powder				5	
Millet		5	5		
Tapioca					9
Maida	10			10	
Vitamin/mineral premix	2	5	5		1
Total	100	100	100	100	100
Form	pellets	balls	balls	balls	balls
Feed conversion rate	2.5-3	5	4.0-4.5	4	2.5
Yield (kg/ha/crop)	1,500	700	700-800	700	500-1,000

* A.P. = Andra Pradesh

Table 16. Farm-made feed formulae adopted by shrimp farmers in India * (%)

Feed ingredients	A.P.	M	T.N.#1	T.N.#2
Fish meal	21	50	16	20
Squilla powder			16	
Shrimp waste	20			
Squid meat				20
Meal/bone meal	20			
Snail meat			20	
Fish oil	1			
Soya oil	1			
Soybean cake	20	15		28
Groundnut cake		10	20	
Sesame cake				5
Corn meal				15
Wheat flour		10		
Rice bran	10	10	16	12
Tapioca	6		10	
Vitamin/mineral premix	1	5	2	
Total	100	100	100	100
Form	n.a.	n.a.	balls	pellets
Feed conversion rate	n.a.	5	3.2	2.0-2.8
Yield (kg/ha/crop)	n.a.	900	90	1,000-1,300

* A.P. = Andra Pradesh; M = Maharastra; T.N. = Tamil Nadu

Table 17. Farm-made feed formulae adopted by Indian shrimp farmers* (%)

Feed ingredients	W.B.#1	W.B.#2	K.A.	K.E.
Fish meal	32	30	20	
Clam meat			5	50
Fish oil				3
Groundnut cake		40		30
Defatted groundnut cake			15	
Mustard cake	33			
Wheat bran		15		
Rice bran	33	15	50	15
Rice powder			10	
Vitamin/mineral premix	2			2
Total	100	100	100	100
Form	n.a.	dry powder	dry powder	balls
Feed conversion rate	2.5	n.a.	3.7	3.5-4.0
Yield (kg/ha/crop)	400	700	680	900-1,000

* W.B. = West Bengal; K.A. = Karnataka; K.E. = Kerala

In the successful commercial carp culture operations of Andhra Pradesh it has been clearly demonstrated that natural pond productivity through fertilization plays an important role in reducing production costs. Since the cost of oil cakes and rice bran has been increasing rapidly, Andhra Pradesh farmers manipulatively use the combinations of these two ingredients based on fish growth rate and pond productivity. During the summer, when the fish growth is rapid, they are provided with a feed mixture consisting 30-40% oil cake and 60-70% rice bran, fed *ad libitum*.

Other ingredients are also used, based on their availability and cost, mostly in addition to rice bran and groundnut cake or in substitution for the latter. However, when the growth of fish is poor, particularly during the monsoon and winter seasons, fish are fed with rice bran alone or mixed with a small percentage of oil cake. When the pond water is rich in plankton, feeding rates are reduced. The protein input through feed is always kept to the lowest possible, while maintaining adequate carbohydrate. Some farmers in this region also use poultry vitamin and mineral mixtures, but their value in fish culture in fertilized systems is not yet quantified.

The use of animal ingredients in grow-out systems is almost non-existent in carp culture. Fish meal is used only in broodstock rearing, at up to 15-20% of the diet. In broodstock diets, along with fish meal, rice bran (20%), broken rice (10-15%), oil cakes (20-25%) and pulses (15-20%) are commonly used. These ingredients are ground, cooked and fed moist. In nursery rearing 1:1 rice bran:oil cake mixtures are most commonly used.

The use of farm-made aquafeeds in shrimp culture is gaining momentum. A number of feed combinations are used in different localities, based on the ingredients locally available. While some farmers just mix and feed the ingredients in dry or wet form, quite a few now cook the ingredients and feed them moist. In some cases, pellets are prepared on the farm on a daily basis, dried and fed. Simple cooking ovens and meat mincers are used to cook the feed and prepare the pellets. Animal sources of feed ingredients are used at 20-50% of the diet, while the rest is contributed by plant sources (Tables 15-17). Several farmers also add vitamins, minerals and fish oil. The feed conversion rates reported by these farm-made feed producers vary between 2.5-5.0:1, with yields ranging from 400-1,500 kg/ha/crop.

ON-FARM FEEDING STRATEGIES

Only two species groups are widely cultured in India on a commercial scale, so detailed descriptions of feeding strategies are only given here on carps and shrimp. Air breathing fishes, such as *Clarias batrachus* and *Heteropneustes fossilis*, are cultured on a small scale in a few localities of the northeastern states, using collected wild seed. Although technology packages for these fishes have been developed, this type of culture has not yet gained good momentum due to shortages of seed. Where they are cultured, they are fed with domestic wastes. Rice bran, alone or mixed with trash fish and groundnut-cake, is also used by some farmers.

Freshwater prawn culture has not yet become commercially successful in India. *Macrobrachium rosenbergii* has been widely researched both for monoculture and for polyculture with carps. Attempts are now being made by the Marine Products Export Development Authority (MPEDA) to promote the culture of this species. Two hatcheries have also been established. Attempts have been made to feed the prawns either with clam, mussel or snail meat, mixed with rice bran and groundnut cake (Silas 1992). So far no results from farmers are available. With the participation of MPEDA, it is expected that freshwater prawn culture will soon succeed.

On-farm feeding strategies in carp culture

Standard recommendations for carp culture involve either three species of Indian major carps (catla, rohu, and mrigal) or six species of carps consisting of both Indian major carps and exotic carps (silver carp, grass carp and common carp). A stocking density of 10,000 seed/ha for the latter and 4,000 seed/ha for the former type of culture is generally advocated. In "six species polyculture" Indian carps are stocked at about 40% of the total. Fish are recommended to be fed at 2-3% body weight with a 1:1 rice bran:oil cake mixture. To obtain the optimum production of 10 t/ha/yr, ponds have to be periodically fertilized with both organic (poultry manure at 10 t/ha/yr or cow dung at 20 t/ha/yr) and inorganic fertilizers (nitrogen and phosphorus fertilizers at 300-500 kg/ha/yr). Standard practices have not been completely adopted by farmers due to the non-availability of all varieties of seed and the poor market demand of exotic carps in some areas. Hence, these culture practices are often modified to suit the local requirements and availability of inputs.

Fish are mainly fed rice bran mixed with a small amount of oil cakes of various origins in an uncooked form. The feeding rate varies from 1-3% and fish are fed by broadcasting once a day, generally in the morning. National average pond production of farms assisted by the Fish Farmers Development Agencies has been reported to be 1,895 kg/ha/yr (Srivastava *et al.* 1990). A survey of these farms has identified seed and feed availability as the major constraints faced by farmers. Poor availability of feed ingredients in adequate quantities and their escalating cost have prevented small-scale farmers adopting regular feeding.

The most spectacular success in Indian aquaculture has been the development of the carp culture industry by the farmers of Andhra Pradesh during the last 8-10 years. Nearly 50,000 ha of paddy fields have been converted into fish ponds and Indian major carps are grown in the whole area. The culture of carps was started on an experimental scale by a few farmers around a natural lake employing the recommended standard technology packages during the early 1980's. However, adoption of the technology packages did not yield adequate returns on investment, due to various constraints. Most of the production from this area is marketed in the northeastern states, the main market being in West Bengal. Silver carp showed poor keeping quality after long distance transport of more than 30 hours and common carp posed the problem of burrowing and breeding in the ponds and the bursting of their bellies during transport. Grass carp, although having good market demand, voraciously feeds on the food provided to other species, therefore it became unpopular with farmers.

Through experience, and the application of scientific skills, farmers modified the culture system, not only to obtain high production, but also the most profitable return on investment. The modified technology consists of stocking only two species of Indian major carps, namely catla and rohu, the latter being dominant (80%) because of its high market demand. However, about a quarter of the farmers resort to three species culture, still with rohu as the dominant species (70%) followed by catla (20%) and mrigal (10%). Ponds are usually larger, with the minimum of 1 ha. These large ponds are stocked with 6-12 month-old stunted fish of 100-150 g, instead of fry or fingerlings as practised in the standard system. Since carps are known to grow faster during their second year, this phenomenon is exploited by the farmers. Stocking density is kept below 5,000/ha.

Ponds are generally fertilized with both poultry manure and inorganic fertilizers at almost double the dose recommended in the standard technology. Periodic fertilization, based on the colour of the water, is normal; the volume of fertilizers is increased until satisfactory plankton production is obtained. Fish are fed daily with supplementary feed, consisting of defatted rice bran alone or a 7:3 mixture of defatted rice bran and oil cakes (Figure 3), depending on the growth rate of the fish. Feed is provided in an uncooked form. In order to avoid feed wastage, farmers have developed a bag-feeding method. In this, polyethylene inorganic fertilizer bags are trimmed to hold 15-20 kg of feed and two to three rows of small holes are made at the bottom. The feed mixture is placed into these bags, which are then suspended on bamboo poles in various locations in the pond (Figure 4). About 20-25 bamboo poles are fixed in every hectare of water area. Fish, which were already trained to feed on such feeding bags during the stunting process, consume the feed within 1-2 hours after the feeding bags are fixed to the bamboo poles. If the feed is not consumed, farmers suspect some disease or water quality problems and resort to immediate curative measures.

Figure 3. Mixing of rice bran and oil cake before placing them in feeding bags

Figure 4. Fixing the feed bags to bamboo poles placed in different areas of the pond

Feeding bags are removed daily and washed and dried before reuse. The feeding rate is regulated more on the basis of fish growth rather than on its size. During the fast growing season they are fed *ad libitum*, but during the slow growing period the feed quantity is kept to a minimum. Feed conversion ratios vary from 2.0-4.5:1, depending on the management practices adopted, to maintain good water quality. The introduction of feeding bags has also helped farmers to apply medication effectively through feed. Some also aerate their ponds to increase the growth of carps. While water exchange is not a routine practice, evaporation losses are replaced. Further, when the fish show signs of distress, the water is exchanged immediately.

Fish are generally cultured for a period ranging from 9 to 12 months, by which time most would have reached at least 1.5 kg. On average, rohu and mrigal would have exceeded 1.7 kg and catla 2.5 kg. A recent detailed survey has shown that farmers obtain an overall net production of about 5,300 kg/ha/yr. A considerable portion obtain more than 8,000 kg/ha/yr. One farmer reached a record net production of 14,620 kg/ha/yr using only three species of Indian carps. In general, variation in the overall production between the two and three species systems was not statistically significant. The survey also identified a number of constraints. Parasitic and bacterial diseases have become very common and farmers employ a wide range of pesticides and antibiotics to overcome them (Rao *et al.* 1991). There is an urgent need to educate farmers on the adverse effects of the excessive use of such practices.

The developments which have taken place in carp culture in Andhra Pradesh have triggered the interest of other states. Similar development is now taking place in Punjab, the state which has achieved a phenomenal success in the green revolution. Wheat bran, which is more commonly available there, is commonly used instead of rice bran. It is often supplemented with potato, which is a major source of carbohydrate. The average production obtained in Punjab is 3,000-4,000 kg/ha/yr and it is anticipated that these enterprising farmers will make similar progress to those in Andhra Pradesh.

On-farm feeding strategies in shrimp farming

The high demand for shrimp in international markets and the need for foreign exchange have prompted the Indian government to support this sector on a massive scale. MPEDA is acting as the lead agency in the promotion of shrimp culture and many schemes have been formulated and implemented to augment cultured shrimp production. Many states have already allocated land for shrimp culture and rapid development is expected, with the participation of several major entrepreneurs and industrial companies, in addition to small- and medium-scale farmers. Three major types of shrimp culture systems are seen in the country as described below.

The traditional or extensive method has been in practice for many years in West Bengal, Kerala, Karnataka and Goa. Shrimp entering with the tide are trapped and harvested after 5-6 months. There is no feeding or fertilization involved, except that paddy stumps left behind after the rice harvest serve as a principal source of detritus. Shrimp largely depend on the natural food in the incoming tidal water. Production from such systems is less than 500 kg/ha/yr.

In semi-intensive systems, selective stocking, liming, fertilization and, to the extent possible, artificial feed is provided. Seed is stocked at 30,000-100,000/ha, depending on the species. *Penaeus monodon* is stocked at 30,000-40,000/ha, while *P. indicus* at 50,000-100,000 per ha. The culture period varies between 4 and 6 months. Production in this system ranges from 500 -1,000 kg/ha/crop. Minimal water exchange is also provided depending on the facilities and the location of the farm. Most of the newly developed shrimp culture area belongs to this category.

The intensive system has been adopted by medium and large scale farmers and it involves, in addition those practices adopted in the more extensive systems, the provision of good quality supplementary feed, higher stocking densities of 100,000-500,000/ha, regular exchange of water and aeration. Production varies between 1,500-3,000 kg/ha with some progressive farmers even achieving up to 13,000 kg/ha.

The actual area under each culture system is not known. A major portion of the artificial feeds used in the semi-intensive and intensive systems are produced on-farm. In general most of the successful farmers prepare the ponds, apply lime, depending on the soil pH, and apply organic manure at a rate of 500-1,500 kg/ha, inorganic fertilizers consisting of both nitrogen (20-40 kg/ha) and phosphorus (10-25 kg/ha), before stocking postlarval shrimp. Some farmers re-fertilize later, to increase the availability of natural food.

Supplementary feed is provided in various amounts. Most farmers feed at 10% of body wt/day at the beginning and gradually reduce to 3% as the shrimp grow. In a few cases the feeding level starts at 25%, gradually reducing to 3% towards the end of the culture period. A wide variety of ingredients (Tables 15-17) is used to make farm-made shrimp feeds but, in almost all cases, fish meal forms an important component. Ground feed ingredients are mixed, either manually or mechanically, depending on the scale of operation.

Many farmers in Andhra Pradesh cook the ingredients, make small balls of it (Figure 5) and place these in various places in the pond. Some feed nearly 60% of the cooked feed through broadcasting and place the remaining quantity in trays made of wood and mosquito netting, or in earthen pots. Feed is also given in an uncooked form, simply mixing the ground feed ingredients with a small quantity of water and making small balls out of it. Some farmers use meat mincers to prepare farm-made pellets successfully (Figure 6). The pellets are generally sun-dried, though driers have been employed experimentally.

Figure 5. Preparation of feed balls for shrimp

Figure 6. Mechanized preparation of farm-made shrimp feed

Feeding frequencies vary between two and six times daily. However, most farmers feed twice a day, in the morning and the evening. The cost of farm-made feed varies between US$ 0.13-0.19/kg depending on the quality of feed ingredients. Most farmers use vitamins and minerals at a 2% level, which alone constitutes about 10-15% of the feed cost. Most vitamin and mineral premixes used are designed for poultry. The production obtained with farm-made feeds does not vary greatly from that achieved by the use of domestically produced commercial feeds. In fact, it is for this reason that farmers increasingly opt for producing farm-made feeds, which saves them considerable cost. In some instances farmers have been able to obtain higher production than with bought-in feeds, due to their own strict control of ingredient quality, and of their processing system.

In Karnataka, farmers use factory-made starter feeds for the first month and then switch to farm-made feeds. Production from this system is around 800-1,000 kg/ha/crop.

Shrimp farming on a semi-intensive and intensive scale is developing rapidly in Andhra Pradesh and Tamil Nadu, compared to other coastal states. In both states, pump-fed ponds are common. In Tamil Nadu, *Penaeus indicus* is the most widely cultured species, while in Andhra Pradesh, *P. monodon* is dominant. The Victory farm, located in Tuticorin, Tamil Nadu, with a water area of 27 ha, is reported to routinely obtain a production of 13 t/ha/crop in 4 months with *P. indicus*. A stocking density of 400,000-500,000/ha, a minimum of 30% water exchange and the use of quality feed are the major factors accounting for such high production. Other intensive farms in India produce about 2-4 t/ha/crop using either imported feeds or farm-made feeds. In addition to feed, these successful farms place great emphasis on water quality management.

MAJOR INSTITUTIONS INVOLVED IN AQUAFEED RESEARCH AND DEVELOPMENT

The Indian Council of Agricultural Research (ICAR) is the nodal agency for all research work pertaining to agricultural sciences, including fisheries. There are two major institutions under its direct supervision, namely the Central Institute of Freshwater Aquaculture, located in the state of Orissa and the Central Institute of Brackishwater Aquaculture, located in Tamil Nadu. Both are actively engaged in research in fish nutrition. Besides these institutes a number of Fisheries Colleges, part of the state Agricultural Universities, are also heavily involved in this type of research and in training. In addition to these agricultural institutions, many other universities have taken an active role in nutritional research (Annex 2). Most of the institutions have been more active in research and training on freshwater fish species than on shrimp. With the anticipated development of the shrimp culture industry, many are likely to divert their research focus to the brackishwater sector.

The Marine Products Export Development Authority (MPEDA), based in Cochin, Kerala, but with offices in all the maritime states, is playing an important role in the development of brackishwater aquaculture. Since late 1991, this agency has also been promoting freshwater prawn culture. The FAO-executed Bay of Bengal Programme has been extending research and developmental assistance for coastal aquaculture and their reports (*e.g.* Wood *et al.* 1989) have greatly assisted the development of the domestic shrimp feed industry.

CURRENT NATIONAL REGULATIONS

There are no regulations in force formulated specifically for the promotion of the aquafeed industry or the quality control of aquafeeds. Existing regulations can be effectively used in quality control. Under existing foreign exchange policy, feed can be imported duty-free by those units which are 100% export oriented, while others have to pay 25%.

CURRENT PROBLEMS AND CONSTRAINTS

Several problems face the aquafeed industry, both at the commercial and farm levels. Although the nutritional requirements of the majority of the cultured species in India have been investigated, farmers are unable to use such information, since the nutritional suitability of the raw materials readily available to them have not been studied in depth. Similarly, appropriate inclusion levels in diets, methods of processing and the economics of using such ingredients in aquaculture operations are poorly known (Keshavanath 1992). Carp farmers still rely heavily on rice bran alone since economics do not permit the use of other ingredients. Shrimp farmers face a problem in choosing available plant protein ingredients, since recommendations on their inclusion level, based on their composition are not available. Although much technical manpower is available in the country, the expertise needs upgrading to solve the basic nutritional problems faced by farmers. Due to lack of such expertise and adequate funding support to address these problems on a long term basis, the nutrition sector has suffered, like other sectors of aquaculture.

Many predict that animal protein ingredients, particularly fish meal, will be in short supply since there is a high demand for them from the livestock sector. Many other sources of marine protein have high export value and are not readily available at an affordable cost to local farmers and feed manufacturers. Plant protein ingredients cannot fill the gap unless their deficiencies (such as anti-nutritional factors) are removed. Low water stability and high fibre content of those commercial shrimp feeds which are available are major causes for concern. Technological developments to increase water stability using locally available ingredients and to reduce the fibre content of plant protein ingredients would greatly assist the development of the domestic aquafeed industry. Some ingredients are not available in all parts of the country. Hence, technological development should consider the local availability of resources, since long-distance transport would not only be uneconomical, but is also technically unacceptable to farmers.

TRENDS IN AQUAFEED MANUFACTURING AND USE

Carp culture will continue to rely upon plant ingredients, mainly rice bran, due to the growth of shrimp farming. The latter is expected to use almost all the available animal protein, and the majority of the high quality plant protein, ingredients. Farm-made feed manufacturing and use will be critically important in carp culture since the use of commercial feeds would reduce the income of farmers due to the low market value of fish. Government estimates indicate that, by the turn of the century, at least another 500,000 ha will be brought under aquaculture. Even if the feed demand is calculated at a minimum of 2 t/ha, this additional culture area would require an additional 1 million t of feed. More realistic estimates have been made for the shrimp culture industry. By the year 2000 about 100,000 ha is expected to be under shrimp culture and a total of 0.2 million t of feed would be required (Sunderraj *et al.* 1992). MPEDA has proposed support for the establishment of 10 feed mills, with a capacity of 10 t each, during the Eighth Five Year Plan (1992-97). Several feed mills are expected to be established by the private sector, jointly with foreign investment. Since the cost of commercial feeds is a deterrent for small-scale farmers, farm-made feeds are likely to continue to play a significant role in aquaculture in India.

CONCLUSIONS

Much information on the basic nutritional requirements of some of the important Indian indigenous cultured species has been generated. However, these results require further verification and study in actual farming environments. Most of the work carried out so far does not take into consideration the effect of other pond inputs on the nutritional requirements of the cultured species. Feed being an expensive input, alternative pathways need to be investigated to reduce its cost without compromising growth rates. Issues raised by New (1989), Tacon (1992), and De Silva and Davy (1992) require serious consideration under Indian conditions. Solutions to these practical nutritional problems, which are already experienced by the farmers, would pave the way for the faster expansion of aquaculture.

Raw material availability and chemical composition requires thorough study in order to develop effective aquacultural programmes. Since animal protein ingredients are in short supply, there is a need to develop techniques for the effective utilization of plant proteins. Since many of the plant ingredients contain anti-nutritional factors, technologies are required to overcome such problems. Non-conventional feed ingredients have to be considered for use in aquafeeds to overcome the short supply of traditional ingredients and to reduce production costs.

Machinery suitable for the commercial manufacturing of high quality feeds or for farm-made feed production is not easily available in India. Not only is it necessary to develop simple techniques for farm-made feed production, but feed production machinery which can be manufactured locally should be designed. Research on energy-saving devices for cooking and drying farm-made feeds would help farmers reduce production costs.

Technological development, based on locally available resources and the creation of an effective extension mechanism for farmers, requires priority consideration. The experience of carp farmers in Andhra Pradesh clearly indicates that their problems remain unresolved, while new technological packages are advocated which are not suitable to their farming systems. In fact, there are no serious takers to solve their problems. Since technical development does not clearly address actual problems, the existing extension systems have been largely ineffective.

LIST OF REFERENCES

Ahamad Ali, S. and K.S. Mohamed. 1982. Utilization of prawn waste and mantis shrimp for compounding feeds for the culture of penaeid prawns, p. 615-616. *In* Proceedings of the Symposium on Harvest and Post-harvest Technology of Fish. Society of Fisheries Technologists (India), Cochin, India.

Anon. 1983. First 138 case studies of composite fish culture in India. Bulletin No.23. Central Inland Fisheries Research Institute, Barrackpore, West Bengal, India. 153 p.

Becker, E.W. and L.V. Venkataraman. 1984. Production and utilisation of blue-green alga *Spirulina* in India. Biomass 4:105-125.

Chow, K.W. 1982. India - carp nutrition research at the Freshwater Aquaculture Research and Training Centre, Dhauli : establishment of a nutrition laboratory and initiation of a diet development programme for carp culture. Project: IND/75/031, Field Document 4. FAO, Rome, Italy. 33 p.

Dehadrai, P.V. 1992. Aquaculture development in India - Government policies. Lecture presented at the Summer Institute on Aquaculture and Environment. Central Institute of Fisheries Education, Bombay. (Ms.)

De Silva, S.S. and F.B. Davy. 1992. Fish nutrition research for semi-intensive culture system in Asia. Asian Fisheries Science 5:129-144.

Jayaram, M.G. and H.P.C. Shetty. 1980. Studies on the growth rate of catla, rohu, mrigal and common carp on different formulated diets. Mysore Journal of Agricultural Sciences 14: 598-606.

Keshavanath, P. 1992. Nutritional studies on Indian major carps - a review. Consultancy project report, International Development Research Centre, Canada. 72 p.

Meyers, S.P. 1988. Development of coastal aquaculture : a report prepared for the Government of India brackishwater aquaculture programme. Project IND/85/059, FAO, Rome, Italy. 35 p.

New, M.B. 1989. Formulated aquaculture feeds in Asia: some thoughts on comparative economics, industrial potential, problems and research needs in relation to the small-scale farmer, p. 19-30. *In* Report of the Workshop on Shrimp and Finfish Feed Development, 25-29 October, Johor Bahru, Malaysia; ASEAN/UNDP/FAO Regional Small-Scale Coastal Fisheries Development Project, Manila, Philippines.

Rao, K.G., C.V. Mohan and D. Seenappa. 1991. The use of chemotherapeutic agents in fish culture in India, p. 503-513. *In* M. Shariff, R.P. Subhasinghe and J.R. Arthur (eds.) Diseases in Asian aquaculture I. Fish Health Section, Asian Fisheries Society, Manila, Philippines.

Silas, E.G. 1980. Status of prawn culture in India and strategy for its future development, p. 209-218. *In* Proceedings of the First National Symposium on Shrimp Farming. MPEDA, Cochin, India.

Silas, E.G. 1992. Freshwater prawns. Proceedings of the National Symposium on Freshwater Prawns (*Macrobrachium* spp.). Kerala Agricultural University, Trissur, India. 272 p.

Singh, B.N. 1991. Nutrition and feed development strategies for aquaculture in India. Journal of the Inland Fisheries Society of India 23:99-112.

Sinha, V.R.P., K.K. Ghosh and N.K. Thakur. 1991. Manpower requirement of fisheries development in India. Journal of the Inland Fisheries Society of India 23:9-19.

Srivastava, U.K., B.H. Dholakia, S.S. Rao and S. Vathsala. 1990. Evaluation of Fish Farmers' Development Agencies' programme for freshwater aquaculture. Indian Institute of Management, Ahmadabad, India. 252 p.

Sukumaran, N., Md. Kaleemar Rahman and M. Devaraj. 1990. Status report on shrimp feed: formulation, manufacture and marketing. Fisheries College, Tamil Nadu Veterinary and Animal Husbandry Science University, Tuticorin, India. 64 p.

Sunderaraj, V., M. Devaraj and Prince M.J. Jeyaseelan. 1992. Prawn farming in India, p. 51-58. *In* I.C. Liao, C.Z. Shyu and N.H. Chao (eds.) Aquaculture in Asia. Proceedings of the 1990 APO Symposium on Aquaculture. TFRI Conference Proceedings 1., Taiwan Fisheries Research Institute, Keelung, Taiwan, R.O.C.

Tacon, A.G.J. 1992. Aquaculture nutrition and feeding in developing countries: a practical approach to research and development. Paper presented at the IVth International Symposium on Nutrition and Feeding in Fish, 24-27 June, Biarritz, France.

Tripathi, S.D. 1992. Status of freshwater prawn fishery in India, p. 42-49. *In* E.G. Silas (ed.) Freshwater Prawns. Proceedings of the National Symposium on Freshwater Prawns (*Macrobrachium* spp.), Kerala Agricultural University, Trissur, Kerala.

Veerina, S.S., M.C. Nandeesha and S.S. De Silva. 1992. Production *versus* economics - a case study of Indian carp culture. Paper presented at the Third Asian Fisheries Forum, 26-30 October, Singapore.

Veerina, S.S., M.C. Nandeesha, K.G. Rao and S.S. De Silva. 1993. Status and technology of Indian major carp farming in Andhra Pradesh, India. Special Publication No. 9, Asian Fisheries Society, Indian Branch. Mangalore, India. 76 p.

Viswakumar, M. 1992. Technical and economic considerations for shrimp culture in Andhra Pradesh. Fishing Chimes, June, 1992: 30-40.

Wood, J., J. Coulter and I. Rajendran. 1989. India's expanding prawn culture industry - where will the raw materials come from? p. 257-268. *In* V.R.P. Sinha and H.C. Srivastava (eds.) Aquaculture Productivity. Oxford and IBH Ltd., New Delhi.

Annex 1. Names and addresses of fish and shrimp feed mills in India

Name and address	Type of feed
M/s. Tata Oil Mills Co., Ltd. 18 & 22, Elaiya Mudai Street Tondierpet, Madras - 600 081 Tamil Nadu	Shrimp
M/s. Him Feeds (India) Ltd. 2-A, Shanker Market Connaught Circus New Delhi - 110 001	Shrimp and fish
M/s. Higashimaru Feeds (India) Ltd. Nima House, IV Main Road Willingdon Island, Cochin - 682 003 Kerala Tel.: 6440, 6071, 6171 Fax.: 34-0130, 34-0133 Cable: AMALGAM COCHIN	Shrimp
M/s. Jalasampada Feeds 9-3-1, Morrispet Tenali - 522 202 Andhra Pradesh Tel.: 4160, 4242	Shrimp
Mysore Feeds Ltd. Post Box 7884, 283-284, Old Tharagupet Bangalore - 560 053 Karnataka Tel.: (91-812)71-471,72-215 Cable: MYFEEDS	Shrimp and fish (production stopped temporarily)
Pioneer Aqua Farms Ltd. 23-A,Pereira Street Tuticorin - 628 001 Tamil Nadu Tel.: 24-826 Tlx.: 0434-219 TIMP IN	Shrimp
M/s. Karnataka State Agro Corn Products Ltd. P.B.No.2479 Hebbal, Bangalore - 560 024 Karnataka Tel.: (91-812) 330-314, 330-281 Cable: AGROCORN	Shrimp

Annex 1. (cont.)

M/s. Tayo Matsya Super Feeds Ltd. Continental Chambers 15A, Hemanta Basu Sarani Street Calcutta - 700 001 West Bengal Tel.: (91-33) 281-035, 288-057, 284-557 Tlx.: 021-2300 KSOR IN Cable: KISORCAL	Shrimp
M/s. Godrej Agrovet Ltd. 1-B Industrial Estate Ambatur, Madras - 600 098 Tamil Nadu Tel.: (91-44) 65-4831/34 Cable: FEEDGODREJ	Shrimp and fish
M/s. Sakthi Soyas Ltd. 180, Race Course Road Coimbatore - 641 018 Tamil Nadu Tel.: (91-422) 27-329, 27-392 Fax.: (91-422) 25-574 Tlx.: 0855-333 SSL IN Cable: SAKTHISOYA	Shrimp
M/s.Lakshmi Agro Products Shed No.16, Phase III Autonager, Vijayawada - 520 007 Andhra Pradesh Tel.: 53-286, 473-240, 475-511	Shrimp and fish
M/s. Lipton India Ltd. Enikepadu, Vijayawada - 521 108 Andhra Pradesh Tel.: 84-462, 52-995	Fish
M/s. Nav Maharashtra Chakan Oil Mills 42/43, Shanker Sheth Road Pune - 411 037 Maharashtra Tel.: (91-212) 650-406/8 Fax.: (91-212) 650-411 Tlx.: 0145-349 NMC IN Cable: SHENGA	Fish

Annex 1. (cont.)

M/s. Provimi Products Pvt. Ltd. 42, Bharathi Street Gobichettipalayam - 638 452 Tamil Nadu Tel.: (91- 4285) 24-93, 25-93, 20-59 Fax.: (91- 4285) 26-77 Tlx.: 08501-2677 PROV IN Cable: PROVIMI	Shrimp
M/s. B. K. Agencies P.B. No. 929 Autonager, Vijayawada - 520 007 Andhra Pradesh	Shrimp and fish
M/s. Ma Laxmi Fertilizer Suit No. 303, 31-A, S.P. Mukherjee Road Calcutta - 700 025 West Bengal	Shrimp and fish
M/s. Aar Gee Formulated Gangapatnam Village Indukurpet Mandal, Nellore District Andhra Pradesh	Shrimp
M/s. Shymala Feeds B3-B4 Industrial Estate Nellore Andhra Pradesh	Shrimp

Annex 2. Names and addresses of major research institutions involved in aquafeed research, titles of major research projects and source of funding

---

Annamalai University, Centre of Advanced Studies in Marine Biology, Parangipettai - 608 502, Tamil Nadu

(a) Nutritional evaluation of Indian *Artemia* nauplii for the larvae of *Penaeus* spp. and *Macrobrachium* spp. Funded by the International Foundation for Science, Sweden (1987-1989)
(b) Larviculture of commercially important Portunid crabs from Port Novo coast. Funded by the Indian Council of Agricultural Research, New Delhi (1991-1994)
(c) Efficacy of different formulated feeds on the growth of Indian white prawn *(Penaeus indicus)*. Annamalai University project (1990)

Research workers:

Dr. T. Kannupandi (larval nutrition and hatchery technology of shrimp)
Dr. S. Ajmalkhan (diet development for shrimp)
Dr. S. Sethuramalingam (diet development for shrimp)

Central Institute for Freshwater Aquaculture (CIFA)
Kausalyagang, Bhubaneswar - 751 002, Orissa

(a) Feed formulation and diet development for carps, catfish and prawn. Funded by ICAR, New Delhi (1983-1995)
(b) Investigations on the digestibility of nutrients and formulation of low cost feed formulae for carps. Funded by ICAR, New Delhi (1990-1995)
(c) Studies on the digestive physiology of carps. Funded by ICAR, New Delhi (1988-1990)
(d) Effects of oxygen, carbon-dioxide and temperature on the metabolism and assimilation of feed in certain Indian and Chinese carps. Funded by ICAR, New Delhi (1975-1978)

Research workers:

Dr. K.M. Das
Dr. S.N. Mohanty
Dr. P.K. Mukhopadhyay
Mr.P.V.Rangacharyulu
Dr.B.N.Singh
Mr.S.K.Swain

(All the above scientists are working mainly with carps; some work has also been carried out on *Clarias batrachus* and *Macrobrachium* spp. nutrition)

Annex 2. (cont.)

Central Institute for Brackishwater Aquaculture (CIBA)
12, Leith Castle Street, Santhome, Madras - 600 028, Tamil Nadu

Complete list is of projects is not available. Several papers have been published on the nutritional requirements, larval and grow-out feed development. Currently a major project entitled "Studies on quantitative requirement of essential amino acids and fatty acids for prawn (*Penaeus monodon*) and use of additives in grow-out feeds for improving feed efficiency and growth promotion" is in operation as a collaborative project with Central Institute of Fishery Technology, Cochin. This project is funded by the Department of Biotechnology (DBT), New Delhi.

Research workers:

Dr. K. Alagarswami (nutritional requirements of shrimp)
Dr. Syed Ahmed Ali (nutritional requirements of shrimp and diet development)
Dr. S. Paul Raj (nutritional requirements of shrimp and brackishwater finfishes)

Central Institute of Fisheries Education
P.B. No. 7392, Andheri (West), Bombay - 400 058

Most publications from this institute relate to nutritional requirements of carps and diet development for fishes. Recently work has been initiated on shrimps nutrition. List of projects in operation is not available.

Research workers:

Dr. K.K. Jain (digestive physiology of fish and shrimp)
Mr. K. Venugopal (amino acid requirement of carps)
Prof. P.S.R.K. Sinha (protein requirement of penaeid shrimp)

Central Institute of Fishery Technology (CIFT)
Matsyapuri P.O., Cochin - 682 029, Kerala

Studies on quantitative requirements of essential amino acids and fatty acids for *Penaeus monodon* and use of additives in grow-out feeds for improving feed efficiency and growth promotion. Ongoing collaborative project with CIBA and funded by DBT, New Delhi (1991-1994)

Research workers:

Dr. K. Gopakumar
Dr. K.G. Ramachandran Nair
Mr.P.Madhavan
Mr.A.G.Radhakrishnan

(All the above scientists work on the nutritional requirements of shrimp and the use of additives in shrimp feed)

Annex 2. (cont.)

Cochin University of Science and Technology
School of Marine Sciences, Cochin - 682 016

Studies on the synthesis and nutritional evaluation of shrimp feeds

Research workers:

Dr. N. Chandramohanakumar (diet development for shrimp)
Mrs. B. Ushakumari

Central Food Technological Research Institute (CFTRI),
Mysore - 570 013

(a) Recycling of feed grade wastes of the food processing industry for intensive animal production. Funded by the Council of Scientific and Industrial Research (CSIR), New Delhi (1986-1988)
(b) Conversion of organic wastes of agro-industries for the production of low-cost animal feed. Funded by CSIR, New Delhi (1988-1990)
(c) Development of technology for the utilization of food processing wastes and low-cost by-products as a source of protein in the preparation of animal feeds. Funded by DBT, New Delhi (1991-1994)

Research workers:

Mr. N.P. Dani
Mr. V.S. Kabade
Mr. N.S. Mahendrakar

(All the above scientists are working on the utilization of various agricultural wastes in fish feed formulation)

Annex 2. (cont.)

College of Fisheries, University of Agricultural Sciences
Mangalore - 575 002, Karnataka.

(a) Optimization of nutrient inputs in carp culture. Funded by ICAR, New Delhi (1991-1993)
(b) Mixed feeding schedules in carp culture. Funded by the International Development Research Centre (IDRC), Canada (1990-1991)
(c) Formulation and evaluation of nutritionally balanced diets for carps. Funded by IFS Sweden (1988-1994)
(d) Broodstock and larval nutrition of carps. Funded by Zilla Parishad, Shimoga, Karnataka (1991-1994)
(e) Amino acid requirement of Indian major carps. Funded by IFS, Sweden (1992-1995)
(f) Studies on the nutritional requirements of and feed formulation for cultivable carps. Funded by the US Department of Agriculture, US Embassy, New Delhi (1987-1991)

Research workers:

Dr. K.V. Devaraj
Dr. Joseph K. Manissery
Dr. P. Keshavanath
Mr. Keshavappa
Mr. K.V. Mohire
Dr. M.C. Nandeesha
Dr. D. Seenappa
Mr. H. Shivananda Murthy
Dr. T.J. Varghese

(All the above scientists work on different aspects of carp nutrition)

College of Fisheries, Kerala Agricultural University
Panangad, Cochin - 682 506, Kerala

(a) Studies on the basic nutritional requirements of cultivable fish and shellfish
(b) Feed formulation for commercially important shrimp species
(c) Nutritional studies on ornamental fishes

Research workers:

Dr. P.M. Mathew (nutrition of ornamental fishes)
Dr. P.M. Sherief (nutritional requirements of and diet development for fish and shrimp)

Annex 2. (cont.)

College of Fisheries, Tamil Nadu Veterinary and Animal Sciences University, Tuticorin - 628 008, Tamil Nadu

(a) Utilization of squid and shrimp wastes in shrimp feeds. Funded by DBT, New Delhi (1990-1992)
(b) Effects of incorporation of crystalline amino acids on the growth performance of white prawn, *Penaeus indicus*
(c) Studies on the quantitative requirement of methionine and the effects of its supplementation in the diets of *Penaeus monodon*

Research workers:
Mrs. G. Indira Jasmine (use of wastes in shrimp feed)
Dr. N. Sukumaran (nutritional requirements of shrimp)

Department of Limnology and Fisheries, Rajasthan Agricultural University, Udaipur Campus, Udaipur - 313 001, Rajasthan

Dietary vitamin requirements of fingerlings of selected Indian major carps with respect to growth, feed conversion and deficiency symptoms. Funded by ICAR, New Delhi (1991-1994)

Research workers:
Dr. A.K. Gupta (vitamin requirement of fishes)
Dr. V.K. Rajabanshi (diet development for fishes)

Department of Zoology, Aligarh Muslim University Aligarh - 202 002, Uttar Pradesh

(a) Investigation on the nutrition of some cultivable finfish species and development of cost-effective feed formulae. Funded by USDA, US Embassy, New Delhi (1985-1991)
(b) Studies on the effect of anti-nutritional substances in commonly used feed ingredients of plant origin on the growth, feed conversion efficiency and health of the Indian major carps. Funded by ICAR, New Delhi (1991-1994)

Research workers:
Dr. A.K. Jaffri (nutritional requirements and antinutritional factors in freshwater fish)
Dr. Mukhtar A. Khan

Annex 2. (cont.)

Department of Zoology, Punjab University, Chandigarh.

Role of minerals in fish reproduction: trace mineral nutrients in the regulation of seasonal reproductive cyclicity. Funded by USDA, US Embassy, New Delhi (1991-1994)

Research workers:
  Dr. K.C. Kanwar (mineral nutrition in carps)
  Dr. Usha Kanwar
  Dr. M.S. Johal

Department of Zoology, Punjab Agricultural University
Ludhiana - 403 004, Punjab

Economic aspects of fish biomass based and fish biomass/pond ecology based feeding strategies for carp. Funded by IFS, Sweden (1991-1994)

Research workers:
  Dr. H.S Sehgal (nutritional requirement of fishes)
  Dr. H.S. Toor

Department of Zoology, Visva-Bharathi University
Santiniketan - 731 235, West Bengal

(a) Efficacy of some artificial diets on the kinetics of food intake, growth promotion and associated physiological and biochemical changes in the air-breathing fishes. Funded by CSIR, New Delhi (1985-1988)
(b) Evaluation of the nutritional role of some aquatic weeds and possibilities of their utilization as feed in carp culture. Funded by the University Grants Commission, New Delhi (1988-1991)
(c) Nutrition of Indian major carps. Funded by IDRC, Canada (1990-1991)

Research worker:
  Dr. Arun Kumar Ray (carp and catfish nutrition and diet development)

Department of Zoology, Nagarjuna University
Nagarjunanagar, Andhra Pradesh.

Formulation of supplementary feeds for better management of Indian major carp fry. Funded by ICAR, New Delhi (1987-1990)

Research worker:
  Dr. S.V. Sharma (larval nutrition)

Annex 2. (cont.)

Department of Zoology, University of Delhi, Delhi

Feeding ecology of the larvae of aquaculturally important species. Funded by ICAR, New Delhi (1987-1990)

Research worker:
Dr. T.R. Rao (feeding behaviour and feeding ecology of carps)

Department of Biochemistry, University College of Science
35 Ballygunge Circular Road, University of Calcutta
Calcutta - 700 019, West Bengal

Nutritional studies and digestive physiology of air-breathing fish under different ecological conditions. Funded by IFS, Sweden (1976-1982)

Research worker:
Dr. Sudip K. Banerjee (nutritional and enzymatic studies in catfishes)

Department of Zoology, St.Xavier's College
Palayankottai, Tamil Nadu.

Nutrient requirements of young and adult *Channa marulius* and other *Channa* species in relation to their bioenergetics. Funded by ICAR, New Delhi (1991-1993)

Research worker:
Dr. M.A. Haniffa (bioenergetics of carnivorous fishes)

Department of Zoology, University of Madras
Madras, Tamil Nadu

Studies on the nitrogen metabolism of commercially important cultivable penaeid shrimp. Funded by ICAR, New Delhi (1991-1993)

Research worker:
Dr. G. Durairaj (nutritional requirements of penaeid shrimp)

Annex 2. (cont.)

National Research Centre on Coldwater Fishes
Haldwani, Uttar Pradesh

Estimation of the nutritional requirements of golden mahseer (*Tor putitora*)

Research worker:

Dr. K.L. Sehgal (nutritional requirements of coldwater fishes and diet development for them)

School of Biological Sciences, Madurai Kamaraj University
Madurai - 625 021, Tamil Nadu

(a) Nutritional requirements of the giant freshwater prawn (*Macrobrachium rosenbergii*). Funded by DBT, New Delhi (1991-1993)

(b) Formulation of suitable low-cost diets for *Macrobrachium* spp. using chicken intestines as major ingredient. Funded by IFS, Sweden (1989-1991)

Research workers:

Dr. T.J. Pandian
Dr. Peter Marian
Dr. S. Mathavan

(All the above scientists work on bioenergetics of fishes and *Macrobrachium* spp. nutrition)

AQUAFEEDS AND FEEDING STRATEGIES IN INDONESIA

Iin S. Djunaidah

Brackishwater Aquaculture Development Centre
P.O. Box 1, Jepara, Central Java
Indonesia

DJUNAIDAH, I.S. 1993. Aquafeeds and feeding strategies in Indonesia, p. 255-281. *In* M.B. New, A.G.J. Tacon and I. Csavas (eds.) Farm-made aquafeeds. Proceedings of the FAO/AADCP Regional Expert Consultation on Farm-Made Aquafeeds, 14-18 December 1992, Bangkok, Thailand. FAO-RAPA/AADCP, Bangkok, Thailand, 434 p.

INTRODUCTION

For centuries Indonesian people have been familiar with fish and shrimp culture, but production was low because wild fry were used. Some more advanced rearing systems of freshwater aquaculture commenced in 1972 with the introduction of running water ponds, raceways, cages, floating net cages, and pen culture (Jangkaru 1981). Of the freshwater species cultured, common carp is the most popular in the local markets, while red tilapia has a promising future in international markets.

Artificial feeds (pellets) are applied by farmers operating running water ponds and floating net cages to grow carp, while rice bran or soybean cake is used by traditional pond farmers. For running water ponds where production is 50-60 kg/crop and FCR is 2.5:1, the feed cost is US$ 0.64/kg* of fish. For floating net cages with a production level of 1,250 kg/unit/crop (10.2 kg/m^3) and an FCR of approximately 2.0:1, the cost of feed will be around US$ 0.51/kg of fish produced. Unlike common carp, for which aquaculture is well established, efforts to culture tilapia are just developing, utilizing the floating net cage method. Production reaches 1,500 kg/unit/crop and, with an FCR of 2.0:1, the feed cost is also around US$ 0.51/kg of fish produced.

In brackishwater aquaculture, shrimp has become the primary commodity because of its high export value. Since 1980, many Indonesian fish farmers have changed from milkfish to shrimp culture. Shrimp culture was encouraged after 1980, in conjunction with a prohibition on the use of trawling to catch shrimp.

---

* Exchange rate US$ 1 = Rp 1,943

Shrimp farming has great potential, because the availability of land is estimated to be 800,000 ha, but the farm areas so far established total only 200,885 ha (Anon. 1992). Most are still using traditional farming techniques.

The production of traditional shrimp farms is very low, around 150 kg/ha/crop. Efforts to increase pond productivity are carried out through various means, including the use of artificial feed. Production in semi-intensive shrimp ponds reaches about 1.2 t/ha/crop, with an FCR of 1.3:1; the feed cost is US$ 1.74/kg of shrimp. Intensive ponds average about 5,000 kg/ha/crop with an FCR of 1.7:1, and the cost of feed is around US$ 2.27/kg of shrimp produced.

AVAILABILITY OF FEED INGREDIENTS

The livestock feed industry in Indonesia developed earlier than the aquafeed industry but each uses similar ingredients, which may be classified into major ingredients and additives. Indonesian natural resources provide considerable quantities of feed ingredients. However, the amounts available are not always in balance with requirements, on account of imperfect processing or handling techniques, or due to conflicts of interest with their demand for human food. Some of the major raw materials available in Indonesia are discussed below.

Shrimp head meal

Shrimp heads are one of the industrial wastes of shrimp processing plants, especially from shrimp which are exported "headless". Other wastes from these shrimp processing plants are in solid and liquid form. On average, shrimp are exported in frozen form; 95% is "headless" (Ariyani 1989). Exports of frozen shrimp reached 47,710 t in the first 6 months of 1992. Assuming that a shrimp head is 35 % of the total weight, over 24,000 t of shrimp head were theoretically available, providing (at 20% of the wet weight) a potential local supply of 4,880 t of shrimp head meal (814 t per month).

Fish meal

Fish meal originates from spoiled fish, unsold fish from markets and fish processing factories, the by-products of capture fisheries, and the wastes of fish and shrimp processors.

Fish meal resources which have not been optimally exploited include the shrimp by-catch (Martosubroto and Nurzali 1985). The unutilized fisheries by-catch is up to 78 % from the seas of Eastern Indonesia, and up to 25 % from Western Indonesia. Of these, 889,235 t are estimated to be potential fish meal resources (Anon. 1989). This huge quantity gives an idea of the potential for the development of the Indonesian fish meal industry without affecting the supply available for human consumption.

So far, fish meal demand cannot be totally met through domestic production and some has to be imported. The amount of fish meal imported in 1991 was 25,700 t (Hastuti, pers. comm.). The national use of fish meal is quite high. Estimates of fish meal requirements in 1988, for instance, were 135,500 t for livestock (Poespodihardjo 1985), 44,000 t for shrimp (Poernomo 1985) and

Table 1. Demand, domestic production and imports of fish meal in Indonesia (1,000 t)

	1984	1985	1986	1987	1988	Growth rate (%)
Demand	87.8	98.1	110.0	122.8	141.3	12.6
Production	1.9	2.4	2.9	3.8	5.8	32.2
Imports	85.9	95.7	107.1	119.0	135.5	12.1

around 3,200 t for freshwater fish culture. In future it is expected that the demand will be greater, due to the expansion in aquaculture and livestock production in Indonesia (Table 1).

The increasing demand for fish meal for animal feed production has stimulated government policies designed to encourage private companies to establish fish meal factories. Up to 1989, 28 fish meal factories with a production

Table 2. Number and capacity of fish meal plants in Indonesia in 1989

Province	Number of plants	Production capacity (t/yr)
East Java	7	24,340
Bali	11	37,800
East Kalimantan	1	720
South Sulawesi	2	7,680
Southeast Sulawesi	2	206
Maluku	4	23,520
Irian Jaya	1	5,500
Total	28	99,766

capacity of nearly 100,000 t/year have been established in several provinces in Indonesia (Table 2). A list of fish meal producers is given in Annex 1.

Soybean meal

Data obtained during the 5-year period 1984-89 show that increased demand was not met by increased soybean meal production (Table 3). As a plant protein source, soybean meal has proven reliable and capable of substituting animal protein sources such as fish meal. For centuries, soybean has had a high value for Indonesian people, since it can be processed into "tempe" (fermented soybean), "tahu" (soybean extract) and other derivative products.

There are still conflicts of interest in the utilization of this commodity, between its importance for human consumption and the demand for livestock and aquafeeds. Home industries which produce various products are now well established and scattered all over Indonesia. One by-product of soybean processing which is a favorable material for feed, is soybean cake which has a crude protein content of around 21 %.

Table 3. Demand, domestic production and imports of soybean meal in Indonesia

Year	Demand (t)	Production (t)	Imports (t)
1984	1,142,000	769,384	400,478
1985	1,143,000	869,818	330,119
1986	1,602,000	1,226,727	342,721
1987	1,401,000	1,160,963	287,000
1988	1,417,000	1,270,418	465,839

Leaf meal

The leaf meals "lamtoro" (*Leucaena* sp.) and gliricidia (*Gliricidia sepium*) are potentially suitable for feeds. Leaf meal from *Leucaena* sp. is well known in Indonesia and some other countries, while *Gliricidia* leaves (a leguminous plant) are also much used.

From the nutritional point of view, *Gliricidia* leaves are as favourable as *Leucaena* leaves. The crude protein content is quite high, around 23%. In Indonesia, these two plants are frequently used as shelter plants, fences, and supporting plants, and the leaves are used as feed for livestock, fish and shrimp. Data on the production of the two plants is shown in Table 4.

Table 4. **Estimated production of *Leucaena* and *Gliricidia* in, Indonesia and their nutrient content (kg/ha/yr)**

Item	Production		Nutrients	
	Fresh	Dried	Protein	Crude fibre
Leucaena	256,000	80,384	19,212	14,459
Gliricidia	116,000	24,592	5,853	4,131

Other plant ingredients

Various other plant ingredients can be used for feeds, such as rice bran, corn, cassava, etc. (Table 5). In certain cases the great natural potential of feed materials in Indonesia has been unable to meet the demand for feed raw materials.

Table 5. Actual and forecast production of selected feed ingredients in Indonesia (1,000 t)

Commodity	1989	1990	1991	1992	1993
Rice	44,399	45,362	46,450	47,566	48,707
Corn	6,499	6,676	6,912	7,155	7,409
Soybean	1,379	1,467	1,567	1,689	1,845
Cassava	16,133	16,532	17,549	17,833	18,627
Peanut	639	674	710	747	787
Mung bean	293	310	333	352	371

Proximate analyses of several potential locally available raw materials are shown in Table 6, while the levels of feed ingredient imports in 1991 are shown in Table 7.

Table 6. Proximate analyses of selected Indonesian feed ingredients

Feed ingredient	Proximate composition (%)					
	Moisture	Lipid	Protein	Carbohydrate	Fibre	Ash
Shrimp head meal	17.28	6.65	53.74	n.a.	14.67	n.a.
Shrimp head meal	5.59	5.85	47.30	35.05	n.a.	6.21
Squid head meal	n.a.	2.16	70.19	n.a.	n.a.	n.a.
Tilapia meal	12.04	2.70	43.57	9.00	0.67	n.a.
Leiognatus fish meal	9.12	5.23	67.27	2.37	n.a.	n.a.
Squid meal	11.47	5.21	67.54	2.45	1.46	11.87
Anchovy fish meal	12.42	4.84	58.40	n.a.	n.a.	n.a.
Rice bran	12.47	11.19	10.86	34.73	10.73	n.a.
Coconut meal	8.62	11.86	21.27	34.73	7.67	n.a.
Soy meal, extracted	10.45	5.54	23.55	26.92	26.50	n.a.
Wet soy extract	87.35	2.62	2.47	3.86	2.50	1.20
Wet soybean cake	6.65	25.26	59.20	n.a.	n.a.	n.a.
Soybean meal	8.23	10.57	42.86	n.a.	5.01	n.a.
Peanut cake meal	11.29	13.06	36.99	21.05	6.02	11.59
Starch flour	12.60	0.54	0.41	86.40	n.a.	0.05
Yeast	6.12	n.a.	50.00	38.93	n.a.	n.a.
Peanut leaf meal	9.09	3.22	15.85	18.04	20.94	32.86
Leucaena leaf meal	8.80	5.40	36.82	16.05	18.14	14.79
Leuc. leaf meal, young	8.27	2.27	21.07	57.23	10.48	0.68
Moringa sp. leaf meal	1.12	6.73	22.26	12.56	n.a.	n.a.
Sesbania leaf meal	11.97	4.73	27.54	21.30	14.01	20.45
Dried seaweed	22.90	n.a.	0.82	31.51	n.a.	n.a.
Algae powder	21.44	0.88	11.72	n.a.	47.72	n.a.
Chaetomorpha sp. meal	8.42	0.65	11.89	17.66	12.51	48.87
Enteromorpha sp. meal	14.70	0.60	10.84	21.22	12.03	40.61

Source : Chemical Laboratory, BADC, Jepara, Indonesia

Table 7. **Imports of shrimp feed raw materials in Indonesia in 1991**

Raw materials	Volume (t)	Price (US$/kg)
Fish meal	25,700	0.70-0.90
Squid meal	7,842	0.70-1.20
Soybean meal	4,571	0.40
Wheat gluten	3,342	1.40-1.80
Peanut, extracted	2,809	0.55-0.70
Groundnut, extracted	1,001	n.a.
Corn gluten meal	957	n.a.
Fish/squid oil	554	0.50-2.00
Pellet binder	504	0.3
Liver powder/meal	336	1.3
Premix vitamin/mineral	283	2.00-3.50
Mussel/oyster meal	251	0.50
Shrimp meal/soluble	237	0.55
Meat bone meal	200	n.a.
Total	48,587	

AQUAFEED MANUFACTURING INDUSTRY

Encouraged by the rapid development of fish and shrimp culture, the aquafeed industry has expanded rapidly. For freshwater fish culture, in particular for common carp, feed manufacturing has been established since 1975, whereas shrimp feed production commenced only in 1983, along with the government programme to intensify shrimp culture.

So far, the Indonesian feed industry is still producing grower feeds only; no feed mill produces larval feeds for shrimp hatcheries. According to Martosubroto and Wibisono (1989), the number of companies operating shrimp farms in Indonesia in 1988 was 184, with a farming area of 964 ha (Table 8).

To support the activities of the grow-out shrimp farms, 43 feed mills were established up to 1989 (Table 9) but many are still in the preparation phase and have not yet been commercially operated (Wahyono 1989). A list of shrimp feed producers is given in Annex 2. Several other fish and shrimp feed mills comprise livestock feed manufacturers which have implemented product diversification. Most are producing at less than capacity. Table 10 provides information on the current value of shrimp feed imports while Table 11 records their value.

Table 8. Companies operating intensive shrimp farms in Indonesia in 1988

Province	Number of companies	Area of farms (ha) planned	Area of farms (ha) constructed
Aceh	21	1,225	118
North Sumatra	31	1,416	98
Riau	5	886	-
South Sumatra	7	2,150	90
Bengkulu	8	3,025	76
Lampung	8	1,257	-
West Java	28	3,806	110
Central Java	10	227	50
East Java	14	734	37
Bali	4	289	245
West Nusa Tenggara	7	3,655	-
East Nusa Tenggara	1	1,200	-
South Kalimantan	3	960	3
West Kalimantan	12	1,243	17
North Sulawesi	11	1,789	120
Central Sulawesi	5	1,942	-
South Sulawesi	6	1,431	-
Southeast Sulawesi	1	2,500	-
Maluku	2	495	-
Total	184	30,230	964

Table 9. Number and production capacity of Indonesian shrimp feed mills in 1989

Province	Number of companies	Capacity (1,000 t/yr)
Aceh	1	2.4
North Sumatra	3	79.5
Lampung	2	8.0
West Java	15	990.2
Central Java	4	43.0
East Java	11	178.9
South Sulawesi	3	22.8
Central Sulawesi	2	20.0
North Sulawesi	1	1.0
West Nusa Tenggara	1	6.0
Total	43	1,351.8

Source: Directorate of Feed Industry, 1989

Table 10. Volume of Indonesian shrimp feed imports in 1992

Brand of feed	Country of origin	Annual total (t)	Monthly average (t/month)
Apitoon	Thailand	0	0
Biofeed	Singapore	36	3
Chia Hsin	Taiwan	0	0
Chin Da	Taiwan	143	12
Cin Ta	Taiwan	0	0
Fong Leng	Taiwan	0	0
Fu Lu	Taiwan	144	12
Hanaqua	Taiwan	0	0
Mitsutomo	Taiwan	0	0
Pa Lue	Taiwan	0	0
Sea Horse	Hongkong	0	0
Shinta	Taiwan	0	0
Tai Roun	Taiwan	543	45
Vip	Singapore	0	0
Total		866	72

Monthly import volumes:				
	January	108 t	June	193 t
	February	126 t	July	350 t
	March	53 t	August	36 t

Table 11. Value of Indonesian shrimp feed imports in 1992 (US$)

Brand of feed	Country of origin	Annual total	Monthly average
Apitoon	Thailand	0	0
Biofeed	Singapore	34,440	2,870
Chia Hsin	Taiwan	0	0
Chin Da	Taiwan	139,367	11,614
Cin Ta	Taiwan	0	0
Fong Leng	Taiwan	0	0
Fu Lu	Taiwan	140,166	11,681
Hanaqua	Taiwan	0	0
Mitsutomo	Taiwan	0	0
Pa Lue	Taiwan	0	0
Sea Horse	Hongkong	0	0
Shinta	Taiwan	0	0
Tai Roun	Taiwan	523,717	43,643
Vip	Singapore	0	0
Total		837,690	69,808

Monthly import volumes:				
	January	106,511	June	192,380
	February	122,334	July	331,337
	March	55,428	August	29,700

The shrimp hatchery business is also developing. Martosubroto and Wibisono (1989) reported that, up to 1989, 113 hatchery companies have been recorded. In the last two years, hundreds of backyard hatchery units have also been developed. Unfortunately, this rapid development has not been instantly followed by adequate domestic feed supply. So far, 90 % of the larval feed required is still imported.

ON-FARM FEED FORMULATION AND MANUFACTURING

Currently the majority of fish farmers applying intensive technology prefer to use commercial feeds rather than to make their own feed. This is because the former are considered more practical, whereas making on-farm feeds requires a lot of time and the purchase of ingredients is relatively difficult.

Some freshwater fish farmers, however, still prepare feed for themselves. The feed is made by means of simple equipment, such as manual meat mincers. They use easily available local raw materials, such as rice bran, cassava leaves, soybean cake, *etc*. These on-farm feeds are usually used for common carp, gourami and red tilapia.

Fish farmers in Indonesia formulate their feeds using raw materials which are readily available and meet the nutrient requirements. Formulated feeds made by researchers are better than those made by farmers, because they use feed additives. For research purposes purified diets are often used.

ON-FARM FEEDING STRATEGIES

Farmers operating traditional ponds usually rely on natural food and household waste products. Sometimes the farmers themselves make simple artificial feeds, such as mixtures of fine rice bran and soybean cake. Commercial feeds given in intensive freshwater aquaculture are fed at about 3 % of biomass per day. For common carp and red tilapia, feed is broadcast over the surface of the water.

Commercial feeds are applied to both intensive and semi-intensive tiger shrimp culture. At present, feeding regimes in semi-intensive culture are almost the same as those in intensive ones, except for feeding frequency (Table 12). The different characteristics of intensive and semi-intensive culture are shown in Table 13.

Table 12. **Feeding regime for *Penaeus monodon***

Period of rearing (day)	Intensive Size (g)	Intensive Feed dosage (%)	Intensive Frequency (/day)	Semi-intensive Size (g)	Semi-intensive Feed dosage (%)	Semi-intensive Frequency (/day)
1-7	0.008	100	2	0.008	75	2
8-14	0.05	25	3	0.05	25	2
15-21	0.4	15	3	0.5	14	2
22-28	1.1	10	4	1.3	9	3
29-35	2.1	9	4	2.4	8	3
36-42	3.5	8	4	3.8	7	3
43-49	5.5	6	5	5.9	6	3
50-56	7.2	5	5	7.7	5	4
57-63	9.0	4	5	9.5	4	4
64-70	11.5	4	5	12.0	4	4
71-77	14.5	3.5	6	15.0	3.5	4
78-84	17.0	3.5	6	18.0	3.5	5
85-91	20.0	3.5	6	21.0	3.0	5
92-98	22.5	3.0	6-7	23.5	3.0	5
99-105	25.0	3.0	6-7	27.0	3.0	5
106-112	27.5	3.0	6-7	29.5	2.5	5
113-120	30.5	2.5	6-7	32.5	2.5	5
120<	33.5	2.5	6-7	34.5	2.5	5

Table 13. Characteristics of *Penaeus monodon* culture systems in Indonesia

Characteristics	Intensive	Semi-intensive	Notes
Size of pond (m^2)	5,000-8,000	5,000-10,000	
Stocking density (pc/m^2)	30-35	10-13	
Fertilization	once in 5-7 days	once in 5-7 days	applied before stocking according to the density of the natural food
Water depth (cm)	120-160	80-110	
Water exchange (%/day)	30-40	10-15	
Paddlewheel aerators (pc)	4-8	none	
Feeding control (%)	1-2.5	1-2	% placed in feeding trays
Feed conversion rate (FCR)	1.6-2:1	1.1-1.3:1	

MAJOR INSTITUTIONS INVOLVED IN AQUAFEED RESEARCH AND DEVELOPMENT

Several institutions are involved in fish and shrimp feed research and development. Besides those controlled by the Directorate General of Fisheries, other institutions work in this field, such as the Indonesian Science Institution and some universities (Annex 3).

CURRENT NATIONAL REGULATIONS

The following Indonesian government policies apply to the feed industry:

A decree of the Minister of Industry No. 37/M/SK/3/1992 which refers to the establishment of Indonesian standards for the shrimp feed industry. The principle of this decree covers quality standards (Table 14), sampling methods, formulation, and labeling. Standards for labeling include the trade mark, the nature of the feed, net weight, ingredients and nutritive content.

Table 14. Shrimp feed quality standards* in Indonesia

Formulation criteria	Unit	Starter 1		Starter 2		Grower 1		Grower 2		Finisher	
		min	max	min	max	min	max	min	max	min	max
Moisture	%		12		12		12		12		12
Crude protein (Nx6.25)	%	39		38		37		36		35	
Crude lipid	%	4		4		4		4		4	
Crude fibre	%		3		3		3		3		3
Ash	%		17		17		17		17		17
Water stability**	%	90		90		85		85		85	
Free nitrogen	%		0.15		0.15		0.15		0.15		0.15
Microbial contamination											
total bact. count:	col/g		$7.5x10^3$		$7.5x10^3$		$7.5x10^3$		$7.5x10^3$		$7.5x10^3$
fungi:	col/g		50		50		50		50		50
salmonella:	neg.		25 g		25 g		25 g		25 g		25 g
Antibiotics											
oxytetracycline:			n.s.		n.s.		n.s.		n.s.		n.s.

* col = colonies; n.s. = not significant; neg. = negative in
** % remaining after 2 hours in water

A circular letter of the Minister of Agriculture to all governors, dated 4 October 1973, states that the fish meal supply can be obtained from fisheries waste products, and was issued to reinforce efforts to increase village processing industries.

Presidential Decree No. 85, 1982 regulates the use of the shrimp by-catch which has to be rendered to the Fisheries State Enterprise for utilization according to the public interest.

CURRENT PROBLEMS AND CONSTRAINTS

Efforts to develop the aquafeed industry have encountered several constraints, as noted below.

Lack of raw materials

One of the constraints in the aquafeed industry is the supply of raw materials. Up to now, the primary raw materials for shrimp feed, *i.e.* fish meal and soybean meal, are still mostly imported. Local raw materials are inadequate to meet the need, and their quality is sometimes inconsistent. In addition, their price is still relatively high. Actually, Indonesia has the potential to provide adequate amounts of fish meal. However, due to the high price of trash fish in some regions, the price of fish meal is also high. Trash fish prices in Indonesia range from US$ 0.13-0.51/kg. Assuming that 1 kg of trash fish can be processed into 0.25 kg of fish meal, the raw material cost of the fish meal would be US$ 0.51-2.06/kg. Assuming that the feed formula requires 20% of fish meal, this would contribute up to US$ 0.41/kg of feed produced.

Lack of research and development funding

While fish generally consume feed quickly, shrimp eat their feed slowly. Shrimp feed should therefore have a higher water stability than fish feed. Manufacturing feed with good water stability requires certain conditions, such as fine feed particles, high pressure and temperature, as well as moisture during feed processing. Factories should be provided with equipment suitable for producing good quality feed to meet the requirements. As a consequence, high levels of investment are required for equipment. Table 15 shows the requirements of shrimp and fish feed in terms of texture and water stability.

With the promising development of fish and shrimp hatcheries, research and development on artificial larval feeds is urgently required. Physical and chemical criteria for larval feeds differ from grow-out feeds. Thus, in turn, different information, techniques and equipment are also required. So far, there is a lack of facilities for larval feed development. Research funds are always scarce.

Table 15. Requirements for texture and water stability in fish and shrimp feeds

Nature of feed	Texture	Water stability
Fish	fairly refined (0.6 mm)	low (15-20 minutes)
Shrimp	refined (0.4 mm)	high (> 3 hours)

Source: Djajasewaka and Sumpeno (1986)

Lack of larval feed manufacturing

Up to now the aquafeed industry in Indonesia has focused on producing grow-out feeds. The hatchery industry, particularly that for tiger shrimp hatcheries, requires good quality larval feeds, even though demand is less than for grow-out feeds in terms of quantity.

At present, all the larval feeds used in Indonesia are imported from Taiwan, Japan and Europe. Without competition from the local feed industry, the price of imported larval feeds will be inflated. While the price of grow-out feeds is around US$ 1.29/kg, the average price of larval feed is US$ 70/kg, fifty-two times as much.

TRENDS IN AQUAFEED MANUFACTURING AND USE

Opportunities to develop the fish and shrimp feed industry are technically determined by the rate of development of aquaculture itself. A significant non-technical factor is the negotiation common between feed producers and fish farmers or fisheries entrepreneurs to settle the price of feed. The price of feed should be proportional to the feed quality and its ability to raise productivity.

Diversification of feed products

Diversification of cultured species will affect the availability of feed. In the last two years, red tilapia has become a local as well as export commodity. Besides that, the government is trying to increase the number of species cultured in freshwater, brackishwater and marine farms (including seabass and groupers). It is thus crucial to develop feeds for every species of fish or shrimp reared. At present, artificial feeds for red tilapia, milkfish, catfish and shrimp are produced. New species will require new feed formulations.

Possibilities for farm-made feed manufacturing

A cursory study shows that feed companies are now capable of responding quickly to the needs of fish farmers and entrepreneurs. Once the culture of a certain species is introduced, most of the feed companies make feeds for it available. Their response does not only concern production, but also the provision of services to customers. Good cooperation between fish farmers and feed producers will stimulate increased production.

Provided that cooperation between feed producers and consumers is improved and that feed prices becomes proportional to quality, all fish farmers are expected to become dependent upon the feed industry. To some extent, this will bias the larger-scale farmers against producing their own feed.

Projected growth of aquafeed manufacturing and use

The fish and shrimp feed industry has great potential for expansion, which is closely related to the development of fish and shrimp culture itself. The growth of fish culture shows a promising annual increase; in the case of pond and cage culture, production is increasing at around 12% and 44% respectively (Anon. 1990).

Further increases in production are likely, particularly in shrimp culture. This is supported by intensification, the rate of which is around 10% per year, and by the expansion of the area under shrimp farming. Accordingly, the requirement for feed will also increase. However, the increase in the amount of feed required will not exactly correspond to the scale of intensification, since the latter also includes semi-intensive systems. So far the FCR in intensive pond culture is about 1.6:1, while in semi-intensive pond culture it is 2.0:1. The average production in intensive ponds is about 5 t/ha/yr. In semi-intensive ponds, it is around 1.2 t/ha/yr. The potential for the use of aquafeeds for shrimp culture, as estimated by Hardjolukito in 1989, is shown in Table 16 and Figures 1 and 2.

The development of aquafeeds for other species also has good potential, due to the increase in the number of species cultured in both brackish and freshwater. As noted before, according to fisheries statistics for 1990, the production of freshwater aquaculture in cages is increasing rapidly, by 44.3% per year. This trend is likely to continue. Finally, common carp and red tilapia will require higher quality feed than that currently available.

Table 16. Projection of shrimp farm area, production and feed requirement in Indonesia

	1987	1988	1989	1990	1991
Pond area (ha)					
- total area	**249,000**	**257,466**	**266,220**	**275,271**	**284,631**
- intensive	3,886	4,275	4,702	5,172	5,690
- semi-intensive	33,964	37,360	41,097	45,206	49,727
- extensive	211,150	215,831	220,421	224,893	229,214
Production (t)					
- total production	**121,591**	**128,820**	**136,604**	**144,993**	**154,042**
- intensive	19,430	21,373	23,510	25,861	28,448
- semi intensive	38,816	42,698	46,968	51,664	56,830
- extensive	63,345	64,749	66,126	67,468	68,764
Feed requirement (t)					
- total requirement	**71,847**	**79,032**	**86,935**	**95,628**	**105,191**
- intensive	33,031	36,334	39,968	43,964	48,361
- semi-intensive	38,816	42,698	46,967	51,664	56,830
- extensive	0	0	0	0	0

	1992	1993	1994	1995
Pond area (ha)				
- total area	**294,308**	**304,315**	**314,661**	**325,360**
- intensive	6,259	6,884	7,573	8,330
- semi-intensive	54,699	60,170	66,186	72,805
- extensive	233,350	237,261	240,902	244,225
Production (t)				
- total production	**163,811**	**174,365**	**185,776**	**198,123**
- intensive	31,292	34,422	37,864	41,650
- semi intensive	62,514	68,765	75,641	83,206
- extensive	70,005	71,178	72,271	73,267
Feed requirement (t)				
- total requirement	**115,711**	**127,281**	**140,010**	**154,010**
- intensive	53,197	58,516	64,368	70,805
- semi-intensive	62,514	68,765	75,642	83,205
- extensive	0	0	0	0

Assumptions:

Intensive ponds - private investment programme, 5 t/ha/yr, 1.6:1 FCR
Semi-intensive ponds - intensification programme, 1.2 t/ha/yr, 1.0:1 FCR
Extensive ponds - 0.03 t/ha/yr

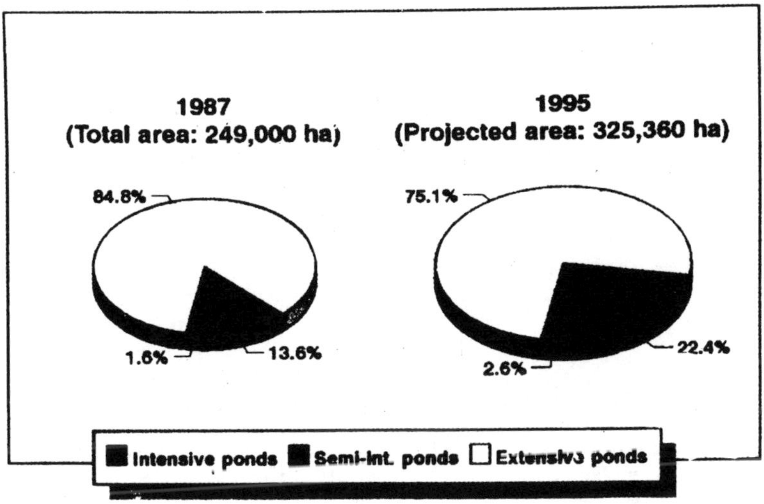

Figure 1. Share of shrimp culture systems in Indonesia in 1987 and projection for 1995

Figure 2. Projected shrimp farm area, production and feed requirement in Indonesia

LIST OF REFERENCES

Anon. 1985 a. Kemungkinan Pengalokasian Sebagian Sumberdaya Hayati Perikanan Untuk Bahan Baku Tepung Ikan. Rapat Teknis Tepung Ikan, Jakarta, 28-29 May 1985. Badan Penelitian dan Pengembangan Pertanian.

Anon. 1985 b. Permasalahan Produksi dan Konsumsi Tepung Ikan. Rapat Teknis Tepung Ikan, Jakarta, 28-29 May 1985. Badan Penelitian dan Pengembangan Pertanian.

Anon. 1989. Status dan Perkembangan Industri Pakan, p. 1-20. *In* Temu Karya Ilmiah Penelitian Menuju Program Swasembada Pakan Ikan Budidaya. Jakarta, 30-31 August 1989. Pusat Penelitian dan Pengembangan Perikanan. Badan Penelitian dan Pengembangan Pertanian.

Anon. 1990. Statistik Perikanan Indonesia. Direktorat Jenderal Perikanan. 74 p.

Anon. 1992. Buletin Warta Mina No. 68. Direktorat Jenderal Perikanan, Jakarta. 43 p.

Ariyani, F. 1989. Pemanfaatan Limbang Udang Sebagai Komponen Pakan Udang, p. 96-106. *In* Temu Karya Ilmiah Menuju Swasembada Pakan Ikan Budidaya. Jakarta, 30-31 August 1989. Pusat Penelitian dan Pengembangan Perikanan. Badan Penelitian dan Pengembangan Pertanian.

Bastari, T. 1989. Status dan Proyeksi Bahan Baku Pakan Ikan. Temu Karya Ilmiah, p. 6-10. *In* Penelitian Menuju Swasembada Pakan Ikan Budidaya. Jakarta, 30-31 August 1989. Pusat Penelitian dan Pengembangan Perikanan. Badan Penelitian dan Pengembangan Pertanian.

Djajasewaka, H. and S. Ningrum. 1985. Kualitas dan Kuantitas Tepung Ikan dalam Ransum Ikan. Rapat Teknik Tepung Ikan. Jakarta, 28-29 May 1985, Badan Penelitian dan Pengembangan Pertanian.

Djajasewaka, H. and P. Sumpeno. 1986. Peranan Pakan Dalam Budidaya Ikan dan Udang, Seminar Hari Pangan Sedunia ke VI. Jakarta, 16 October 1986.

Hardjolukito, S. 1989 a. Prospek Pengembangan Pabrik Pakan Udang Dalam Persangan Dengan Pakan Impor Pada Situasi Pasaaran Udang yang menurun, p. 88-94. *In* Prosiding Lokakarya Industri Budidaya Udang. Balai Budidaya Air Payau Jepara.

Hardjolukito, S. 1989 b. Potensi Pengem bangan Usaha Pabrik Pakan Udang dan Peranan Pola TIR dalam distribusinya. Lokakarya Efisiensi Penggunaan Pakan Udang. Direktorat Jenderal Perikanan.

Jangkaru, Z. 1981. Intensive Fish Culture in Indonesia. Indonesian Agricultural Research and Development Journal, Vol. III, 4:95-98.

Martosubroto, P. and N. dan Nurzali. 1985. Sumberdaya Perikanan dan Industri Tepung Ikan. Rapat Teknis Tepung Ikan, Jakarta, 28-29 May 1985. Badan Penelitian dan Pengembangan Pertanian.

Martosubroto, P. and R. Wibisono. 1989. Menuju Industri Budidaya Udang Berwawasan Lingkungan, p. 116-122. *In* Prosiding Lokakarya Industri Budidaya Udang. Balai Budidaya Air Payau

Mathius, I.W. 1992. Penggunaan hijauan glicinidia sebagai pakan pengganti hijauan lomtoro. Jurn. Penelitian dan Pengembangan Pertanian, Vol. X, 1:1-7.

Poespodiharno, S. 1985. Peronasalahan Produksi dan Konsumsi Tepung Ikan. Rapat Technis Tepong Ikan. Jakarta, 28-29 May, 1985. Pusat Penelitan dan Pengenbangan Perikanan. Badan Penelitian dan Pengenbangan Pertanian.

Poernomo, A. 1985. Persyaratan Pakan Untuk Budidaya Pantai. Rapat Teknis Tepung Ikan. Jakarta, 28-29 May 1985. Pusat Penelitian dan Pengembangan Perikanan. Badan Penelitian dan Pengembangan Pertanian.

Wahyono, U. 1989. Status Produksi Udang Indonesia. Lokakarya Industri Udang. Jepara, 25-28 September 1989.

Annex 1. List of fish meal producers in Indonesia in 1992

Name of company	Office address	Factory location	Capacity (t/yr)
Arumbai Arika Teknik	Jl. Karet No.147 (atas) Surabaya	Anbon, Malukn	120
Abuhasdha & Co.	Ji. Kyai Caringin No. 148 Jakarta	Ternate, Maluku	5,040
Ariyo Sentosa Wahana	Jl. Ratulangi No. 132 Ujung Pandang	Ujung Pandang	6,000
Binstra Trading Co.	Jl. Bangka No. 4/I Mampang Prapatan Jakarta Selatan	Kodya Padang	3,300
Bali Kota Baru Fish Meal Mfg. Co.	Jl. Gunung Sahari No. 29 Jakarta	Kuala Karang Kalimantan Barat	12,000
Banda Sea Tuna Fisheries	Jl. Mangga Besar Raya No. 68A, Jakarta	Ternate, Maluku	
Bina Kemakmuran Jaya	Jl. K.H. Hasyim Ashari No 28A, Jakarta	Kendari, Sulawesi	2,000
Sinar Mahakam Company	Jl. Dr.Tehupriry SK 42/33 Ambon, Maluku	Maluku	360
Merdeka Jaya	Jl. Kesehatan No. 27 Jakarta	Jayapura, Irian Jaya	5,500
CV. Bumi Indo	Jl. Tirta Suam No. 1 Mojokerto, Jawa Timur	Mojokerto, Jawa Timur	12,000
CV. Duta Samudera Jaya	Jl. Imam Bonjol No. 94 Denpasar	Badung, Bali	3,000
CV. Inkud (Bali)	Jl. Letjen. S. Parman Kav. 80, Jakarta	Badung, Bali	7,800
Sumber Tuna Cakalang	Jl. Cideng Barat No. 16 Jakarta	Lembar, Sulawesi Tenggara	120

Annex 1. (cont.)

Pelagis Samudera Lestari	Jl. Ps. Minggu Raya 99A-B Jakarta	Pelabuhan Perikanan Jakarata	
Sami Rahayu Bali	Jl. Gatotkaca No. 170 Bali	Pengambengan Negara Bali	3,840
Cakalang Pitu	Jl. Jenderal Sudirman 57 Jakarta	Perairan Sulteng dan	86
Bali Raya	Jl. Benoa Sanggaran Denpasar	Badung, Bali	10,000
Samudera Rejeki Murni	Jl. Bungur Besar Raya No. 39 Jakarta	Tanjung Leidong, Tanjung Balai, Asahan Sumatera Utara	3,000
Third World Fisheries	Jl. Amanagappa No. 3 Ujung Padang	Maros, Sulawesi Selatan	
Pasific Fish Meal			900
Slamet Jadi			1,800
Hasil Laut Sibolga			396
Altantic Utama Fishery Industry			2,500
Halpindo			1,800
Praban Sari Pasific			1,200
Tri Bakti I			900
Tri Bakti II			334
Muncar			900
Indo Bali		Tegal Badeng, Negara, Bali	1,500
Bali Maya Permai		Tegal Badeng, Negara, Bali	3,600
Bali Indah		Blambangan, Negara	3,600

Annex 1. (cont.)

Sarana Nelayan	Teluk Bagan, Samarinda	720
Deho Canning Co.	Jl.dr.Sutomo 11	60
Suria Irian Jaya	Sorong, Irian Jaya	300
Bangka Pioneer	Kp. Merawang Kendo Baturasa, Bangka	600
Central Jawa Cold Storage	Jl. Citandui,	1,800
Fish Mando Fishery	Kamp. Munçar- Baturasa	7,500
Bangka Pioneer Industry	Kp. Merawang Kendo Baturasa, Bangka	7,920
Prabawa Sari Pasific	Desa Penipahan, Kec. Kubu Bengkalis Riau	2,400
Indonesia Fishing	Muncar, Jatim	10,000
UD. Santosa	Muncar, Jatim	5,000
Blambangan Raya	Muncar, Jatim	
UD. Sumber Jala	Muncar, Jatim	
Mega Muncar	Muncar, Jatim	
NV. Muncar	Muncar, Jatim	
Enggeng Metem Camp	Samarinda, Kaltim	
Daya Guna Samudera	Dobo, P. Aru	
Maya Muncar	Muncar, Jatim	5,000
Total		134,896

Annex 2. List of shrimp feed producers in Indonesia in 1992

Name of company (Shrimp feed brand) Head office address	 Company address	Installed capacity (t/yr)	Actual production (t/yr)
1. NORTH SUMATERA			
PT. Mabar Feedmill (Mabar Feed) Jl. Rumah Potong Hewan 44 Desa Mabar, Medan Tel. : (62-61) 612-244, 612-252, 612-451 Fax.: (62-61) 512-233 Tlx. : 51967 MABAR IA	 Jl. Belawan Km 9, Medan Tel.:(62-61) 29-759 24-920 Tlx.: 51967 MABAR IA	 18,000	 6,000
PT. Charoen Pokphand Indonesia Jl. Medan Tanjung Morawa Km 8.5, Medan P.O.Box 306/MDN Tel. : (62-61) 27-724/27-790 Fax.: (62-61) 28-999 Tlx. : 51729 POKPHAND MDN	(CP Bintang, Irawan, Novo) Jl. Medan Tanjung Morawa Km 8.5, Medan P.O.Box 306/MDN Tel.: (62-61) 27-724/27-790 Fax.: (62-61) 28-999 Tlx.: 51729 POKPHAND MDN		
2. WEST JAVA			
PT. Winayaharjautami (Aquafeed) Blok Caringin 9, Cigondewah Bandung Tel. : (62-22) 631-066 634-929 Fax. : (62-22) 611-694 634-898	 Blok Caringin 9, Cigondewah Bandung Tel.: (62-22) 631-066 634-929 Fax.: (62-22) 611 694 634-898	 12,000	 1,000
Pega Pakan Sejati (Pega Pakan) Jl. Kebon Kacang Raya 1 Jakarta Tel. : (62-21) 321-208 335-677, 336-772 Fax.: (62-21) 513-041 Tlx. : 62338 PEGA IA	 Jl. Industri II Blok J, No.5, Kawasan Industri Jatake, Tangerang	 7,200	 1,000
PT. Nugraha Mitra Jaya (Raja Udang) Jl. P Jayakarta 121/11 Jakarta Tel. : (62-21) 639-8366 659-539	 Jl. Batuceper Dalam I Tangerang, Jabar	 1,000	 750

Annex 2. (cont.)

Company / Office	Factory		
PT. President Feed Indonesia (President Feed) Wisma Argo Manunggal Lt 9 Jl. Gatot Subroto Kav. 22 Jakarta Tel. : (62-21) 514-314 Fax. : (62-21) 514-857 Tlx. : 514317 FUMIRA IA	Jl. Raya Pengas Dengklok Km 2-3, Desa Tunggak Jati Karawang Tel.: (62-267) 81-088, 81-103 Fax.: (62-267) 81-089	25,000	4,000
PT. Bestari Indo Prima (Bestari) Gedung Galva Lantai V Jl. Hayam Wuruk 27 Jakarta 10120 Tel. : (62-21) 385-4994 Fax. : (62-21) 380-4908 380-5886 Tlx.: 42764 GT IA	Jl. Raya Pasar Kemis Km 4,1 Kp. Gelam Ds. Kutajaya Tangerang Tel.: (62-82) 132-238	36,000	
PT. Shinta Prima Feedmill (Percobaan) Jl. Sulaiman 25, Slipi, Jakarta Barat Tel. : (62-21) 548-0959 548-0372, 548-1379 Tlx. : 45941 MECOSIA Cable: SHINTAFEED, Jakarta	Desa Narogong, Cilengsi Bogor, Jabar	7,500	
PT. Central Pangan Lestari (Bintang, Irawan, CP) Jl. Parangtritis Raya 1 Jakarta 14430 Tel. : (62-21) 686-537/39, Fax.: (62-21) 686-538	Jl. Raya Krawang-Cikampek Km 17, Desa Purwasari Cikampek P.O.Box 46/CKP		
Pabrik Pakan Udang TIR (TIR Feed) Gdg. Hanurata Graha Lt. 6 Jl.Kebon Sirih 67-69 Jakarta Tel. : (62-21) 305-3243	Proyek TIR, Karawang	2,000	500
PT. Grobest Indomakmur (Chuen Shin) Jl. Daan Mogot 119 B/6 Jakarta Tel. : (62-21) 560-3732 566-5922/25	Jl. Industri VI/6 Kawasan Industri Jataka, Tangerang Jabar	28,800	6,000
PT. Kartika Americia Aquasys (Kartika) Jl. Sadar 1, No. 1-5 Jakarta 10130 Tel. : (62-21) 380-3464 380-3494 Fax. : (62-21) 380-3536 380-7199	Jl. Mundu Pesisir 32 Km 4,2, Cirebon Tel.: (62-231) 27-433 27-466 Fax.: (62-231) 26-137	10,800	

Annex 2. (cont.)

PT. Duta Persada Union (Dupper-U) Jl. Raya Banten Komplek Sapta Marga 147, Serang, Banten Tel.: (62-254) 81-422	Jl. Raya Banten Komplek Sapte Marga 147, Serang, Banten Tel.: (62-254) 81-422	4,800	1,000
PT. Bina Pemuda Industri Krawang Jl. A. Yani/Bypass 17 Krawang	Krawang, Jabar		
PT. Inter Niaga Makmur (Super Pelet) Jl. Raya Kemayoran Gempol 81, Kemayoran Jakarta	Kuningan, Jabar		
PT. Hanaqua Feedmill Indonesia (Hanaqua) Delta Building Blok C-12 Jl. Suryopranoto 1-9 Jakarta Tel.: (62-21) 854-0402	Jl. Ds.Curug Kulon Tangerang	21,000	
PT. Mitra Manggalindo Jl. Cengkeh 8B-10 Jadarta Tel.: (62-21) 690-1577	Muara Baru Ujung Blk. N Kav. 3-4, Penjaringan Jakarta Utara		
PT. Cargill Indonesia (Cargill) P.O.Box 129/Kby-B Kebayoran Baru, Jakarta	Ds. Tanjung Udik Gunung Putri, Bogor	12,000	
PT. Panggung Interprises Jl. Jembatan Tiga Barat Blok E/12 Muara Karang, Pluit Jakarta Tel.: (62-21) 668-1021 Fax.: (62-21) 660-2689 Tlx.: 41266 NATEK IA	Tanjung Pasir, Tangerang	9,600	
PT. Metro Inti Sejahtera Kampung Pondok Ungu Kelurahan Medan Satria Bekasi Km 29	Kampung Pondok Ungu Kelurahan Medan Satria Bekasi Km 29	6,000	
PT. Nagarona Langkah Maju (Nagarona) Jl. Orpa 23A Jakarta Tel.: (62-21) 675-906 676-761			

3. CENTRAL JAVA

PT. Pulau Madu (Tiger Brand) Jl. Brigjen Sudiarto 198/443 Semarang Tel.: (62-24) 27-759	Kabluk, Semarang		
PT. Ika Putra Feedmill Jl. KHM. Mansyur 87 Pekalongan, Semarang			

4. EAST JAVA

PT. Bamaindo Foodstuff (Bama) Andhika Plaza Building Blok B 1-2 Jl. Simpang Dukuh 38-40 Surabaya	Jl. Raya Gedangan Desa Tebel Sidoarjo Tel.: (62-3129) 41-467, 41-968	15,000	4,000
PT. Central Proteina Prima (Bintang, Irawan, CP) Jl. Parangtritis Raya 1 Jakarta Tel.: (62-21) 686-537/39	Jl. Dupak Rukun Tromol Pos 5018, Surabaya Tel.: (62-31) 21-973/5	48,000	2,500
PT. Wirifa Sakti (Galina) Jl. Rungkut Industri III/48B Surabaya	Jl. Rungkut Industri III/48B Surabaya	1,200	100
PT. Matahari Sakti (Feng Li) Jl. Margo Mulyo Industri I Kav. A 10-13 Tandes Commersial Estate Surabaya Tel.: (62-31) 333-983 334-014 Fax.: (62-31) 510-733	Jl. Margo Mulyo Industri Kav. A 10-13 Tandes Commersial Estate Surabaya Tel.: (62-31) 333-983 334-014 Fax.: (62-31) 510-733	7,500	1,500
PT. Windu Fairly Int. (Fairly) Andhika Plaza Building Blok B13 Jl. Simpang Dukuh 39-40 P.O.Box 76, Surabaya Tel.: (62-31) 511-662 Fax.: (62-31) 514-112	Desa Kedungrejo P.O.Box 6 Muncar Banyuwangi Tel.: (62-333) 21-184 Fax.: (62-333) 21-184	9,000	6,000
PT. Golden Greatwall Ind. (Ta Chen) Ds. Cangkir, Driyorejo Gresik Km 21.5, Jatim Tel.: (62-31) 833-3569 833-3579	Ds. Cangkir, Driyorejo Gresik Km 21.5, Jatim Tel.: (62-31) 833-3569 833-3579	12,000	5,400

Annex 2. (cont.)

Company / Office	Farm location		
PT. Karka Nutri Industri (Karka). Jl. Raya Darmo 23-25 Surabaya Tel.: (62-31) 571-371 576-538 Fax.: (62-31) 572-318 Tlx.: 32155-33107	Jl. Raya Pembangunan Km31 Tanggul Angin Sidoarjo Jatim	48,000	12,60C
PT. Matandi Makmur Sugih Waras, Jenuk, Tuban Jatim Jatim	Sugih Waras, Jenuk, Tuban		
PT. Suri Tani Pemuka (Comfeed) Jl. Basuki Rahmat 129 Po. Box 197, Surabaya Tel.: (62-31) 43-395	Jl. Raya Biduıan Km 3.5 Sidoarjo Tel.: (62-3129) 21-961/64	19,000	8,000
5. SOUTH SUMATERA			
PT. Prayasa Mina Tirta Jl. Majapahit 34/2 Jakarta	P. Bangka, Sumatra	2,000	
PT. Bangka Pioneer Jl. Kapt. A. Surahman 37-39 Pangkal Pinang Tel.: 22-256			
6. LAMPUNG			
PT. Jaka Utama Kraftfutter (Viterna) JL. Hayam Wuruk 47 Jadarta Pusat Tel.: 588-2085 548-2085	Jl. Raya Kalianda Km 12 Serengsem, Bandar Lampung Tel.: 31-181/82		
7. JAKARTA			
BinaSwadaya (Prori) Koperasi Pegawai Negri Dinas Perikanan DKIJ Jakarta	Jl. Gunung Sahari III/7 Jakarta Pusat Tel.: 420-4402		
8. SOUTH SULAWESI			
PT. Bukaka (Bukaka 66) Jl. HOS Cokroaminoto Ujung Pandang			

Annex 3. Major institutions involved in aquafeed research and development in Indonesia

Institution	Activities Research	Activities Development	Area of study	Address
Indonesian Science Institution	X		freshwater and brackishwater	
Universities	X		freshwater and brackishwater	
Freshwater Aquaculture Development Center		X	freshwater aquaculture	Jl. Selabintana 17 Sukabumi
Brackishwater Aquaculture Development Center		X	brackishwater aquaculture	Jl. Pemandian Kartini P.O. Box 1 Jepara
Marine Culture Development Center		X	marine aquaculture	P.O. Box 74 Teluk Betung Bandar Lampung
Freshwater Fisheries Research Center	X		freshwater aquaculture	Jl. Sempur No. 1 P.O. Box 51 Bogor
Coastal Aquaculture Research Center	X		brackishwater aquaculture	Alliri Tengah Village Maros

AQUAFEEDS AND FEEDING STRATEGIES IN MALAYSIA

Che Musa Che Utama

National Prawn Fry Production and Research Centre
Kampung Pulau Sayak, 08500 Kota Kuala Muda, Kedah, Malaysia

UTAMA, C.M. 1993. Aquafeeds and feeding strategies in Malaysia, p.282-296. *In* M.B. New, A.G.J. Tacon and I. Csavas (eds.) Farm-made aquafeeds. Proceedings of the FAO/ AADCP Regional Expert Consultation on Farm-Made Aquafeeds, 14-18 December 1992, Bangkok, Thailand. FAO-RAPA/AADCP, Bangkok, Thailand, 434 p.

INTRODUCTION

The aquaculture industry in Malaysia is basically categorized into freshwater and brackishwater culture. Freshwater culture sites consist of freshwater ponds, freshwater cages and ex-mining pools. On the other hand, brackishwater culture includes brackishwater ponds, brackishwater cages, cockle, mussel and oyster culture (Figure 1). The area of freshwater and brackishwater aquaculture in 1990 is shown in Table 1.

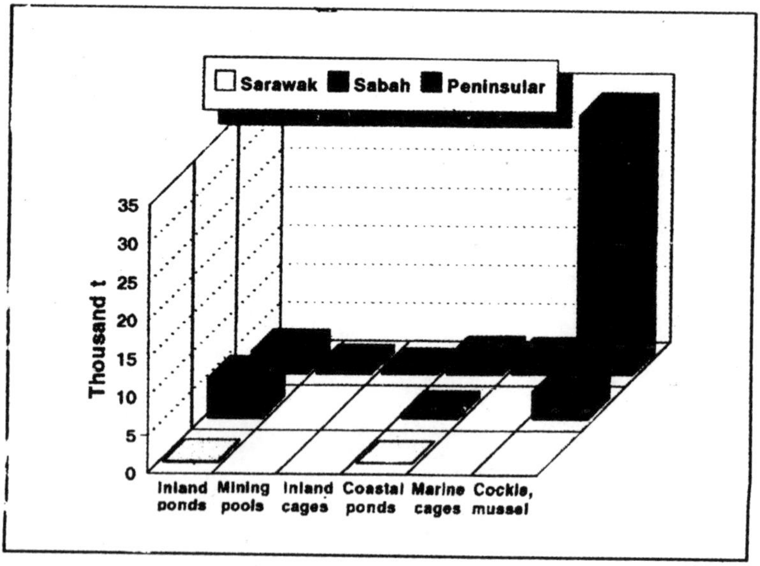

Figure 1. Estimated aquaculture production in Malaysia in 1990

Table 1. Estimated area of ponds, disused mining pools and cages in Malaysia in 1990

Region	Freshwater ponds (ha)	Freshwater ex mining pools (ha)	Freshwater cages (m²)	Brackishwater ponds (ha)	Brackishwater cages (m²)
Peninsular Malaysia	3,015	641	21,413	1,120	242,971
Sarawak*	211	0	0	0	0
Sabah	1,023	0	0	432	0
Total	4,249	641	21,413	1,552	242,971

* ponds constructed in 1990

Source: Department of Fisheries, Malaysia (1991).

Total aquaculture production in Malaysia was about 52,300 t in 1990 (Anon. 1991). Cockle production, at 37,514 t, accounts for 71%, freshwater fish culture production, estimated at 9,690 t, accounts for 19% and brackishwater/marine finfish and shrimp culture production, estimated at 5,100 t, accounts for 10% of total annual aquaculture production. The production, broken down by culture systems and by species, is shown in Tables 2 and 3, respectively, and illustrated in Figure 2.

Table 2. Estimated aquaculture production in Malaysia in 1990 by culture system (t)

Region	Freshwater ponds	Freshwater mining pools	Freshwater cages	Brackishwater ponds	Brackishwater cages	Brackishwater cockle & mussel
Peninsular Malaysia	2,900	905	516	2,195	2,055	33,874
Sarawak	169	0	0	50	0	0
Sabah	5,200	0	0	800	0	3,640
Total	8,269	905	516	3,045	2,055	37,514

Source: Department of Fisheries, Malaysia (1991)

Table 3. Aquaculture production in Malaysia in 1990 by species (t)

Species	Production
Local carp	1,846
Chinese and common carp	3,154
Freshwater prawns	137
Tilapia	1,144
Catfish	198
Sand gobi	419
Miscellaneous*	2,792
Inland sub-total	9,690
Seabass	1,953
Grouper	144
Mangrove snapper	421
Banana shrimp	156
Black tiger shrimp	2,184
Mud crab	190
Miscellaneous**	52
Coastal sub-total	5,100
Grand total	14,790

* includes snakehead, river catfish and gouramy

** includes snapper, rabbitfish and horse mackerel

Source: Department of Fisheries, Malaysia (1991)

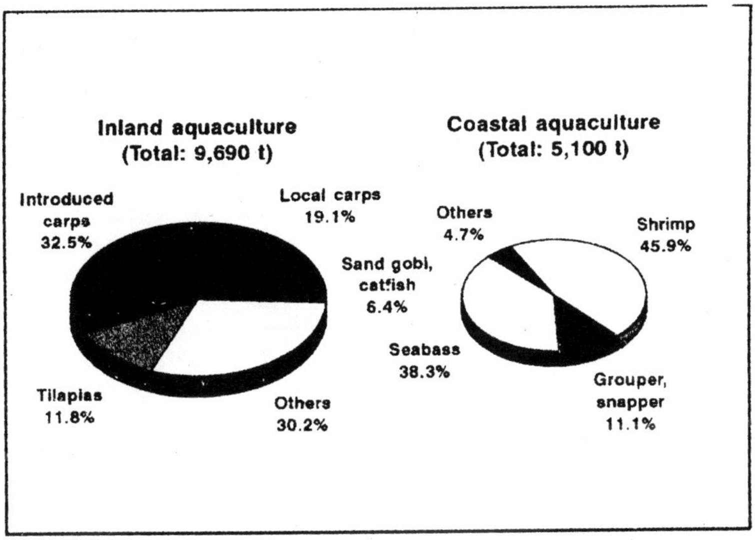

Figure 2. Cultured finfish and crustacean production in Malaysia by major species groups in 1990

It is recognized that formulated feeds are required to increase the productivity of cultured species. In Malaysia, utilization of formulated feed is common in the culture of marine shrimp, freshwater prawns, tilapia and catfish. Based on trials conducted in farmers' ponds, the cost of feed for culturing freshwater prawns ranged from 50% to 75% of operational costs (Yaakob 1991). From the information gathered from tiger shrimp farms in the northern region the cost of feed is estimated at around 50% to 60% of the production cost. In the case of tilapia culture in cages in the central region, the feed cost is more than 40% of the production cost.

AVAILABILITY OF FEED INGREDIENTS

When aquafeeds were first produced, they consisted of ingredients commonly used in animal feeds. Some of these ingredients were not suitable for the production of aquafeeds or were poorly digested by the cultured species. Since aquaculture species are more sensitive to the quality of ingredients and have higher nutritional requirements, most of the locally produced ingredients are not suitable for aquafeed production.

Most local feed mills, with their long experience in animal feed production, can source ingredients globally. However, for the sake of freshness, local or regional supplies are preferred. The main feed ingredients imported into Malaysia in 1990 are shown in Table 4, while the proximate composition of some of the available feed ingredients is shown in Table 5.

Table 4. Type and quantity of imported feed ingredients in Malaysia in 1990

Ingredient	Quantity (t)	Country of origin (% of total)
Fish meal	23,627	Thailand (67), Chile (9)
Invertebrate meal	483	Indonesia (51), Thailand (15)
Meat meal	604	France (47), New Zealand (37)
Soybean meal	137,300	China (88)
Corn	112,000	Thailand (46), China (22)
Wheat pollard/bran	102,200	
Rice bran	15,000	Indonesia (71)
Corn bran	6,500	Argentina (80)
Groundnut meal	46,100	India (70), China (23)

Source: Statistic Department of Malaysia (1991)

Table 5. Proximate composition of feed ingredients in Malaysia

Ingredients	No. of samples analyzed	Moisture (%)	Crude protein	Crude fat	Ash
			(% of dry matter)		
Fish meal (Malaysian)	4	9.9	57.6	7.5	32.0
Fish meal (Thai)	1	13.0	64.8	8.3	28.7
Fish meal (Peruvian)	1	12.7	70.7	8.5	16.8
Fish meal (home-made)	3	6.2	64.3	6.3	27.0
Shrimp head meal	1	8.3	29.1	3.3	58.0
Shrimp meal	2	13.6	70.1	4.1	24.7
Meat and bone meal	1	9.8	62.4	12.8	21.8
Anchovy head meal	3	12.5	59.6	8.3	34.7
Squid head meal	3	10.2	78.3	10.0	8.4
Paste-shrimp meal	2	8.5	67.4	4.7	21.6
Soybean meal	4	10.2	46.1	2.5	6.6
Corn meal	2	12.6	9.0	7.5	1.9
Wheat pollard	1	10.4	15.5	4.1	5.5
Rice bran	2	10.1	15.4	17.9	8.5
Baker's yeast	1	9.1	44.0	1.4	5.0
Brewer's yeast	1	5.1	59.4	1.1	8.3

Source: NAPFRE's Laboratory.

About 311,000 t of trash fish was available in the country in 1990. Trash fish was used for feeding cultured finfish and crustaceans (like mud crab) and for manufacturing fish meal.

AQUAFEED MANUFACTURING INDUSTRY

Shrimp

More than a dozen brands of formulated larval feeds are available in the market, but almost all are imported. However, they are generally being used to supplement natural food.

Almost all marine shrimp farmers use dry formulated feeds for the grow-out phase. Only three out of the more than 10 brands of shrimp feeds available are produced locally. Most imported brands are from Taiwan and Thailand. At least 50% of farmers indicate a preference for imported feeds. One of the reasons given is the higher probability of inconsistent quality in local feeds, especially when the demand is high, while imported feeds are only slightly more expensive.

In freshwater prawn culture, only the large farms use dry commercial feeds while the smaller ones still depend on supplementary feeds (such as chicken entrails) in their grow-out phase. The main reason is that this type of feed is only processed by feed mills if farmers order a considerable quantity; it is not a stock item.

Finfish

At present, the rearing of finfish larvae totally relies on live food (such as rotifers, brine shrimp and cladocerans) depending on whether they are marine or freshwater species. Nursery and grow-out phases are fed with trash fish in the case of marine species. For freshwater species, dry formulated feeds are available but they are only widely used in tilapia and catfish culture, due to economic reasons.

General

The proximate composition of some of the aquafeeds available in the country is shown in Table 6.

Table 6. **Proximate composition of commercial aquafeeds in Malaysia**

Species/stage	Moisture	Crude protein	Crude fat	Ash
Feed brand	(%)	(% of dry matter)		
***Penaeus monodon* - larvae**				
Nippai B.P	3.8	50.4	42.3	5.4
Nippai A.S	7.8	65.8	5.4	16.4
Topal	5.6	48.0	13.2	5.4
Yeaster BMC	2.5	49.0	31.1	5.5
Yeaster TP-1	6.4	61.9	9.1	15.2
Shrimp-gen	1.0	35.7	26.4	9.5
Frippak CAR #1	2.2	50.0	17.3	19.8
Frippak CD #2	7.0	51.1	18.5	20.7
Kyowa-Hakko A-250	5.0	52.6	24.7	11.1
Higashimaru #0	8.7	53.7	5.2	18.3
President HP #1	5.0	47.7	26.8	8.1
Bio-science #0	7.0	49.4	4.0	21.5
Hatchfry Encapsulon	6.8	51.7	15.4	12.1
Gold Coin	4.5	42.0	15.2	16.1
Shar Mio Pao	6.1	42.6	5.5	28.1
Gin Shar Miu	8.0	41.6	2.5	33.3
ZM	4.3	50.9	8.6	8.7
EPAC	3.7	48.9	14.5	5.0
***Penaeus monodon* - postlarvae**				
Gold Coin	9.8	41.8	5.0	13.6
Nippai	9.4	60.9	9.7	18.3
Lucky Star	10.4	46.8	7.9	15.4
President	13.9	45.1	7.3	22.3
Chuen Chin	7.8	40.3	5.5	19.3
Higashimaru	9.4	45.0	5.7	15.0
CP	7.7	43.8	7.4	13.4
***Penaeus monodon* - grower**				
Gim Hin	8.3	44.4	7.8	13.3
Gold Coin	11.6	43.1	7.6	15.1
Fuw-Sow	9.8	43.1	9.7	12.2
Lucky star	11.3	41.1	5.0	12.7
VIP	9.4	48.7	9.5	18.9
Victory	10.5	42.0	8.5	16.0
Marine Feed #5	10.7	44.2	9.4	15.5
Leong Huat	8.9	43.6	6.1	13.6

Table 6. (cont.)

Macrobrachium rosenbergii **- larvae**				
ACAL	7.1	48.9	30.8	12.4
SUTIMAL	12.6	55.6	9.2	18.8
Macrobrachium rosenbergii **- grower**				
Gold Coin	8.2	35.6	4.7	12.8
Tilapia - grower				
Gold Coin	8.4	20.1	2.1	6.9
DSM	7.1	24.3	4.2	7.6
Joy Fish	6.7	23.5	1.6	6.3

Source: NAPFRE's Laboratory

ON-FARM FEED FORMULATION AND MANUFACTURING

At present freshwater fish culture is in a transitional phase where farmers are shifting from the traditional extensive culture systems, dependent on pond fertilization alone, to semi-intensive systems requiring supplementary feeding and some intensive culture systems requiring complete feeds (Figure 3). General observations indicate that the majority of the farmers practice semi-intensive culture. Among the commonly utilized supplementary feeds are chicken viscera, Napier grass, culled eggs, rice bran, bread crusts, noodle waste, soybean waste and the like (Figure 4). Farms which have developed on a commercial scale normally use factory-made feeds.

Figure 3. Cage culture of tilapia in an abandoned tin mining pit in Selangor, Malaysia

Figure 4. Soybean processing waste used as supplementary feed in Selangor, Malaysia

Currently marine shrimp culture, with initial stocking densities of at least $10/m^2$, depends on factory-made aquafeeds. For finfish culture, trash fish is still the main feed.

Formulation and manufacturing of farm-made feeds is primarily carried out by government farms and research institutions. Typical formulae of farm-made feeds used for fresh water fish and marine shrimp are shown in Tables 7 and 8, respectively.

Table 7. Composition of experimental feed for tilapia in Malaysia

Ingredients	%
Fish meal	20.0
Soybean meal	29.2
Copra cake	19.8
Maize	2.0
Rice bran	27.5
Vitamin mix	0.5
Mineral mix	1.0
Total	100.0

Source: Pathmasothy (1989)

Table 8. Composition of experimental feed for tiger shrimp in Malaysia

Ingredients	%
White fish meal	40.0
Shrimp head meal	15.0
Soybean meal	20.0
Corn gluten	5.0
Wheat flour	10.0
Fish oil	1.0
Starch	2.0
Basfin	0.3
Vitamin and mineral mix	6.7
Total	100.0

Source: Department of Fisheries, Malaysia (1990).

ON-FARM FEEDING STRATEGIES

Feed management is the most critical factor in determining the profitability of fish and shrimp farms, because of the high related input costs. The feeding behavior of cultured species dictates the feeding strategy. In freshwater fish culture floating feed is commonly used, especially for the culture of catfish and tilapia. Fish are fed to satiation at each feeding time, usually more than twice a day. When sinking pellets are used for feeding, the fish are normally fed two or three times a day at rates of 6% of body weight in the first month reducing to 2% in the last month of culture.

Marine shrimp are fed several times per day with a major portion of the daily feed allotment given at night. It is common practice to feed more to young juveniles than to adults. The feeding rate used is normally between 3-10% of body weight. However, the actual amount of feed given is judged by the appetite of shrimp. The main problem encountered in feed estimation is to assess survival rate which is difficult due to the benthic, cannibalistic and evasive nature of shrimp. However, farmers can estimate survival rate according to growth stage or by means of regular sampling at several spots in the pond. The unpredictability of survival underlines the need to monitor feed by consumption rather than to feed a calculated amount. Feed consumption is normally monitored by the use of feeding trays. Generally, six to eight feeding trays are used per hectare with a minimum of four trays per pond regardless of pond size. Approximately 1-4% of the total feed provided per meal is given on feeding trays. Feed consumption on the feeding trays is monitored one or two hours after feeding, depending on the age of shrimp.

In the cage culture of marine fish, the fish are normally fed to satiation twice a day for seabass and once a day for groupers. The trash fish used is either chopped, or ground with a mincer.

MAJOR INSTITUTIONS INVOLVED IN AQUAFEED RESEARCH AND DEVELOPMENT

Research and development of aquafeeds is one of the responsibilities of the Department of Fisheries. The Department acknowledged the importance of formulated feeds when it established the feed section of the Fisheries Research Institute (FRI) in Penang in 1976. Since the establishment of the research branch of the Brackishwater Aquaculture Research Centre (BARC) in Johor in 1979, the feed section has been expanded to cover pond growout feeds for marine fish and shrimp. When the National Prawn Fry Production and Research Centre (NAPFRE) in Kedah was established in 1987, its feed section concentrated on larval and postlarval feeds for shrimp and other crustaceans (Figures 5 and 6). Research on freshwater fish feeds began in 1975 at the Freshwater Fish Research Centre (FFRC) in Malacca. Several universities also carry out aquafeed research, including the Universiti Pertanian Malaysia. The mailing addresses of the above institutes are given in Annex 1.

Figure 5. Preparation of farm-made feed at the National Prawn Fry Production and Research Centre (Kedah, Malaysia)

Figure 6. Feeding trials in indoor tanks at the National Prawn Fry Production and Research Centre (Kedah, Malaysia)

CURRENT NATIONAL REGULATIONS

There are no set standards or regulations imposed on local aquafeed mills. Their feeds are basically formulated on biological and economic considerations. There is, however, some concern regarding the use of certain antibiotics in aquafeeds. A 13% government tax is imposed on imported feeds to encourage the development of local aquafeeds. However, local feed mills also have to pay taxes on imported feed ingredients.

CURRENT PROBLEMS AND CONSTRAINTS

Small and dispersed markets

Total annual production of freshwater fish was around 9,700 t in 1990. Based on experience in the production of freshwater fish, the apparent feed conversion (AFCR) is estimated at 2.5. Thus 24,300 t of feed would be sufficient to feed all the cultured freshwater fish in the country, assuming that all fish were given formulated feeds alone. Pond experiments have shown that, in the case of black tiger shrimp and seabass, an AFCR of 2.5 can also be obtained with local feeds. Therefore another 12,800 t of feed would be sufficient to produce the 1990 production (5,100 t) of brackishwater/marine finfish and shrimp. The total feed for aquaculture species calculated in this arbitrary way would amount to 37,100 t/yr, or 119 t/day (for 312 working days per year).

Feed mills, for greater economy, are constructed to operate on a large scale. The minimum batch size for many mills is 2 t and the built-in capacity is often 20 to 40 t per hour. Consequently, local feed mills have to resort to competing in foreign markets, the local market currently being very small and dispersed.

High quality raw material supply

Since cultured aquatic species are more sensitive to the quality of raw feed ingredients than other livestock and have higher nutritional requirements, only high quality raw materials are used in aquafeeds. Most have to be imported, either because they are not available locally or the quality of locally produced ingredients cannot meet the standards. Due to this great dependence on imports, feed prices are subject to the global supply and demand of ingredients.

Lack of precise data on nutritional requirements

Although approximate specifications for aquafeeds are available, precise data on the nutritional requirements of several cultured species are still lacking. This is reflected by the difference in performance of feeds with apparently the same nutritional specifications.

TRENDS IN AQUAFEED MANUFACTURING AND USE

There is considerable potential for aquaculture development in Malaysia, especially for marine shrimp farming. Such development, together with the trend towards increasing intensities of various culture systems, would lead to a greater demand for commercial feeds. The future for the feed milling industry is hence bright and there is considerable room for expansion.

LIST OF REFERENCES

Anon. 1991. Annual Fisheries Statistics 1990. Department of Fisheries, Malaysia. 172 p.

Pathmasothy, S. 1989. The culture of all-male, all-female and mixed-sex batches of Nile tilapia in static pond, 9 p. *In* A.A. Jothy (ed.) Proceedings of Fisheries Research Seminar, 1988. Department of Fisheries, Kuala Lumpur.

Yaakob, A. 1991. Studies on monoculture of freshwater prawn: effectiveness and viability. Paper presented at the IRPA Seminar (Socio-economic Sector), UUM, Kedah, Malaysia.

Annex 1. Addresses of Malaysian institutions involved in aquafeed research

Fisheries Research Institute
11800 Glugor
Penang

Brackishwater Aquaculture Research Centre
Fisheries Department of Malaysia
81550 Gelang Patah
Johor

Freshwater Fish Research Centre
Fisheries Department of Malaysia
Batu Berendam
75350 Melaka

National Prawn Fry Production and Research Centre
Kg. Pulau Sayak,
08500 Kota Kuala Muda
Kedah

Universiti Pertanian Malaysia
Faculty of Fisheries and Marine Science
Department of Aquaculture
43400 Serdang
Selangor

AQUAFEEDS AND FEEDING STRATEGIES IN NEPAL

Madhab B. Pantha

Fisheries Development Division, Department of Agriculture,
Harihar Bhawan, Pulchowk,
Kathmandu, Nepal

PANTHA, M.B. 1993. Aquafeeds and feeding strategies in Nepal, p. 297-316. *In* M.B. New, A.G.J. Tacon and I. Csavas (eds.) Farm-made aquafeeds. Proceedings of the FAO/AADCP Regional Expert Consultation on Farm-Made Aquafeeds, 14-18 December 1992, Bangkok, Thailand. FAO-RAPA/AADCP, Bangkok, Thailand, 434 p.

INTRODUCTION

The kingdom of Nepal is situated along the southern slopes of the Himalayas. It is a landlocked, mountainous country in the South Asia region, lying between longitude 80° 4" to 80° 12" east and between latitude 26° 22" to 30° 27" north. The total land area of the country is 147,180 km^2 of which two-thirds consists of hills and high mountains and one-third by the southern plain, the Terai. As shown in Table 1, agriculture, forest, pasture and water areas represent 18%, 38%, 13% and 3%, respectively. The climatic variation from north to south is very dramatic, in response to changes in altitude ranging from arctic, through alpine, to sub-tropical and tropical climates.

Table 1. Land use in Nepal

Land use type	Total area (1,000 ha)	Share (%)
Agriculture	2,653.3	18.0
Forest	5,533.4	37.6
Snow-covered	2,246.2	15.3
Pasture	1,978.5	13.4
Water	400.0	2.7
Settlements and roads	103.3	0.7
Others (wasteland, barren land and slopes)	1,803.3	12.3
Total	14,718.0	100.0

Source: CBS (1992)

60-80% of the annual precipitation falls during the monsoon season (June to September); rainfall depends on the southeast winds. The average annual rainfall is about 1,600 mm. An extreme maximum temperature of 44°C was recorded at Dhangadhi in 1983. The population of the country is about 18.5 million, with a growth rate of 2.08% per annum (CBS 1992).

Nepal is predominantly agricultural. Nearly 90% of the population is involved in agriculture, which includes livestock, forestry and fisheries, accounting for about 56% of G.D.P. in 1990/91 (CBS 1992). *Per caput* animal protein consumption was 7.5 kg/yr in 1990, out of which fisheries contributed 809 g/yr (10.8%). Crops represented 61% of the agricultural GDP, while livestock contributed 27%, forestry 11% and fisheries only about 1%. The GDP *per caput* is estimated to be about US$ 170 in 1988/89. Agriculture accounts for about 75% of exports and it absorbs more than 90% of the labour force.

The total nutrients available *per caput* per day, in terms of calories, protein and fat in 1990/91 were 2,288 kcal, 61.5 g and 30.4 g respectively. The contribution of fish and fish products was extremely low (Table 2).

Table 2. Nutritive values of available food in 1990/91 in Nepal (*per caput* per day)

Food items	Calories		Protein		Fat	
	(kcal)	(%)	(g)	(%)	(g)	(%)
Crop based	2,131.0	93.0	52.5	85.0	16.1	53.0
Livestock based	157.0	7.0	9.0	15.0	14.3	47.0
of which						
- meat	39.0	25.0	3.4	38.0	5.8	41.0
- fish	2.0	1.3	0.4	5.0	0.1	0.5
- others	116.0	73.7	5.2	57.0	8.4	58.5
Total	2,288.0	100.0	61.5	100.0	30.4	100.0

Source: MSD (1992)

Fish consumption, fisheries and aquaculture

The *per caput* availability of fresh fish was 0.86 kg/yr and of dried/processed fish 0.01 kg/yr in 1990/91 (Table 3). Exports of fresh fish have declined from an estimated 3,000 t in the mid eighties to a few hundred tons annually, probably due to the overall effect of fish disease outbreaks (EUS) and increasing domestic demand.

Table 3. Availability of fresh and processed fish in 1990/91 in Nepal

Commodity	Production (t)	Imports (t)	Exports (t)	Domestic supply (t)	*Per caput* availability (kg/yr)	(g/day)
Fresh fish	15,595	121	11	15,705	0.86	2.36
Dried, processed fish	-	212	-	212	0.01	0.03

Source: MSD (1992)

The demand for fish is very high and is met mainly by domestic production but partly by importing fish from neighbouring countries. Fish exports to India declined from 290 t in 1986/87 (APROSC 1989) to about 11 t in 1990/91 (MSD 1992).

The fisheries sub-sector provides a supplementary source of high-valuc but cheap animal protein. It also helps to create additional job opportunities and provides a higher income per unit area. In addition it helps to save foreign currency by substituting imports of fish and facilitating fish exports. Finally, the fisheries sub-sector helps in the utilization and exploitation of national land and water resources while maintaining ecological balance.

The main sources of water in Nepal are the small seasonal and large perennial rivers, of which there are more than 6,000. The glaciers and melting snow of the Himalayas constantly supplies water to the three major river systems, namely the Koshi, Gandaki and Karnali. Annual surface run-off from these rivers, including the catchment located in Tibet, amounts to 200,000 million m^3. The total water surface area in the country is 4,000 km^2, 2.7% of the total land area (Table 1). Natural water bodies (rivers, lakes and reservoirs) constitute about 0.4 million ha (Table 4). There are about 5,000 ha of village ponds and about 325,000 ha of irrigated paddy fields.

Table 4. Water resources of Nepal

Type of water resource	Estimated area (1,000 ha)	Projected area (1,000 ha)
Natural waters	401.4	
of which		
- rivers	395.0	
- lakes	5.0	
- reservoirs	1.4	78.0
Ponds	5.0	14.0
Irrigated rice fields	325.0	700.0 *
Total	731.4	792.0

* additional area estimated by Sharma (1992)

Nepal, with an average width of 193 km and a difference in elevation of more than 8,800 m, has, within some river systems, habitats ranging from cold to warm water, and a corresponding range in fish species. Since almost all rivers flow into the Ganga river system in India, this also enriches the fish fauna of Nepalese rivers, although there are some non-migratory fish species. 172 indigenous fish species in the various water systems were reported by Rajbanshi (1982) and 176 by Terashima (1984). These occur from a few hundred metres elevation to 4,000 m above sea level. Besides these, ten warm and coldwater fish and one prawn species have been introduced into Nepal since 1956. Exotic fish species recommended for warm water aquaculture are common carp (*Cyprinus carpio*), silver carp (*Hypophthalmichthys molitrix*), bighead carp (*Aristichthys nobilis*), and grass carp (*Ctenopharyngodon idella*), along with three high-value indigenous fish species, namely rohu (*Labeo rohita*), mrigal (*Cirrhinus mrigala*) and catla (*Catla catla*).

Exotic fish species currently being cultured in various Fisheries Research Centres under experimental conditions are rainbow trout (*Oncorhynchus mykiss*), Nile tilapia (*Oreochromis niloticus*), Nasice carp (*Cyprinus carpio* - a Yugoslavian strain), and silver barb (*Puntius gonionotus*). Some exotic fish species introduced earlier but not now existing were amago (*Oncorhynchus rhoduras*) from Japan and brown trout (*Salmo trutta*) from the U.K. Introduction of the giant freshwater prawn (*Macrobrachium rosenbergii*) is made occasionally, for experimental purposes, either from Thailand or India. *Tilapia mossambica* was introduced by the fish farmers themselves, from India, without recommendation. Efforts are also being made towards the commercial production of three indigenous high-value coldwater fish species, namely asla (*Schizothorax* spp.), katle (*Acrossocheilus hexagonalepis*) and mahseer (*Tor tor*, *T. putitora*) (Pradhan 1987).

Fish production in Nepal is restricted to inland water bodies, mainly through capture fisheries in rivers, lakes, reservoirs, swamps and irrigated paddy fields and fish culture in traditional village ponds and new fish ponds. Traditionally, farmers used to allow flood water, which is a source of wild fish seed, to flow into their village ponds and paddy fields or stock wild seed collected from the rivers, resulting in an extensive fish production of less than 0.5 t/ha/yr.

Systematic aquaculture began in the late 1950's with the introduction of the exotic common carp. By 1960, fingerlings of common carp were successfully produced and distributed to fish farmers. Chinese carps (silver, bighead, and grass carps) were introduced after 1965, when induced breeding techniques were also adopted for the indigenous fish rohu, mrigal and catla (Pantha 1979). Fish polyculture, using all the seven fish species, started after 1980, with the launching of the First and Second Aquaculture Development Projects (Pantha 1991).

There are 13 Fisheries Development Centres in the country, out of which eight are in the Terai and are responsible for technology generation for warm water fish culture, providing support services, including training, to fish farmers. The rest are in the mountain regions for research on cage culture, coldwater fish species and indigenous fish species.

The fish production statistics of Nepal show that from 1979/80 to 1991/92 the fish culture area increased from 890 ha to 4,664 ha (424%) and aquaculture production rose from 1,150 t to 9,296 t (708%) (Table 5). Average capture fisheries production and the *per caput* fish consumption increased in the same period by 140% and 136%, respectively.

Table 5. Fish culture area and fish production in Nepal

Description	1979/80	1986/87	1990/91	1991/92	Increment 79/80-91/92 (%)
Fish culture area (ha)	890	4,070	5,180	4,664	424
Fish production, total (t) of which	3,350	10,715	15,595	16,516	393
- captured (t)	2,200	5,284	5,281	5,281	140
- cultured in ponds (t)	1,150	5,434	10,314	9,296	708
- cultured in other waters (t)	n.a.	n.a.	n.a.	1,939	
Fingerling distribution (million)	4.17	18.16	40.63	46.90	1,024
Pond yield (t/ha)	1.29	1.34	1.99	1.99	54
Fish consumption (g/*caput*/yr)	330	625	819	895	171

Source: FDD Annual Report 1991/92 (unpublished)

Seven fish species (all carps) are recommended by the government for polyculture. Stocking densities range from 7,000- 15,000/ha, depending on the intensity of production, but stocking is reduced to 5,000-10,000/ha when using large fingerlings of 2-3" size. The total pond area, total fish production and average yield for each species is shown in Table 6. The total fish production of the country coming from capture and culture operations in different water bodies was 16,516 t in 1991/92 (Table 7). The percentage of total fish production from different water bodies for 1991/92 is given in Figures 1 and 2. Stocking rates, fish mortality, average weight of harvested fish and the yield of each species is given in Table 8.

Table 6. Water surface area, fish production and average yield by species in 1991/92 in Nepal

Fish species	Water surface area (ha)	Production (t)	Average yield (t/ha)
Common carp	997	1,916	1.92
Silver carp	949	1,825	1.92
Bighead carp	285	548	1.92
Grass carp	475	913	1.92
Rohu	759	1,460	1.92
Mrigal	522	1,175	2.25
Catla	677	1,459	2.16
Total	4,664	9,296	1.99

Table 7. Fish production data of Nepal in 1991/92

Description	Cultured		Captured		Total	
	(t)	(%)	(t)	(%)	(t)	(%)
Fishponds	9,296	56.3	-	-	9,296	56.3
Cages/enclosures	62	0.4	-	-	62	0.4
Ricefields	13	0.1	1,219	7.4	1,232	7.5
Lakes/reservoirs	30	0.2	75	0.4	105	0.6
Rivers	-	-	3,950	23.9	3,950	23.9
Swamps	1,834	11.1	37	0.2	1,871	11.3
Total	11,235	68.1	5,281	31.9	16,516	100.0

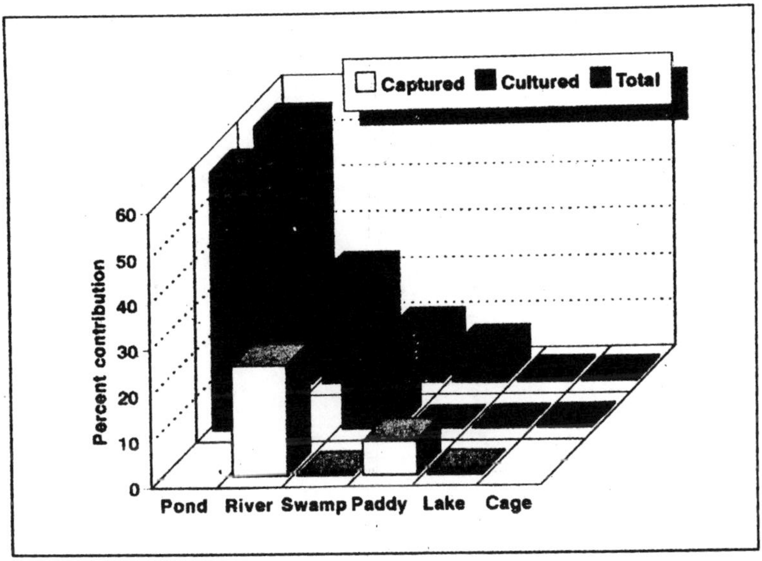

Figure 1. Fish production in Nepal in 1991/92 (total production 16,516 t)

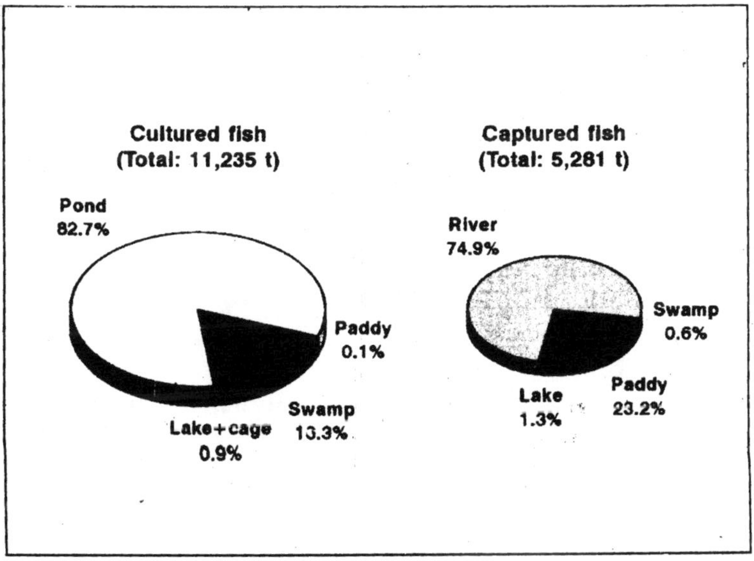

Figure 2. Cultured and captured fish production in Nepal in 1991/92

Table 8. **Aquaculture production data by fish species in 1991/92 in Nepal**

Fish species	Species composition (%) *	Stocking density (pc/ha)	Mortality (%) **	Harvested no. (pc/ha)	Average weight (kg/pc) ***	Production (t/ha)
Cyprinus carpio (Cc)	21	1,861	40	1,117	0.304	0.339
Hypophthalmichthys molitrix (Sc)	20	1,770	42	1,027	0.466	0.479
Aristichthys nobilis (Bhc)	6	531	42	308	0.679	0.209
Ctenopharyngodon idella (Gc)	10	885	38	549	0.563	0.309
Labeo rohita (R)	16	1,417	51	694	0.308	0.214
Cirrhinus mrigala (M)	16	1,417	45	779	0.298	0.232
Catla catla (K)	11	974	48	506	0.413	0.209
Total (average)	100	8,855	(44)	4,980	(0.400)	1.991

* species composition differs from government recommendations, which favour a higher ratio of Chinese carps to Indian major carps

** fish seed is stocked at 1-2" size, causing high mortality

*** culture period is 8 months long

From the data given in the Tables 6 and 8, which are based on a field survey of private fish farmers conducted by the Department of Food and Agricultural Marketing Service in 1991, it is obvious that the farmers are stocking common carp at 21%, Chinese carps at 36% and Indian major carps at 43%, while the government recommendation is 25% common carp, 55% Chinese carps and 20% Indian major carps. This is mainly because the farmers get higher market prices and better keeping quality during marketing for Indian major carps than for Chinese carps. Two groups of carps are cultivated, which have different feeding habits, namely:

- species that thrive best on mixed diets composed of artificial and natural feed (common carp, grass carp, catla, rohu and mrigal); and
- those that feed mostly on the natural food present in the pond (silver carp and bighead carp).

Even before 1980, fish farming was considered to be a profession only for the fishermen caste or for farmers of low status. The government has given first priority to exploiting the waters of village ponds by adopting low cost fish culture technologies. Fish culture practices started first with the monoculture of common carp, which was followed by polyculture, in which emphasis was given to those species which thrive best on natural plankton food.

It is important to note that village ponds were built for religious and social purposes. In them villagers bathe and wash their clothes and utensils, while their animals are stalled on the pond dikes. These ponds were constructed in low lying areas. In the rainy season all the run-off of the villages drains into them, renewing the nutrient load annually. Therefore village ponds are very fertile and fish farmers generally do not apply any supplementary feeds or manure.

Based on the total feed and manure used in the private sector (Tables 9 and 10) and the total production (Table 5), feeding costs are only US$ 65/t of fish produced. This is mainly due to the fact that a large proportion of the fish production comes from village ponds. Most of the feeding and manuring is done by farmers who constructed their ponds after 1980. Therefore the national average cost of feeding and manuring is low, compared to that in other countries. In the government sector the feed and manuring cost of fingerling production is US$ 1.42/ 1,000 fingerlings (US$ 1.27 for feed and US$ 0.15 for fertilizers). In the case of market-sized fish production under the outreach programme, the cost of feed and manuring is US$ 242/t (US$ 180 for feed and US$ 62 for fertilizers). Since the average farm-gate value of fish in Nepal is US$ 0.825/kg, the value of the national farmed production is nearly US$ 7.7 million. National feed and manure costs are less than 10% of this figure.

Table 9. Amount of major fish feed ingredients used in the private sector of Nepal in 1991/92

Ingredients	Quantity (t)	Price* (US$/t)	Total cost (1,000 US$)
Rice bran	2,546	113.4	288.7
Wheat bran	511	103.1	52.7
Maize	301	134.0	40.3
Mustard oil cake	520	82.5	42.9
Total	3,878		424.6

* US$ 1.00 = NRp 48.50

Sources : MSD (1992); FDD Technical Report (unpublished)

Table 10. Amounts of fertilizers and manure used in the private sector of Nepal in 1991/92

Types	Quantity (t)	Price* (US$/t)	Total cost (1,000 US$)
Cow dung	1,200	5.2	6.2
Duck manure	100	20.6	2.1
Chicken manure	100	25.8	2.6
Fertilizer	2,245	84.1	188.8
Total	3,645		199.7

* US$ 1.00 = NRp 48.50

Sources: MSD (1992), FDD Technical Report (unpublished)

In Nepal only 18% of the agricultural land is under irrigation; thus most of the consumable food is produced on rain-fed agricultural land. The Terai, with land suitable for crop production, receives 2,250-2,500 mm rainfall annually; rainfall is much higher than in the hilly (average 1,500 mm) or the mountainous areas (about 300 mm) (CBS 1991). The Terai produces more than 95% of pond cultured fish and about 65% of the capture fisheries from natural water bodies (FDD 1991).

The fish farmers of the Terai have more experience in fish culture and most of the hatchery and nursery operators provide feed, especially to brood fish, fry and fingerlings. Only about 30% of the fish farmers feed grow-out fish.

AVAILABILITY OF FEED INGREDIENTS

Out of many available feed ingredients, including agricultural by-products, about ten are used for feeding fish. The most commonly used are rice bran, wheat flour/bran, maize and oil cakes. The estimated total quantity of these feed ingredients is about 280,000 t (Table 11) and only about 1.4% is used for fish. Other feed ingredients, which are used in smaller quantities (about 125 t in total) are soybean cake, fish meal, meat meal, bone meal, and sesame oil cake. About 50 t of broken rice is also used for fish feed. The proximate analysis of these feed ingredients is given in Table 12.

Table 11. Feed ingredients used for fish feed in 1991/92 in Nepal

Feed ingredients	Available (t)	Used for fish feed (t)	(%)
Major ingredients			
rice bran	20,200	2,546	12.6
wheat flour/bran	28,302	511	1.8
corn	180,504	301	0.2
mustard oil cake	48,230	520	1.1
Sub-total	277,236	3,878	1.4
Minor ingredients			
soybean cake		25	
fish meal		30	
meat meal		10	
bone meal		25	
sesame oil cake		10	
broken rice		50	
Sub-total		150	
Total		4,028	

Table 12. Proximate composition of fish feed ingredients in Nepal (% of dry matter)

Feed Ingredients	Crude protein (%)	Ether extract (%)	Crude .fibre (%)	Ash (%)	Ca (%)	P (%)
Broken rice	7.5	0.5	0.2	0.5	0.1	1.0
Rice bran	13.7	17.9	6.8	12.0	0.1	1.4
Wheat flour	11.6	2.0	1.2	1.0	0.03	0.2
Wheat bran	13.5	4.5	13.0	6.0	0.1	1.3
Corn	10.2	4.5	2.4	1.4	0.03	0.3
Mustard oil cake	25.0	2.1	7.0	15.0	n.a.	n.a.
Sesame oil cake	44.0	7.9	5.7	11.2	2.0	1.4
Soybean cake	29.0	19.0	4.8	5.0	0.3	0.6
Fish meal	35.0	3.5	1.5	21.0	2.1	4.5
Meat meal	54.0	8.2	2.4	25.0	7.9	3.9
Bone meal	11.0	9.2	1.6	80.2	31.0	12.9

AQUAFEED MANUFACTURING INDUSTRY

Although there are more than 30 feed mills in Nepal, with production capacities of 1-10 t/day, all except one produce only animal and poultry feeds. The Animal Feed Production and Development Committee (AFPDC), Hetauda, is operated by the public sector and is responsible for producing fish feeds, as well as cattle and poultry feeds. The total amount of fish feed No.1 (22% protein) and No.3 (16% protein) supplied to government fish farms in 1991/92 was 14 t and 95 t, respectively (Table 13). The AFPDC imported 931 t of nine types of feed ingredients and feed additives in 1991/92 from India (Table 14), out of which only some ingredients were used for the manufacturing of fish feeds.

Table 13. Factory made fish feeds used in public sector fish farms in 1991/92 in Nepal

Feed Ingredients	Fish Feed No. 1 (%)	Fish Feed No. 3 (%)
Maize	20.0	20.0
Wheat	30.0	30.0
Broken rice	5.0	-
Defatted oil cake	-	25.0
Soybean cake	15.5	10.0
Soybean, roasted	15.0	-
Fish meal	10.0	5.0
Bone meal	2.0	2.0
Mustard oil cake	-	6.0
Poultry meal	2.0	2.0
Salt	0.5	-
Rovimix A-D-3 (g/t)	(100)	(100)
Ruby-B (g/t)	(200)	(200)
Total	100.0	100.0
Protein (%)	22	16
Energy level (kcal/kg)	3,000	3,000
Total quantity (t)	14	95
Price (US$/t)	227	170
Total cost (US$)	3,155	16,144

Table 14. Total quantity of feed ing edients imported by AFPDC (Hetauda, Nepal) in 1990/91*

Feed ingredients	Quantity (t)	Price (US$/t)	Protein (%)
Soybean cake, extracted	120	247	42.0
Fish meal	100	247	28.0
Groundnut cake	140	184	42.0
Sesame oil cake	186	142	30.0
Sunflower cake	90	433	30.0
Mustard cake, extracted	239	67	28.0
Salt	32	98	-
Mineral mix (cattle)	11	435	-
Mineral mix (poultry)	13	289	-
Total	931		

* a proportion only was used for fish feeds

Although this is the only feed-mill manufacturing fish feed (about 109 t) about 4,000 t of farm-made feeds are made from locally available feed ingredients. As mentioned earlier, considering the technical and economic capabilities of Nepalese fish farmers, fish farming started mostly with plankton feeders, mainly in polyculture. Therefore little emphasis was given to the use of fish feeds, except for the provision of good quality or complete feeds in the early nursery phase, up to 25-30 days. Supplementary feeding is now recommended for the grow-out phase.

For the past two years, the Fisheries Research Centres have been producing pellets for experimental purposes. Almost all feeds are prepared in a dry mash form. Before use the feed is soaked in water and feed balls formed. These are either thrown to fixed feeding places or placed in bamboo baskets, which are dipped into the pond. In the case of cage culture, common carp and grass carp breeders are fed with pelleted diets. Fish fry/fingerlings are fed with crumbled feeds. Pelleted feeds are also given to coldwater riverine species.

Most existing feed mills are established in the Kathmandu Valley. The feed demand of the Terai is met by feed dealers operating in most of the towns. As fish feed demand increases, there should be few problems in producing and supplying fish feed to the fish farmers of the Terai. Unlike ingredients, compounded feeds are not imported from neighbouring countries.

ON-FARM FEED FORMULATION AND MANUFACTURING

Feed formulae used by fish farmers

Farmers use various mixtures of feed ingredients in feed formulations, examples of which are given in Table 15. Feed type #1 is used mainly by hatchery/nursery operators. Feed types #2 and #3 are used for rearing fry to advanced size fingerlings and broodfish, while feed types #4 and #5 are mostly used for grow-out. Fish farmers are only providing supplementary feed; therefore most are not interested in high quality feeds. However, it is increasingly recognized by hatchery/nursery operators, and the most advanced fish farmers, that good quality feeds would increase their fish production.

Table 15. Feed formulae* used by fish farmers in Nepal (%)

Feed ingredient	#1	#2	#3	#4	#5	Remark
Rice bran	-	35	50	80	60	Land grasses are fed to grass carp
Wheat flour	50	25	20	-	10	
Maize	-	10	10	-	5	
Mustard oil cake	-	10	20	20	25	
Soybean meal	50	15	-	-	-	
Fish meal	-	5	-	-	-	
Total	100	100	100	100	100	

* #1 for hatcheries and nurseries; #2 and #3 for rearing fry to advanced sized fingerlings and broodfish; #4 and #5 for grow-out

Feed formulation for high density verification trials

The characteristics of low density polyculture are shown in Table 16, while a low-cost feed for herbivorous fish species, feeding at low levels of the food chain, is shown in Table 17. Fingerling size is 10-25 g. The stocking pattern and management of high density carp polyculture verification trials is shown in Table 18. The fingerlings size is 20-25 g. About 50% of the production from this system is produced by feeding.

Table 16. Stocking rate and density for polyculture in Nepal

Fish species	Stocking rate(%)	Stocking Density (pc/ha)	Remark
Common earp	25	2,000	Feeding for 50% of total stock; 750 kg/ha of fish is produced from feed
Silver carp	35	2,800	
Bighead carp	15	1,200	
Grass carp	2	160	
Rohu	15	1,200	
Mrigal	8	640	
Total	100	8,000	

Table 17. Low-cost feed for herbivorous fish in Nepal

Feed ingredients	(%)	Cost (US$/t)	FCR	Yield (t/ha)
Rice bran	80			
Mustard oil cake	20			
Total	100	103	3:1	1.5

Table 18. Stocking rate and density for high density polyculture in Nepal

Fish species	Stocking rate (%)	Stocking density (pc/ha)	Remarks
Common carp	20	2,000	Commercial feed (17% protein) calculated for common carp and mrigal only; 250-day growing season; FCR 1.7:1; US$ 169/t feed cost; and about 3 t/ha yield
Silver carp	35	3,500	
Bighead carp	15	1,500	
Grass carp	10	1,000	
Rohu	10	1,000	
Mrigal	10	1,000	
Total	100	10,000	

Source: FDD (1991)

Three indigenous (asla, *Tor* spp. and *Acrossocheilus hexagonalepis*) and one exotic coldwater fish species are also important for commercial purposes and for stocking natural waters in Nepal. The latter is rainbow trout (*Oncorhynchus mykiss*), which is under experimental culture in the Fisheries Research Centre at Trisuli. This centre, which has facilities for pellet feed production and a laboratory for feed analysis, uses high protein diets, the composition of which are given in Table 19.

Fish farmers do not use machinery for fish feed preparation. They purchase various raw materials or use their own by-products, *e.g.* rice bran, oil cake or roasted soybean meal, which are mixed manually. Generally farmers prepare sufficient feed for 15-30 days at a time.

Table 19. Feed formulae for coldwater riverine fish species in Nepal (%)

Ingredients	Diet A*	Diet B*	Diet C*
Soybean, whole	35.00	35.00	-
Fish meal	37.00	20.00	13.60
Wheat	6.75	12.00	13.60
Maize	6.75	12.00	24.30
Mustard oil cake	6.75	10.00	24.30
Rice polishings	6.75	11.00	24.20
Vitamins & minerals	1.00	-	-
Total	100.00	100.00	100.00
Crude protein (%)	45	30	20
Price (US$/t)**	814	589	298

* Diet A: For asla (*Schizothorax* spp.) fry and broodstock
Diet B: For *Tor* spp., *Acrossocheilus* and rainbow trout
Diet C: For common carp broodstock and grow-out

** feed prices are high, mainly because whole dried fish is used in the formulae

Source: Fisheries Research Centre, Trisuli

ON-FARM FEEDING STRATEGIES

Hatchlings and early fry are fed with high quality diets *ad libitum*. After the second week the fry are fed 2 to 3 times per day. Advanced size fingerlings, grow-out fish and broodfish are fed at 3.5% of their body weight, twice daily. Feed quantity is adjusted after biomass assessment at monthly intervals. The stocking rates of different age groups of fish are given in Table 20.

Table 20. Stocking rates of different age groups of fish in Nepal

Age group	Size	Stocking rate
Brood fish (kg)	2-3	1,000-1,100 kg/ha
Hatchlings (cm)	0.7-1.5	2 million/ha
Fry (cm)	2.5-3.5	0.5 million/ha
Fingerlings (g)	<10 g	0.25 million/ha
Advanced fingerlings (g)	>10 g	8-10 thousand/ha

Source: FDD (1991)

Organic manure (1% N, 0.7% P_2O_5) is applied to broodstock ponds at 2,000 t/ha as a basal dose during pond preparation. Inorganic fertilizers are used at a 1:2 N:P ratio, *i.e.*, 917 kg of urea and about 1,900 kg of TSP. One-sixth of this is applied as a basal dose and the remainder is split into 23 doses at 15-day intervals.

For nursery ponds, organic manure is applied in the form of decomposed cow dung, poultry, duck, and pig manure or bio-gas effluent at the rate of 100 kg/week. Urea and TSP are applied at the rate of 2 kg/week and 4 kg/week, respectively, for 4 weeks.

In low density polyculture grow-out ponds, stocked with fish which feed at the lowest level of the food chain, organic manure is applied as a basal dose at the rate of 2,500 kg/ha. Inorganic fertilization, in the form of urea and TSP at rates of 467 and 963 kg/ha, respectively, is split into 20 doses; the first (basal) dose is 1/6 of the total and the remainder is split into 19 doses applied every fifteen days.

In high density carp polyculture verification trials organic manure is applied at the rate of 3 t/ha as a basal dose. Successive doses are applied at the rate of 2.5 t/ha/month for 7 months. Inorganic fertilizers are applied at the rate of 50 kg/ha/month for 8 months, with a total of 400 kg/ha N and 800 kg/ha P, *i.e.*, 870 kg/ha urea and 1,740 kg/ha TSP.

Traditional village ponds are the main water bodies for aquaculture in the country. More than 75% are rain-fed. Since these ponds are deep, fish culture is practiced without water exchange, except that the water level is maintained during the summer months when suitable water supplies are available. Since Nepalese extensive fish culture systems practice low levels of manuring and feeding, regular water exchange is not necessary. However, those adopting semi-intensive fish culture systems occasionally practice water exchange (one to two times a year). The cost of water management for 1 m deep water under the Terai climate is about US$ 52-103/yr, depending on soil structure; more water is needed for ponds constructed on light soils.

MAJOR INSTITUTIONS INVOLVED IN AQUAFEED RESEARCH AND DEVELOPMENT

For the last two years the Fisheries Research Centres (FRC's) at Janakpur and Trisuli have been conducting applied research on fish nutrition. After positive results are obtained, the feeding techniques are verified in other Fisheries Development Centres (FDC's) and one or two experiments are also conducted in farmers' ponds in all of the 20 districts of the Terai, under an outreach verification programme, before being recommended as a technological package to fish farmers nationwide.

Nutritional research and fish feed development in Nepal is of recent origin and there are only a few qualified technical personnel. Research facilities are not adequately developed but are planned under the Natural Water Fishery Development Project, Pokhara/Trisuli. At the moment the author of this paper, Mr. M.B. Pantha, Chief, FDD, also being a nutritionist, is helping to conduct nutritional research and fish feed formulation in the Fisheries Research Centres in Janakpur, Trisuli and Pokhara.

Nutritional research is also implemented in the field by other Fisheries Development Officers, namely Mr. R.R. Dhital, Mr. J.D. Bista and Mr. S.K. Wagle. Aquafeed research and development for cultured carp species is being conducted by the National Aquaculture Research Centre, Janakpur, and the Fisheries Research Centre, Pokhara, while the Fisheries Research Centre in Trisuli is involved in feed development for indigenous and exotic coldwater fish species.

CURRENT NATIONAL REGULATIONS

There is no specific regulation for aquafeed manufacturing but there is a feedstuff law enacted in 1976, and feedstuff by-laws of 1984, which give detailed rules and regulations regarding the procurement, storage, marketing and quality control of animal feeds generally. These are only available in the Nepali language.

CURRENT PROBLEMS AND CONSTRAINTS

The current problems and constraints in the manufacturing and use of aquafeeds include a lack of basic facilities and adequately trained manpower for nutritional research and feed development programmes. In addition, the majority of fish ponds are operated by poor small-scale farmers, who may not be able to afford to purchase expensive feeds or to apply the recommended amounts. As a result of the training provided to fish farmers, the high price of fish, and the fact that fish culture is less labour intensive than other agricultural enterprises, aquaculture is getting popular. To increase fish production, farmers are in favour of using locally available feed ingredients or agricultural by-products for fish feeds. Another problem is the lack of adequate infrastructure for fish feed distribution in the main fish production region.

TRENDS IN AQUAFEED MANUFACTURING AND USE

In the near future about 78,000 ha of new reservoirs are expected to be constructed for hydro-power generation. Pond area is expected to expand in the low-lying parts of the paddy field area. Similarly, the area of irrigated rice is expected to increase by 700,000 ha by the end of the 10th Five Year Plan period, *e.g.* by 2005. Visualizing the country's existing water resources and its future potential, there is great scope for expansion, in terms of area and intensification, of fish production in Nepal.

A decade ago, farmers in Nepal believed that it was a waste of money to provide feed and fertilizers in ponds, because they were unable to see the fish feeding on them. Now concepts are changing and almost all the hatchery or nursery operators are feeding their broodfish and fish fry. Owners of new fish ponds also feed and fertilize their ponds regularly to achieve higher fish production. Fishery Research Centres are now giving priority to nutritional research and the development of economic aquafeeds.

Experiments with various aquafeeds in the Fisheries Research Centres, and their outreach verification trials in other FDC's and fish ponds of the private sector, will help to develop improved technological packages for fish farmers. This will create an increasing fish feed demand of at least 10% per year, which can easily be met by established feed mills. Establishing feed mills specifically for aquafeeds would not be viable.

Numerous agricultural by-products have still not been tested for their suitability in fish feed preparation. Nutritional research is necessary to develop economical aquafeeds utilizing these locally available ingredients. The price of inorganic fertilizers has quadrupled recently. Therefore, there is good scope for aquafeeds to increase fish production through semi-intensive methods. By stocking advanced fingerlings (>10 g) and the provision of good quality supplementary feeds in the grow-out ponds the present mortality levels of 35-45% can easily be reduced to less than 20%.

LIST OF REFERENCES

ADB. 1991. Agricultural Research Study. Fisheries and Agricultural Research in Nepal: 1975-2005. A study team report prepared for the National Agricultural Research Centres and ADB (Ts. No. 1182, NEP). 90 p.

APROSC. 1989. ADB Evaluation Division - Impact study of agriculture financing in Nepal. APROSC Agricultural sector strategy of the Eighth Plan (1990-95). 55 p.

CBS.1991. Statistical Year Book of Nepal.Central Bureau of Statistics, Kathmandu,Nepal.15 p.

CBS. 1992. Statistical Pocket Book, Nepal. Central Bureau of Statistics, Kathmandu, Nepal. 12 p.

FDD. 1991. Fisheries Development Division Budgeting Norms for 1991/92. 25 p.

MSD. 1992. Food Balance Sheet of Nepal, 1990/91. Marketing Service Division. 29 p.

Pantha, M.B. 1979. Induced spawning of *Catla catla*, 12 p. *In* Proceedings of the Fourth National Fisheries Seminar, 22-29 November 1979, FDC, Harihar Bhawan, Kathmandu,Nepal.

Pantha, M.B. 1991. History of fish culture in Nepal: past, present situation and future strategies (unpublished). 16 p.

Pradhan, B. 1987. Potential of reservoir fishery development in Asia, 12 p. *In* Proceedings of a Workshop, Kathmandu. IDRC.

Rajbanshi, K.G. 1982. A general bibliography on fish and fisheries of Nepal. Royal Nepal Academy, Kathmandu. 96 p.

Sharma, B.P. 1989. Status of Aquaculture Development in Nepal. 10 p. *In* Aquaculture in Asia. Asian Fisheries Society, Indian Branch, Bombay, India.

Sharma, K.R. 1992. Some thoughts on irrigation sector. Gorkhapatra, 21 September 1992.

Terashima, A. 1984. Three new species of the Cyprinid genus *Schizothorax* from Lake Rara, North Western Nepal. Japanese Journal of Ichthyology 31(2):122-135.

AQUAFEEDS AND FEEDING STRATEGIES IN THE PHILIPPINES

Felicitas Piedad-Pascual

Aquaculture Specialists, Inc.,
30, Fajardo Extension Jaro, Iloilo City, 5000, Philippines

PASCUAL, F.P. 1993. Aquafeeds and feeding strategies in the Philippines, p. 317-353. *In* M.B. New, A.G.J. Tacon and I. Csavas (eds.) Farm-made aquafeeds. Proceedings of the FAO/AADCP Regional Expert Consultation on Farm-Made Aquafeeds, 14-18 December 1992, Bangkok, Thailand. FAO-RAPA/AADCP, Bangkok, Thailand, 434 p.

INTRODUCTION

The Philippines has some 7,100 islands and a total coastline of 17,460 km. There are ten major lakes with a total area of 186,072 ha. Of the 237,438 ha of existing fishponds, 14,531 ha are freshwater and 222,907 ha are brackishwater ponds (BFAR 1991). All of these mean a substantial fisheries sector, as seen by the total number of people it employs and its production. In 1991, the sector employed 990,872 persons: 258,480 in aquaculture, 675,677 in municipal fisheries and 56,715 in commercial fisheries. Total fish production was 2,599,000 t of which 27% (692,401 t) came from aquaculture (Figure 1), which was valued at US$ 908 million (Figure 2). The remaining 1,147,000 t and 760,000 t were produced from municipal and commercial fisheries, respectively (Figure 1).

Aquaculture gained importance in the last two decades to meet the growing demand for fish protein (Camacho and Lagua 1988). For instance, the *per capita* consumption of fish and other marine products in 1987 was 40 kg/yr (Table 1). This is more than the total consumption of fresh meat (9.9 kg/yr), offal (1.5 kg/yr), processed meat (2.2 kg/yr), dairy products (15.7 kg/yr), poultry meat (3.3 kg/yr) and eggs (3.7 kg/yr) (BFAR 1991). Among the cultured species, milkfish has the highest *per capita* consumption (2 kg/yr) followed by tilapia and shrimp (1 kg/yr) (Table 1).

The major species produced through aquaculture in 1991 were seaweeds (41% of the total production of 629,401 t), milkfish (33.8%), tilapia (11.0%) and shrimp and prawns (7.4%). Mussels and other finfishes (seabass, grouper) comprised 2.5% and 4.2%, respectively (Table 2).

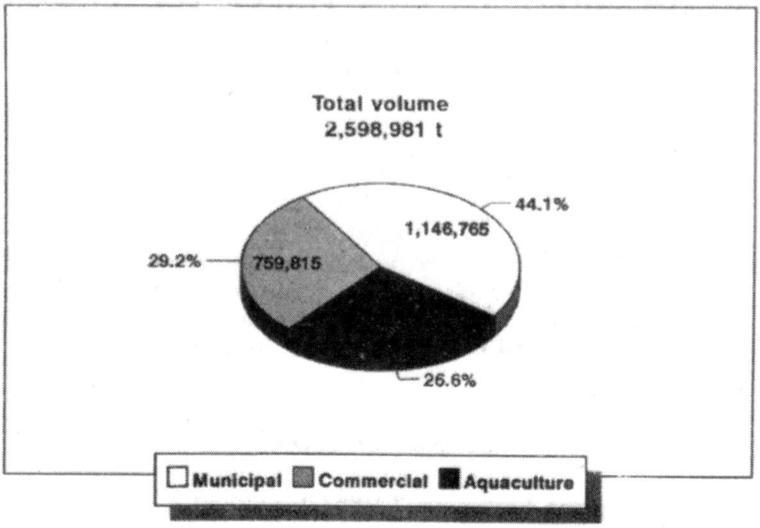

Figure 1. Volume of fish production in the Philippines in 1991

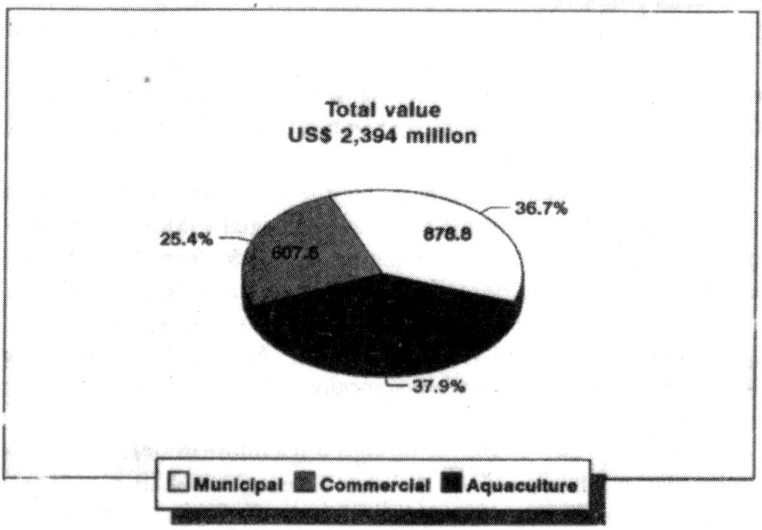

Figure 2. Value of fish production in the Philippines in 1991

Table 1. *Per caput* consumption of fish and other products in the Philippines in 1987

Food group	Per capita consumption (kg/yr)
Fresh fish	
frigate tuna (tulingan)	4
milkfish (bangus)	2
round scad (galunggong)	4
anchovy (dilis, buo)	1
Indian mackerel (alumahan)	1
sardines (tamban)	1
fusilier (dalagang bukid)	1
slipmouth (sapsap)	n
threadfin bream (bisugo)	1
tilapia	1
others	9
Sub-total	25
Dried fish	4
Processed fish	
bagoong	1
patis	n
canned fish (sardines, mackerel, etc.)	1
smoked fish (all tinapa type)	1
others	n
Sub-total*	4
Crustaceans and molluscs	
shrimp (all types)	1
others (fresh)	5
dried and processed	n
bagoong	1
Sub-total	7
Grand total	40

* does not total due to rounding up; n = negligible

Source: Fisheries Policy and Economics Division, Department of Agriculture, Bureau of Fisheries and Aquatic Resources, June 14, 1990

Table 2. **Aquaculture production in the Philippines by species and by culture system**

Culture system	Species	Production (t) 1990	1991
Fish pens	milkfish	19,004	20,449
	tilapia	3,948	4,092
	carp	1,427	-
	others		1,390
	Sub-total	24,379	25,931
Fish cages	tilapia	18,220	21,048
	carp	2,711	2,891
	Sub-total	20,931	23,939
Brackishwater ponds	milkfish	191,878	213,674
	black tiger shrimp	47,591	45,740
	others*	28,345	31,861
	Subtotal	267,814	291,275
Freshwater ponds	tilapia	35,169	37,358
	carp	642	616
	mudfish	5	-
	Sub-total	35,816	37,974
Mariculture	oysters	13,485	12,154
	mussels	17,515	17,345
	seaweeds	291,176	283,783
	Sub-total	322,176	313,282
Grand total		671,116	692,401

* others in 1990 included tilapia (18,805 t); endeavour shrimp (5,619 t); grouper (2,363 t); seabass (799 t) and white shrimp (779 t)

Source: BFAR (1991; 1992)

Though aquaculture contributes less than other sources of fish production, its output is increasing, from 392,348 t in 1982 to 692,401 t in 1991 (Figure 3). The total production by type of aquaculture system in 1991 was as follows: 291,275 t from brackishwater ponds, 37,974 t from freshwater fish ponds and 25,931 t, 23,939 t, and 313,282 t from fish pens, fish cages and mariculture, respectively (Table 2).

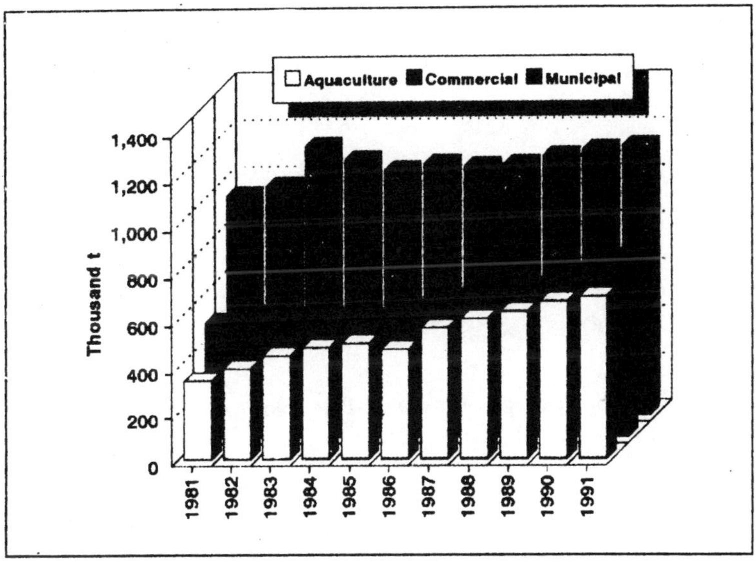

Figure 3. Fish production in the Philippines by sector, from 1981 to 1991

AVAILABILITY OF FEED INGREDIENTS

Fish meal is the main animal protein source in most fish feeds. In 1990, 34% of the fishery imports was fish meal (BFAR 1991). Table 3 shows total feed ingredient imports from 1987 to 1991 while Table 4 gives the amount of some feed ingredients produced locally.

According to Woiwode *et al.* (1992), fish meal and soybean meal comprise 35% and 15% of shrimp diets, respectively. Assuming that 35% of the 46,000 t annually produced commercial feed is fish meal and 15% is soybean meal then 16,100 t and 6,900 t of fish meal and soya meal, respectively, went into shrimp feeds.

Table 3. Some imported feed ingredients in the Philippines (t)

Ingredients	1987	1988	1989	1990	1991
Meat and bone meal	19,629	14,884	13,198	1,885	20,433
Fish meal	16,656	28,166	21,520	5,126	15,503
Soybean meal	1,748	34,135	24,684	8,577	184,488
Blood meal			128	140	405
Feed supplements	7,016	1,024	7,302	916	5,975
Whey powder		120	3,354		3,019
Feed grade milk			2,626	18	5,248
Wheat bran/flour/gluten	101	252	325	250	1,005
Squid meal			143	130	211
Teaseed cake			31		90
Starch residue					72
Thai alpha starch				15	
Feed flavor			652	10	

Source: Animal Feed Standards Division, Bureau of Animal Industry, Department of Agriculture

Table 4. Production of feed ingredients in the Philippines (t)

Ingredients	1988	1989	1990	1991
Fish meal	3,823	5,183	6,879	8,272
Copra meal	30,522	23,583	31,437	18,229
Corn and byproducts	17,225	17,553	19,641	19,247
Cassava	100			
Bone meal		116	513	865
Pollard				1,913
Yeast	14	274	298	360
Ipil-ipil meal	435			424
Feed additives, supplements	4,514	1,507	1,669	1,822
Minerals	16,653	8,534		7,460

Source: Animal Feed Standards Division, Bureau of Animal Industry, Department of Agriculture

Table 5. Proximate composition of some feed ingredients in the Philippines

Feed ingredient	No. of samples	Moisture (%)	Crude protein	Crude fat	Crude fibre	N-free extract	Ash
			(% dry matter)				
Sweet potato leaf meal	3	4.67	29.18	4.68	11.58	41.00	13.56
Cassava leaf meal		3.83	27.56	7.66	11.73	45.65	7.40
Copra meal	4	7.38	20.32	17.45	8.18	44.98	9.07
Najas graminea (digman)	3	12.51	18.50	1.87	18.08	39.04	22.51
Annelids	3	4.75	44.85	20.93	2.34	22.37	9.51
Fish meal (anchovy)	2	5.97	73.73	4.88	0.54	5.77	15.08
Fish meal (lizard fish)		6.43	76.20	4.40	0.42	2.71	16.27
Shrimp meal (kuaket)		8.30	22.50	0.97	8.92	8.84	58.77
Fish meal (Peruvian)	15	7.32	68.32	6.51	0.85	6.25	18.07
Fish meal (slipmouth)*		5.86	65.95	11.96	0.41	4.51	17.17
Fish meal (slipmouth)**		5.69	77.23	6.98	0.66	4.96	10.17
Fish meal (sirum-sirum)		5.68	74.71	7.04	0.38	0.00	17.87
Fish meal (herring)*		7.05	78.40	5.55	0.63	0.50	14.92
Fish meal (herring)**		6.93	87.54	4.81	0.58	0.11	6.96
Fish meal (trash)	7	6.00	76.82	6.02	1.23	3.00	12.93
Fish meal (tuna)*	2	5.29	73.69	12.11	0.56	2.20	11.44
Fish meal (tuna)**		6.52	86.51	6.70	0.62	0.51	5.66
Fish meal (white)	5	6.37	68.68	6.78	0.64	3.74	20.16
Flour (wheat) ***	16	12.02	14.22	1.10	0.84	83.16	0.68
Frog meat	2	7.57	62.47	9.91	1.16	4.77	21.69
Gluten (corn)	3	7.65	62.83	7.53	2.08	25.95	1.61
Ipil-ipil leaf meal	6	8.82	25.11	8.97	9.51	42.93	13.48
Kangkong leaf meal	2	5.26	29.57	4.24	11.98	38.75	15.46
Ruppia maritima (kusay-kusay)		8.08	15.12	3.82	19.78	46.22	15.06
Lab-lab	5	4.43	8.18	1.02	4.81	16.94	69.05
Chaetomorpha (lumot)	3	5.17	15.26	2.18	15.07	36.10	31.39
Meat and bone meal	10	4.99	45.02	8.60	2.05	7.01	37.32
Mussel	21	4.92	64.97	8.09	1.49	14.53	10.92
Oyster meal	9	5.00	54.42	9.15	3.79	20.82	11.82
Rice bran	21	10.00	13.96	13.82	6.19	55.77	10.26
Sargassum		9.00	9.75	0.75	8.83	40.29	40.38
Shrimp head meal	6	5.04	52.97	6.47	14.04	0.68	25.84
Shrimp meal	12	6.26	71.79	3.62	3.40	5.08	16.11
Snail (bagungon)		4.74	13.75	1.68	0.97	9.76	73.84
Soybean meal (def.)****	12	9.26	46.84	1.65	7.09	36.87	7.55
Tetraselmis chui	4	5.50	47.85	10.74	2.12	20.19	19.10
Yeast (brewer's)		1.40	56.10	2.14	0.33	31.58	9.85
Squid meal			79.00				

* whole sample; ** edible portion; *** factor = 5.4; **** factor = 5.7

Source: SEAFDEC Centralized Analytical Laboratory

Table 6. Proximate composition of some feed ingredients in the Philippines

Feed ingredient	Moisture	Crude protein	Crude fat	Crude fibre	N-free extract	Ash
	(%)	(% dry matter)				
White fish meal	6.35	70.50	5.99	1.11	2.47	19.93
Peruvian fish meal	6.36	70.51	10.56	1.38	1.07	16.48
Slipmouth fish meal	5.86	65.95	11.96	0.41	4.51	17.17
Tuna fish meal	6.18	77.23	6.98	0.66	4.42	10.71
Herring fish meal	7.05	78.40	5.55	0.63	0.50	14.92
Shrimp meal	4.54	71.82	3.22	3.14	4.91	16.91
Squid meal	7.85	78.69	5.44	1.62	7.52	6.73
Soybean meal	8.54	43.80	2.04	7.33	39.27	7.56
Kangkong meal	5.32	31.90	3.80	11.64	38.62	14.04
Sweet potato meal	4.97	30.77	4.76	13.64	37.03	13.80
Ipil-ipil meal	11.16	27.82	8.84	9.48	44.23	9.63
Acacia meal	4.30	25.28	5.10	24.06	39.63	5.93
Tamarind meal	4.17	15.12	5.62	14.03	53.83	11.40

Source: Penaflorida (1989)

Tuna fish meal is locally produced. It usually contains around 73% protein and can be mixed with imported white fish meal. Other locally produced fish meals are used in finfish feeds.

Shrimp meal and shrimp head meal can be produced locally while squid meal is usually imported. In the Philippines, all parts of the squid are utilized for human consumption, unlike South Korea, where the squid mantle is canned and the by-products are converted into squid meal. Sources of carbohydrates, such as corn, rice bran and flour, are bought locally. However, the price of corn and rice bran fluctuate seasonally.

The proximate composition of some feed ingredients used in experimental diets is shown in Tables 5 and 6.

AQUAFEED MANUFACTURING INDUSTRY

The most expensive component in aquaculture, especially in black tiger shrimp culture, is feed. Feeds account for 50-60% of the variable cost of production (Cruz 1991). Posadas (1988) reported that the average feed cost in extensive shrimp culture systems was US$ 0.17/kg; in modified extensive systems it was US$ 1.70/kg; and in both semi-intensive and intensive systems US$ 2.24/kg. Feed costs represent 8% of the total operating cost of the extensive,

55% of the moderately extensive, 60% of the semi-intensive and 55% of the intensive systems.

According to Woiwode *et al.* (1992), white fish meal and soybean meal comprise 35% and 15%, respectively, of the shrimp feed composition. The cost of directly imported feed ingredients is US$ 0.80/kg while locally purchased materials cost US$ 0.13/kg of feed. Duties on the imported ingredients were about US$ 0.06/kg, while value added tax was US$ 0.04/kg. Operating expenses were US$ 0.19/kg and mark up of US$ 0.04/kg for dealers and US$ 0.08/kg for feed mills resulted in a total of US$ 1.34/kg of feed.

Most, if not all, commercially produced shrimp feeds are complete feeds used for intensive culture. A few are used in semi-intensive culture, while some of those recommended for intensive culture are more effective in the semi-intensive system. San Miguel Corporation is now marketing a supplemental feed (SD-90) containing 30% crude protein for shrimp cultured at a stocking density of 5/m^2 (Table 7).

Agbayani *et al.* (1990) state that, of the total variable cost in the monoculture of mud crab (*Scylla serrata*), 37% goes to trash fish for crabs stocked at 5,000/ha. When the stocking rate is 10,000/ha the feed cost is 48.7%, while it is 57.2% for 15,000-20,000/ha.

In milkfish production, the cost of feeding, assuming an FCR of 1.6:1 is US$ 0.53/kg, 71.5% of variable costs at a stocking density of 16,000/ha (Ocean Feed 1992) or 73% of total production costs at 20,000/ha (Philip Cruz, Feed Pro Corporation, pers. comm.). The protein content of milkfish feeds ranges between 24-31% while for tilapia it is 21-26% (Table 7).

With two croppings per year, some farmers like Yusay (pers. comm.) find milkfish culture more lucrative than shrimp culture. The latter is presently considered a very risky venture in the Philippines. A cost /benefit analysis for increasing milkfish production using various supplemental feeds reported by Sumagaysay *et al.* (1991) showed that increasing the quality and quantity of nutrients, particularly protein, is beneficial if the resulting increase in sales outweighs the increase in the cost of production. They reported that supplemental feeding of a 27.4% protein diet increased profit by 46.9% compared to a diet of 22.4% protein. Their study suggests that supplemental feeding, even in less productive ponds, is advantageous. Furthermore, they showed that the use of rice bran as a single ingredient, a common practice in the Philippines, is not economical.

The proximate composition of some commercial feeds is presented in Tables 7 and 8. The crude protein content of shrimp feeds before 1990 was 41%, or higher, while crude fat varied from 3-6%. After 1990, the crude protein content was reduced to 30-39% for feeds used for intensive culture while the fat content ranged from 2-7%.

Table 7. Proximate analyses of various commercial feeds in the Philippines

Feed type	Moisture	Crude protein	Crude fat	Crude fibre	N-free extract	Ash
	(%)	(% dry matter)				
Shrimp feeds						
B-Meg CE-90						
PL/prestarter**	10.0	38.5	4.0	3.5		13.0
starter**	10.0	38.0	4.0	3.0		13.0
grower**	10.0	37.0	4.0	3.0		14.0
finisher**	10.0	37.0	3.5	4.0		15.0
D_1*	12.0	31.1	6.2			11.0
D_2*	10.6	38.5	6.4			10.4
D_3*	10.2	37.7	7.2			10.2
B-Meg SD-90**						
starter	10.0	30.0	5.0	5.0		10.0
grower	10.0	30.0	5.0	5.0		10.0
finisher	10.0	30.0	5.0	5.0		10.0
Chuen Shin *						
#144, C_1	10.0	36.3	9.1			11.6
#145, C_2	11.2	35.9	7.7			11.4
#146, C_3	8.9	36.6	7.8			11.5
Hoc Po *						
A_1	13.1	39.2	7.6			11.8
A_2	11.6	39.4	7.3			11.8
A_3	10.0	36.5	6.6			11.6
A_4	10.4	35.7	7.1			11.8
A_5	10.0	36.0	6.4			11.7
Ocean Feeds **						
starter	10.0	38.0	6.0	3.0		16.0
grower	10.0	37.0	6.0	3.0		16.0
finisher	10.0	36.0	6.0	3.0		16.0
Tateh **						
starter	10.0	39.0	3.0	4.0		15.0
grower	10.0	38.0	2.0	4.0		15.0
finisher	10.0	37.0	2.0	4.0		15.0

Table 7. (cont.)

Oversea *						
B_1	9.3	41.2	6.2			14.4
B_2	9.6	39.0	6.1			13.3
B_3	9.3	38.7	7.0			14.6
B_4	10.1	38.0	6.0			14.1
B_5	10.1	36.5	6.2			14.7
Tokai *	7.9	35.2	5.3	4.2	41.3	14.0
Tateh*						
E_1	9.9	39.9	6.2			11.6
E_2	10.9	38.8	5.9			12.5
E_3	10.6	39.6	6.3			12.2
Finfish Feeds						
Fuji Triumph floater fish feed (grower) **	10.0	24.0	4.0	7.0		10.0
Fuji Triumph sinker fish feed (grower) **	10.0	24.0	4.0	7.0		14.0
Ocean Feed (milkfish) **						
grower	12.0	28.0	8.0	8.0		
finisher	12.0	27.0	8.0	8.0		
Robina fish feed **						
crumbles	13.0	26.0	4.0	9.0		12.0
SMC-B-Meg (tilapia) **						
starter	13.0	29-32	8.0	7.0		16.0
grower	13.0	29-32	8.0	7.0		10.0
finisher	13.0	27-30	8.0	7.0		10.0
Tateh (tilapia) **	8.2	27.0	5.0	3.6		8.2
Tateh (milkfish)**						
starter	13.0	31.0	8.0	8.0		
grower	13.0	28.0	8.0	8.0		
finisher	13.0	27.0	8.0	8.0		

* analysed by Regulatory Division, Regional Feed Laboratory Region VI, Department of Agriculture, Iloilo City

** as labelled

Table 8. Proximate composition of some commercial shrimp feeds in the Philippines

Shrimp feeds	Moisture (%)	Crude protein (% dry wt.)	Crude fat (% dry wt.)	Crude fibre (% dry wt.)	Nitrogen-free extract (% dry wt.)	Ash (%dry wt.)
SMC Starter	4.24	45.87	3.60	1.43	35.76	13.34
SMC Grower	5.32	45.16	3.20	1.40	35.60	14.64
SMC Finisher	6.39	45.25	3.28	1.10	36.96	13.41
Chuen Shin	9.32	44.16	4.82	4.58	31.21	15.23
B-Meg	6.80	47.40	3.08	3.33	32.25	13.94
Fuji Triumph	5.84	45.12	4.61	1.50	33.85	14.92
Fuji Triumph Finisher	7.51	41.53	5.56	2.62	37.52	12.77
President Feed	8.62	42.99	6.12	2.08	29.85	18.96
Robina Crustacean Pellet	7.79	47.22	6.13	4.48	30.27	11.90
Vitarich PR 35	10.36	47.93	4.46	3.28	29.67	14.66
Vitarich Finisher	8.79	44.29	3.64	3.77	33.51	14.79

Source: Central Analytical Laboratory of SEAFDEC/AQD.

Table 9. Aquafeed production and shrimp feed imports in the Philippines (t)

Year	Fish feed	Shrimp feed	Total
Domestic aquafeed production			
1987	386	10,000.	10,386
1988	3,000	25,000	28,000
1989	5,000	23,000	28,000
1990	7,000	35,000	42,000
1991	12,000	46,000	58,000
Shrimp/prawn feed imports			
1987		2,428	2,428
1988		2,527	2,527
1989		5,199	5,199
1990		1,250	1,250
1991		485	485

Most of the major feed manufacturers (Annex 1) and suppliers of feed manufacturing equipment (Annex 2) are located in or near Metropolitan Manila. Domestic aquafeed production in the past five years showed an increase from 10,400 t in 1987 to 58,000 t in 1991 (Table 9). Shrimp feed production increased from 10,000 t to 46,000 t, while shrimp feed imports decreased from 2,428 t in 1987 to 485 t in 1991 (Table 9).

Although there was a slump in the shrimp industry, the demand for feed continued to increase, both for fish and shrimp culture. Most farmers are not yet convinced that feeding finfish like milkfish is economically viable. A few companies like Robina, Vitarich, and Fuji-Triumph have been selling feeds for milkfish for some time, while Ocean Feeds and Feed Pro Corporations began selling them in 1992. San Miguel and Feed Pro Corporations also sell feed for tilapia. The prevailing prices for shrimp and finfish feeds are given in Table 10.

Hatchery owners rely on imported artificial feeds to supplement natural food for larval culture.

Table 10. Prices of various brands of shrimp and finfish feeds in the Philippines (US$/kg)*

Feed brand**	Pre-starter	Starter	Grower	Finisher
Shrimp feeds				
B-Meg CE 90	1.61	1.44	1.34	1.28
B-Meg SD 90		1.29	1.17	1.11
Fuji	1.53	1.36	1.25	1.21
Hoc-Po		1.28	1.20	1.16
Interfeeds	1.34	1.23	1.18	1.11
Interfeeds			1.15	
Oversea		1.40	1.26	1.18
Oversea		1.32	1.22	
Sin Heng Chan (Singapore)	1.14	1.14	1.14	1.14
Tateh	1.54	1.46	1.38	1.32
Tateh			1.30	
V-Lang	1.44	1.44	1.36	1.28
V-Lang		1.44	1.32	
Ocean Feeds	1.36	1.36	1.24	1.20
Finfish feeds				
Fuji-Triumph				
(milkfish floater)			0.48	0.47
(milkfish sinker)		0.48	0.43	0.39
(tilapia)		0.48	0.40	0.39
Hoc Po (milkfish)				0.50
Ocean Feed (milkfish)			0.52	0.52
Tateh (milkfish)		0.54	0.44	0.43
Tateh (tilapia)	0.46	0.42	0.42	0.42
San Miguel B-Meg (tilapia)		0.47	0.46	0.43

* exchange rate: P 25.00 = US$ 1.00

** see Annex 1 for manufacturers' addresses

ON-FARM FEED FORMULATION AND MANUFACTURING

On-farm feeds are not very popular in the Philippines because farmers find it more convenient to buy feeds from feed companies. Also, 90% of feed ingredients are imported (Villacorta, pers. comm.). In areas where trash fish and other fresh components of the feed are available, storage facilities may not suffice. However, the type of feed produced on the farm depends on the species being cultured. Tilapia and other low cost fish, for example, are used as feed to fatten grouper and mud crab and to culture seabass.

Roy Aguilar, an enterprising farmer from Cadiz, Negros Occidental, had his own locally made equipment for making moist feed for red tilapia. The feed ingredients he used were: fresh swamp cabbage (*Ipomea aquatica*), fish meal, rice bran, bread flour and afsillin, a mineral and vitamin mix marketed for swine and poultry (which contains streptomycin!). Swamp cabbage has a moisture content of 89.7% and a crude protein level of 3.9% (Food and Nutrition Research Institute 1968). All the dry ingredients were mixed with the swamp cabbage and then passed through a meat grinder twice. The moisture from the swamp cabbage was sufficient to wet the resulting mass. The noodle-like extrusions were then uniformly cut and steamed for five minutes and dried in a native clay and brick oven. Mr. Aguilar also used chopped squid to supplement commercial shrimp feeds when he stocked shrimps at 80/m^2.

This fish farm, however, was converted to a resort when the owner's red tilapia venture became unprofitable. Most aquaculture systems, except those for shrimps, still rely on natural pond productivity. However, due to the encouraging results of feeding as shown in the shrimp industry, farmers are beginning to accept commercial feeds for tilapia and milkfish. Seabass, grouper and mud crab are also "fattened" in cages or ponds by feeding trash fish and/or crustacean pellets. The term "trash fish" differs from country to country because what is considered trash fish in Thailand, for example, is not trash fish in the Philippines. Here, trash fish is composed of various species of fish that are often spoiled, hence cannot be used for human consumption.

In the case of milkfish, some farmers use fish feeds only as a finisher for fattening, one month before harvest. They also rely on commercial fish feeds during the rainy season, when lablab growth collapses. Chicken manure is utilized to fertilize both the soil (during land preparation) and water (during the culture period). To avoid off-flavour of the milkfish flesh, manure application is terminated two weeks before the expected harvest.

When seabass, grouper or mud crab are monocultured or fattened, trash fish or tilapia is used as feed. Trash fish or a combination of 60:40 trash fish:fish crumbles is fed to groupers (Juario 1991). Fish are fed twice daily at 10% of biomass. Some cage operators feed the fish to satiation twice a day for the first 2-3 months, and once a day in the next 2-3 months. Two months prior to harvest, the fish is fed only every other day. FCR is 4-5:1 (Juario 1991). Seabass and grouper are also fed trash fish or are polycultured with tilapia.

ON-FARM FEEDING STRATEGIES

To lower feed cost some companies, like San Miguel Corporation, are producing supplementary feeds for shrimps (Table 7). However, fertilization with organic and inorganic fertilizers to increase natural productivity of the ponds is still carried out during pond preparation. A low stocking density of $5/m^2$ is suggested, and many farmers at present do not stock beyond $10/m^2$. Shrimp are not fed until 2 weeks after stocking. Each feed manufacturer has a manual or a brochure, in addition to the technical assistance given to their clients, on proper feed management. Shrimp farmers often blame or suspect feed, fry, water quality, and/or disease as causes for low productivity, or when yields fall short of target (Cruz 1991). However, poor feed management is the culprit in many instances, since overfeeding or underfeeding results in high feed conversion, poor survival, water pollution, diseases, and poor quality of the produce.

In the case of milkfish, supplementary feed is given when lablab growth collapses due to frequent rainfall. Continuous fertilization or "dressing" with chicken manure is also carried out until two or three weeks before harvest.

Groupers and seabass are stocked at low densities ($0.5/m^3$) and fed trash fish for fattening. Kohno *et al.* (1989) used trash fish, mainly *Leiognathus* spp., for growing grouper (*Epinephelus malabaricus*) in cages in ponds previously used for shrimp culture. With a stocking density of $1/m^3$ and feeding fish to satiation at 10% of biomass, they found that feeding once in two days gave the best results.

A similar procedure is used in the fattening of mud crab (*Scylla serrata*). Kuntiyo (1992) demonstrated that crabs can be fattened in cages placed in drainage canals. The use of commercial shrimp pellets (36% protein) at 2% biomass, or trash fish composed mainly of *Leiognathus* spp. at 10% of body weight, was found to be effective in fattening crabs for 5 weeks.

Some farmers do not feed the shrimp immediately after stocking. They wait until two weeks to one month later, or until the shrimp are about 2 g, or when natural productivity of the pond has diminished.

MAJOR INSTITUTIONS INVOLVED IN AQUAFEED RESEARCH AND DEVELOPMENT

Institutions engaged in aquafeed research and development include national and regional centers, cooperating stations, and specialized agencies (Annex 3). Their funding comes from both local and foreign agencies. Foreign funding agencies include the International Foundation for Science, the International Development Research Center of Canada, the American Soybean Association and many others.

An overview of the aquaculture programmes of the Philippines, based on the five-year plan of the National Aquatic Resources Research and Development System (NARRDS), coordinated by the Philippine Council for Aquatic and Marine Research and Development, is reported by Guerrero (1991). One of the major programmes is the diversification and expansion of viable aquaculture industries. Though much research on aquafeeds is being carried out in the national centres, particularly at the University of the Philippines in the Visayas (Annex 4), the bulk is done in specialized agencies such as SEAFDEC/AQD, specifically its Feed Development Section (FDS) (Annex 5).

The FDS is tasked to determine the nutritional requirements of fish, such as milkfish, seabass, groupers, catfish, and shrimp and to develop practical diets for each developmental stage. Carp and tilapia studies are carried out in the Freshwater Station of SEAFDEC/AQD at Tapao Point in Laguna Lake.

In 1988, aquaculture research received US$ 4.44 million (70%) and marine capture research US$ 1.76 (28%) of the total fisheries budget (US$ 6.36 million). Of this tota. fisheries budget, 45% came from the government of the Philippines while the rest came from foreign funding (Lacanilao 1991). Furthermore, of the US$ 2.9 million that the government of the Philippines gave for fisheries research, US$ 1.9 million was given to SEAFDEC/AQD.

Examples of supplemental and experimental feeds

Some examples of the supplemental and experimental feeds developed by institutions in the Philippines are given in Tables 11-16. A 28% protein supplemental diet for tilapia, formulated by the Brackishwater Aquaculture Center of the University of the Philippines in the Visayas (UPV-BAC) is shown in Table 11 (Pascual 1991). Santiago *et al* (1988) demonstrated that *Leucaena* leaf meal in diets for broodstock Nile tilapia (*Oreochromis niloticus*) should not exceed 40% of the diet. The best diet, which gave good fry production and growth, is shown in Table 12.

In milkfish experimental diets, feed cost can be considerably reduced without affecting growth rate and yield (Sumagaysay *et al.* 1991) by several techniques:

- mixing feeding schedules by alternately feeding high and low protein diets (De Silva *et al.* 1984);

- diluting a high protein diet with a high carbohydrate diet (Hepher and Pruginin 1981); and

- feeding diets with only 23.8% protein at 1.75% of body weight during the season when temperature is low and salinity is high (Sumagaysay *et al.* 1991).

Sumagaysay *et al.* (1991) also noted that the most cost-effective protein range in supplemental feeding for milkfish is between 22% to 27% protein. The composition of these supplemental diets is shown in Table 13. Sumagaysay and Chiu-Chern (1991) explored the nutritive value of fibre in supplemental feeds for milkfish. They demonstrated that a low protein/high fibre diet can be an economical way of increasing milkfish production in brackishwater ponds. The composition of the diets they used is shown in Table 14. They demonstrated that a low protein/high fibre diet is a feasible way of lowering the cost of milkfish feed and an economical way of increasing milkfish production in brackishwater ponds. Sumagaysay (1991) noted that milkfish diets are economical when given at a rate of 1.75% of body weight. This rate can be gradually increased as biomass increases. The experimental diet is shown in Table 15.

Santiago *et al.* (1991) demonstrated that by feeding bighead carp (*Aristhicthys nobilis*) with a 40% protein diet, the total weight of eggs per female, total number of eggs per spawning and number of eggs per kg of body weight were higher compared to fish that was not given the experimental diet shown in Table 16. Furthermore, when fry of the fed broodstock was deprived of food for ten days, a lower mortality rate was observed compared to fry that was produced from unfed broodstock.

Table 11. UPV-BAC supplemental diet for tilapia (Philippines)

Feed ingredients	**Composition (%)**
Local fish meal	.7.00
Shrimp meal	14.00
Copra meal	15.73
Ipil-ipil meal	15.73
Soybean meal	12.69
Rice bran	31.35
Coconut oil	1.63
$Ca(H_2PO_4)_2$	0.95
$CaCO_3$	0.92
Total	100.00

Table 12. Nile tilapia experimental broodstock diet (Philippines)

Ingredients	Composition (%)
Fish meal	24.70
Leucaena leaf meal	20.00
Fish/vegetable oil (1:1)	2.60
Vitamin mix*	0.73
Mineral mix*	3.60
Starch	48.37
Total	100.00

* for complete practical diets (National Research Council, 1977)

Source: Santiago *et al.* (1988)

Table 13. Two experimental milkfish diets (Philippines)

Ingredients	27.4% protein diet (%)	22.0% protein diet (%)
Peruvian fish meal	5	4
T[illegible] fish meal	5	-
S[illegible] an meal	35	20
Copra meal	14	33
Wheat pollard	18	-
Rice bran	18	42
Crude coconut oil	1	-
Tricalcium phosphate	2	-
Dicalcium phosphate	-	1
Vitamin mix	2	-
Total	100	100

Source: Sumagaysay *et al.* (1991)

Table 14. Experimental milkfish diets (Philippines)

Ingredients	Diet 1	Diet 2	Diet 3
Rice hulls (%)	55.0	30.0	-
Peruvian fish meal (%)	10.0	10.0	10.0
Soybean (defatted) (%)	10.0	30.4	22.0
Rice bran (%)	3.5	-	46.1
Copra meal (%)	16.5	19.6	11.9
Coconut oil (%)	1.8	9.7	5.3
Dicalcium phosphate (%)	3.2	0.3	4.7
Total	100.0	100.0	100.0

Source: Sumagaysay and Chiu-Chern (1991)

Table 15. Milkfish feed (Philippines)

Ingredients	Composition (%)
Peruvian fish meal	4.00
Soybean meal	27.50
Rice bran	37.00
Copra meal	20.00
Bread flour	2.00
Coconut oil	0.08
Dicalcium phosphate	1.00
Lime	8.42
Total	100.00

Source: Sumagaysay (1991)

Table 16. Experimental diet for bighead carp broodstock (Philippines)

Ingredients	Composition (%)
Copra meal	20.00
Fish meal	10.00
Rice bran	5.20
Soybean meal	28.60
Corn gluten meal	28.60
Corn oil/fish oil (1:1)	4.40
Vitamin/mineral premix	2.00
L-lysine	1.00
DL-methionine	0.20
Total	100.00

Source: Santiago *et al.* (1991)

CURRENT NATIONAL REGULATIONS

The Bureau of Animal Industry (BAI) of the Department of Agriculture has a Regulatory Division whose task is to accredit the chemical laboratories of feed manufacturers and privately owned service companies which analyze aquaculture feeds, pond soil, water, and fertilizer. BAI also closely monitors the quality of aquaculture feeds in the market. Furthermore, BAI accredits feed quality control laboratories and requires the registration of premixes, additives and supplements prior to their use in feed manufacture. Free brochures, test reagents, procedures, and protocol and certificates of analyses have to be registered with the BAI. Guidelines for accreditation are issued by its Regulatory Division. Collection of feed samples is made in the 13 regions of the country. BAI can also suspend any feed establishment for non-compliance or violation of the Feed Control Law.

The prescribed nutrient composition for complete aquaculture feeds is given in Table 17. Standards for supplemental feeds will have to be defined in the light of the present situation wherein supplementary rather than complete feeds are becoming more acceptable to farmers, due to a reduction in stocking density.

Feed mills are obliged to have an accredited laboratory and experimental ponds (Bureau of Animal Industry 1989a) where their feeds can be assessed. There are also cooperatives, such as the Negros Prawn Producers Marketing Cooperative, Inc. (Bacolod, Negros Occidental), that have their own laboratory which monitors the proximate composition, calcium, phosphorous, and aflatoxin levels of commercial aquafeeds. This service laboratory can also analyze soil and water, and examine fish and shrimp for disease.

BAI has the authority to issue import permits for feeds and feed ingredients, valid only for the specific importation applied for and for a period of sixty days from the date of issue. All containers of commercial feeds are required to bear a complete label. This includes information on:

- net weight of contents in metric equivalent;
- name and address of manufacturer;
- brand or trademark;
- proximate composition of the feed:
    - (a) minimum % of crude protein and crude fat;
    - (b) maximum % of crude fibre, ash and moisture;
    - (c) maximum % of minerals (if more than 5%, the maximum percentage of calcium and phosphorus is to be indicated);
    - (d) guaranteed aflatoxin level;
    - (e) recommended stocking density;
    - (f) guaranteed peroxide level;

- registration number of the particular kind of feed;
- accepted or official name of feed;
- for shrimp feeds containing drugs or drug ingredients for the prevention of animal diseases:
    - (a) the name and percentage of drug used;
    - (b) directions for use;
    - (c) warning against use under those conditions in which its use may be dangerous to the health of livestock and man;
    - (d) withdrawal period;
- control number, code number or batch number;
- date of manufacture;
- expiry date;
- specific warehousing conditions and specifications for temperature, humidity, *etc.*

Table 17. Prescribed nutrient standards for aquafeeds in the Philippines

Feed	Moisture (max %)	Crude protein (min %)	Crude fat (min %)	Crude fibre (min %)	Ash (max %)
Shrimp prestarter pellets/crumbles	10	45	4	4	10
Shrimp starter pellets/crumbles	10	40	4	4	10
Shrimp grower pellets/crumbles	10	35	4	4	10
Shrimp finisher pellets/crumbles	10	30	4	4	10
Fish prestarter mash/pellets crumbles/flakes granules/powder	10	45	4	4	10
Fish starter mash/pellets crumbles/flakes granules/powder	10	40	4	4	10
Fish grower mash/pellets crumbles/flakes granules/powder	10	35	4	4	10
Fish finisher mash/pellets crumbles/flakes granules/powder	10	30	4	4	10

Source: Administrative Order No. 84, Series of 1990, Bureau of Animal Industry, Department of Agriculture, Quezon City, Philippines

CURRENT PROBLEMS AND CONSTRAINTS

One major constraint is the lack of information on the nutritional requirements of cultured species. Studies are carried out under controlled laboratory conditions with semi-purified feed ingredients in chemically defined diets. Recently, Borlongan and Coloso (in press) defined the ten essential amino acids required by milkfish. Likewise, Santiago and Lovell (1988) have reported requirements for the ten essential amino acids of Nile tilapia (Table 18).

Table 18. Requirements of milkfish (*Chanos chanos*) and Nile tilapia for 10 essential amino acids

Amino acids	Milkfish*	Nile tilapia**
	(% of dietary protein)	
Arginine	5.2	4.2
Histidine	2.0	1.7
Isoleucine	4.0	3.1
Leucine	5.1	3.4
Lysine	4.0	5.1
Methionine + cystine	3.2	3.2
Phenylalanine + tyrosine	5.2	5.5
Threonine	4.5	3.8
Tryptophan	0.6	1.0
Valine	3.6	2.8

* Borlongan and Coloso (1993 in press)
** Santiago and Lovell (1988)

The essentiality of n-3 rather than n-6 fatty acids for growth and survival of milkfish juveniles at a salinity of 32 ppt has been demonstrated by Borlongan (1992). She also demonstrated that eicosapentaenoic and docosahexaenoic acids have higher growth-promoting effects than linolenic acid. No deficiency signs are exhibited if the diet contains at least 1% linolenic acid and 0.5% to 1% of highly unsaturated (n-3) fatty acids. There is hardly any data, however, on the mineral and vitamin requirements of these species. Very little is known about the pond ecosystem and the contribution of natural food (Akiyama 1991). Furthermore, information on nutrient availability both from the ponds and the diet is needed.

Another major constraint in aquafeed manufacturing is the difficulty in sourcing quality raw materials and the availability of animal and vegetable proteins.. The availability of some ingredients, such as corn and rice bran, is seasonal and prices fluctuate depending on supply and demand. In the Philippines, feed ingredients may be abundant in one area but transporting them to where the demand is high is sometimes not feasible since the distribution network is somewhat inefficient. Facilities for drying cereal products are also not sufficient and sun-drying becomes the usual practice. When it rains suddenly, the grain becomes wet and, because of ineffective storage facilities, it will sprout and becomes mouldy. The grain then becomes unfit for use, having high aflatoxin levels which can cause the death of the cultured species. A list of aflatoxin levels found in some commercial feeds is shown in Table 19.

Table 19. Aflatoxin levels found in feed samples in the Philippines

Feed	Date submitted	Production date	Aflatoxin level* (ppb)
A	26/09/91	31/08/91	>10 but <20
B	19/09/91	16/08/91	50
B	19/09/91	16/08/91	20
B	19/09/91	08/16/91	>10 but <20
B	08/01/92	09/91	>0 but <10
B	08/01/92	12/91	>20 but <50
B	10/01/92	12/91	>0 but <10
B	11/02/92	14/11/91	>50 but <100
C	20/07/91	06/91	not detected
C	20/07/91	06/91	>0 but <10
C	26/09/91	07/91	>0 but <10
D	26/09/91	07/91	>10 but <20
E	26/09/91	07/91	>10 but <20
F	23/10/91	10/91	>10 but <20
G	20/07/91	06/91	not detected
G	26/09/91	7/09/91	>0 but <10
G	10/01/92	12/91	>0 but <10
H	26/09/91	07/91	>0 but <10
H	24/01/92	12/91	>0 but <10
H	11/12/91	11/91	>10 but <20
H	11/12/91	11/91	>10 but <20

* Limitations of the analysis: the aflatoxin level was determined using the aflatest kit of Rhone Poulenc. This is a rapid and non-carcinogenic testing system for aflatoxin but not as sensitive as the TLC (Thin Layer Chromatography) or HPLC (High Performance Liquid Chromatography) methods, Nevertheless, it gives an approximate value of the parameter.

Source : Service Laboratory of Negros Prawn Producers Marketing Cooperative, Inc., 17 February 1992.

There is also a lack of capital and interest rates are high (21% or even more). It is also difficult to borrow money without collateral.

Shortages in research and development funding and of competent scientists are often problems in developing countries like the Philippines. Additionally, some research work is poorly designed.

TRENDS IN AQUAFEED MANUFACTURING AND USE

According to Woiwode *et al.* (1992), the demand for shrimp feeds will increase to 64,398 t in 1997. There will be no problem in meeting the feed demand, because the present capacity of feed mills is sufficient for the next five years.

Out of the 24 registered feed milling companies, 18 are in operation. Annually only about 40,000 to 50,000 t of aquafeeds are produced, indicating a significantly lower production than the installed capacity (Woiwode *et al.* 1992). Feed is shipped from Manila to other regions and, during transport, it may not be stored properly. Furthermore, storage facilities on the farms are far from ideal. Nutrient content, aflatoxin levels and peroxide values of feeds stored at the farms can differ from the values at the point of manufacturing.

Some feed mills plan to put up smaller production units far from Manila, where locally produced raw materials are more readily available. Feed manufacturing plants can be conveniently located near fish and shrimp processing plants as well as rice and corn mills. Some farmers are also considering establishing local feed mills that can produce enough feed for one week's consumption. This would be more convenient in terms of proximity to the ponds and monitoring of feed quality. With the shift from intensive to semi-intensive systems in shrimp culture and to intensive systems in the culture of finfish, such as milkfish, there is also a shift in the production from shrimp to finfish feeds.

More and more locally available feed ingredients are now being utilized in feeds. Whereas white fish meal was once the only kind of fish meal used, brown fish meal, locally produced tuna fish meal, and other locally produced feed ingredients are now being incorporated in order to reduce feed costs. In addition, these are used in supplemental diets, although farmers are also using fresh food such as fish and snails. There are fewer imported compound feeds in the market today. Most feeds, if not all, are produced in the Philippines.

There are about four fish farmers who manufacture feeds on their farms. Others are planning to follow suit. Farm-made aquafeeds will most likely be used as supplemental rather than as complete feeds for fish in the near future.

LIST OF REFERENCES

Agbayani, R.F., D.D. Baliao, G.P.B. Samonte, R.E. Tumaliuan and R.D. Caturao. 1990. Economic feasibility analysis of the monoculture of mud crab (*Scylla serrata*). Aquaculture 91:223-231.

Akiyama, M. 1991. Future considerations for the aquaculture industry, p. 5. *In* D.M. Akiyama and R. Tan (eds.) Proceedings of the Aquaculture Feed Processing and Nutrition Workshop. American Soybean Association, Singapore.

BFAR. 1991. Philippine Fisheries Profile 1990. Fisheries Policy Research and Economics Division. Department of Agriculture, Bureau of Fisheries and Aquatic Resources. Quezon City, Philippines 40 p.

BFAR. 1992. Philippine Fisheries Profile 1991. Fisheries Policy Research and Economics Division. Department of Agriculture, Bureau of Fisheries and Aquatic Resources. Quezon City, Philippines. 38 p.

Borlongan, I.G. 1992. The essential fatty acid requirement of milkfish (*Chanos chanos* Forsskal). Fish Physiology and Biochemistry: 9 (5-6) : 401.

Borlongan, I.G. and R.M. Coloso. 1993. Requirements of juvenile milkfish (*Chanos chanos* Forsskal) for essential amino acids. J. Nutr. 123 (in press).

Bureau of Animal Industry. 1989a. Maintenance of ponds for quality control purposes. Animal Industry Circular No. 1, series of 1989. Department of Agriculture, Republic of the Philippines, Visayas Ave., Diliman, Quezon City, Philippines. 1 p.

Bureau of Animal Industry. 1989b. Accreditation of feed quality control laboratories. Memorandum Circular, series of 1989. Department of Agriculture, Republic of the Philippines, Visayas Ave., Diliman, Quezon City, Philippines. 1 p.

Bureau of Animal Industry. 1989c. Registration of feed additives supplements and premixes. Memorandum Circular, series of 1989. Department of Agriculture, Republic of the Philippines, Visayas Ave., Diliman, Quezon City, Philippines. 2 p.

Bureau of Animal Industry. 1989d. Guidelines on feed sample collection at the port of entry. Memorandum Circular, series of 1989. Department of Agriculture, Republic of the Philippines, Visayas Ave., Diliman, Quezon City, Philippines. 2 p.

Bureau of Animal Industry. 1990. Guidelines on the issuance of accreditation of feed laboratories. Circular No. 1. Department of Agriculture, Republic of the Philippines, Visayas Ave., Diliman, Quezon City, Philippines. 2 p.

Bureau of Animal Industry. 1991. Granting authority to Bureau of Animal Industry to impose import/export processing fee. Administrative Order No. 37, Series of 1991. Department of Agriculture, Republic of the Philippines, Visayas Ave., Diliman, Quezon City, Philippines. 1 p.

Camacho, A.S. and N. M. Lagua. 1988. The Philippine aquaculture industry, p. 91. *In* J.V. Juario and L.V. Benetiz (eds.) Perspectives in aquaculture development in Southeast Asia and Japan. Proceedings of the Seminar on Aquaculture Development in Southeast Asia, Iloilo City, Philippines, 8-12 September 1987.

Cruz, P.S. 1991. Shrimp feeding management: principles and practices. Kabukiran Enterprises, Inc. 440-442 R. Magsaysay Ave., Davao City, Philippines. 57 p.

De Silva, S.S. and M.K. Perera. 1984. Digestibility in *Sarotherodon niloticus* fry: effect of dietary protein level and salinity with further observations on variability in daily digestibility. Aquaculture 83:293-306.

Food and Nutrition Research Institute. 1968. Food composition table recommended for use in the Philippines, p. 21. *In* Handbook No. 1, 3rd ed. National Science Development Board, Manila, Philippines.

Guerrero, R.D. III. 1991. Aquaculture research and development programs in the Philippines: an overview, p. 4-9. *In* R.D. Guerrero III and M.P. Garcia, Jr. (eds.) Advances in finfish and shellfish mariculture. Proceedings of the First Philippine-French Technical Workshop on Advances in Finfish and Shellfish Mariculture, October 24-26, 1990, Los Banos, Laguna, Philippines.

Hepher, B. and Y. Pruginin. 1981. Commercial fish farming. Wiley-Interscience, New York. 216 p.

Juario, J.V. 1991. Status of finfish mariculture in the Philippines, p. 27-36. *In* R.D. Guerrero III and M.P. Garcia, Jr. (eds.) Advances in finfish and shellfish mariculture. Proceedings of the First Philippine-French Technical Workshop on Advances in Finfish and Shellfish Mariculture. October 24-26, 1990, Los Banos, Laguna, Philippines.

Kohno, A., P. Trino, D. Gerochi and M. Duray. 1989. Effects of feeding frequency and amount of feeding on growth of grouper *Epinephelus malabaricus*. Phil. J. Science 118:89-100.

Kuntiyo. 1992. Fattening of mud crab *Scylla serrata* Forsskal in net cages installed in the drain canal of intensive prawn ponds fed with trash fish and prawn pellet. M.S. Fisheries Thesis, University of the Philippines in the Visayas. 60 p.

Lacanilao, F. 1991. Some concerns on the development of fisheries and aquaculture in the Philippines, p. 19. *In* Lundayan: A quarterly publication of the Tambuyog Development Center Vol.3, No. 2, June.

National Research Council. 1977. Nutrient requirements of warmwater fishes. National Academy of Sciences, Washington, D.C., U.S.A. 78 p.

Ocean Feed, 1992. Make the best milkfish and prawn feed. Ocean Feed Mill Corp. 8 p.

Penaflorida, R. 1989. An evaluation of indigenous protein sources as potential component on diet formulation for tiger prawn, *Penaeus monodon*, using essential amino acid index (EAAI). Aquaculture 83:319-330.

Posadas, B.C. 1988. Economic analysis of various prawn farming systems, p. 12-24. *In* Y.N. Chiu, L.M. Santos and R.O. Juliano (eds.) Technical considerations for the management and operation of intensive prawn farms. U.P. Aquaculture Society, Iloilo City, Philippines.

Samonte, G., R.F.A. Agbayani and R.E. Tumaliuan. 1991. Economic feasibility of polyculture of tiger shrimp *Penaeus monodon* with Nile tilapia *Oreochromis niloticus* in brackishwater ponds. Asian Fisheries Science 4:335-343.

Santiago, C.B., M.B. Aldaba and M.A. Laron. 1988. Reproductive performance and growth of Nile tilapia *Oreochromis niloticus* broodstock fed diets containing *Leucaena leucocephala* leaf meal. Aquaculture 70:53-61.

Santiago, C.B. and R.T. Lovell. 1988. Amino acid requirements for growth of Nile tilapia. J. Nutr. 118:1540-1546.

Santiago, C.B., A.S. Camacho and M.A. Laron. 1991. Growth and reproductive performance of bighead carp (*Aristichthys nobilis*) reared with or without feeding in floating cages. Aquaculture 96:109-117.

SEAFDEC Aquaculture Department. 1991. Three-year plan of the Aquaculture Department (1992-1994). Submitted to SEAFDEC Program Committee at its 14th Meeting, September 1991. 19 p.

Sumagaysay, N.S. and Y. Chiu-Chern. 1991. Effects of fiber in supplemental feeds on milkfish (*Chanos chanos* Forsskal) production in brackishwater ponds. Asian Fisheries Science 4:189-199.

Sumagaysay, N.S. 1991. Utilization of feed and rice straw compost for milkfish *(Chanos chanos)* production in brackishwater ponds. J. Appl. Ichthyol. 7:230-237.

Sumagaysay, N.S., F.E. Marquez and Y.N. Chiu-Chern. 1991. Evaluation of different supplemental feeds for milkfish (*Chanos chanos*) reared in brackishwater ponds. Aquaculture 93:177-189.

Pascual, F.P. 1991. Manual on a practical guide to the preparation of home made feeds for aquaculture. ASEAN/UNDP /FAO Regional Small-Scale Coastal Fisheries Development Project, Manila, Philippines. 28 p.

Torres, E.G. (ed.) 1992. UPV Research Abstracts 1980-1987. Published by the Office of Research Coordination, University of the Philippines in the Visayas, Iloilo City 5000, Philippines. 164 p.

Woiwode, J.G., R.O. Juliano, R.F. Agbayani, U.L. Corre, U. Hatch, W.G. Deutsch, G. Gonzales, F. Sunaz and B. Perkins. 1993. Philippine Prawn Industry Policy Study. Final Report prepared by International Center for Aquaculture and Aquatic Environments, Auburn University, Auburn, Alabama 36849, U.S.A. for the Coordinating Council of the Philippines Assistance Program and U.S. Agency for International Development Contract No. DAN-4180-B-00-8009'00 Order No. 3. Philippines Assistance Program Support Project No. 492-0452. 170 p.

Annex 1. Some feed mills in the Philippines, their addresses and installed capacity

Company (feed brand)	Installed capacity (t/hr)*
General Reliance Co. (Aquaprime Feeds) Sta. Clara Estate Bldg. Lacson St., Bacolod City Tel.: (63-34) 82-045	3.05
Azure Pacific Trading, Inc. (Quick Grow) Iloilo Fishing Port Complex, Tanza, Iloilo City 5000 Tel.: (63-33) 72-932 Fax.: (63-33) 70-833	0.38
Manila B-Meg Feed Plant (Tilapia feeds) 658 A, Bonifacio Street Balintawak, Quezon City Tel.: (63-2) 353-071/90; 361-4838 Fax.: (63-2) 817-6502; 810-4015 (c/o Mr. R. Saure)	
Pampanga B-Meg Feed Plant (B-MEG, SD-90) Bo. Moras de la Paz Sto. Nino, San Matias, Pampanga Tel.: (63-45) 615-457 or 613-792 Fax.: (63-45) 612-791	7.00
First El Presidente Mfg., Inc. (Chuen Shin) 228 Ortigas St., Manila Tel.: (63-2) 703-115 Fax.: (63-2) 721-1508	1.20
Hoc Po Feeds Co. (Hoc Po Feeds) 554 Edsa St., Caloocan, Metro Manila Tel.: (63-2) 361-4419; 361-4421	2.00
Interfeeds Philippines, Inc. (Interfeeds) Navotas, Metro Manila Tel.: (63-2) 239-685 Fax.: (63-2) 222-596	1.30

Annex 1. (cont.)

Fuji Triumph Agri-Industrial Co. (Fuji-Triumph) Triumph Building, 1610 Quezon Ave. Quezon City, Metro Manila Tel.: (63-2) 968-688; 952-086; 992-486; 961-1213 Fax.: (63-2) 921-3592	5.00
Ocean Feed Mill Co. (Ocean Feeds) Calumangan, Bago City Tel.: (63-34) 22-462/68 Fax.: (63-34) 28-367	2.56
Oversea Agri-Aqua Co. (Oversea Feeds) 82 Plaridel St., Cebu City Tel.: (63-32) 77-691/98 Fax.: (63-32) 52-426; 721-3599	2.00
PEC Enterprises (President Prawn Feeds) 511 Elcano St., Binondo, Metro Manila Tel.: (63-2) 486-890 Fax.: (63-2) 588-266	
Ram Aqua Marketing Co. (Ram) Banilad, Mandaue City Tel.: (63-32) 82-017 Fax.: (63-32) 462-247; 52-022	
Feedpro Co. (Tateh Feeds) 509 Camba-Jaboneros St., San Nicolas, Metro Manila Tel.: (63-2) 401-368; 401-350; 530-0488; 489-260 Fax.: (63-2) 489-255	
Vitarich Co. (Vitarich) Marilao, Bulacan Tel.: (63-97) 851-864; 347-042; 873-684; 873-817 Fax.: (63-97) 816-7235	2.50
Sing Heng Chan (Well Grow Brand) Carmona, Cavite	2.50

* Woiwode *et al.* (1993)

Annex 2. Names and addresses of manufacturers of feed equipment in the Philippines

Name	Address
Adfran Co.	56 K Macabagdal Cor. Howmart St. Baesa, Quezon City
Jamandre Industries	86 Rizal Street La Paz, Iloilo City

Annex 3. National and regional centers, cooperating stations and specialized agencies engaged in aquaculture research in the Philippines

Institution/address	Responsibility (ongoing projects)
National centers	
University of the Philippines in the Visayas, Institute of Aquaculture Miag-ao, Iloilo	Brackishwater aquaculture (milkfish, shrimp, tilapia, molluscs, *etc.*)
Central Luzon State University Freshwater Aquaculture Center Munoz, Nueva Ecija	Freshwater aquaculture (tilapia genetics, integrated farming systems, *etc.*)
University of the Philippines, Marine Science Institute, Bolinao, Pangasinan and Diliman, Quezon City	Marine aquaculture (giant clams and seaweeds)
Regional centers	
Cagayan State University College of Fisheries Aparri, Cagayan	Brackishwater and marine aquaculture (grouper, tilapia, *etc.*)
Bicol University College of Fisheries Tabaco, Albay	Brackishwater and marine aquaculture (grouper and cockles)

Annex 3. (cont.)

Silliman University Marine Laboratory Dumaguete City	Marine aquaculture (giant clams and finfishes)
Zamboanga State College of Marine Science and Technology Zamboanga City	Brackishwater aquaculture (shrimp and tilapia)
Mindanao State University, Institute of Fisheries Research and Development Naawan, Misamis Oriental	Brackishwater and marine aquaculture (crustaceans and finfishes)
Cooperating stations	
Mariano Marcos State University College of Fisheries Currimao, Ilocos Norte	Marine aquaculture (seaweeds)
Don Mariano Marcos Memorial State University, College of Fisheries Sto. Tomas, La Union	Brackishwater and marine aquaculture (grouper, snapper, *etc.*)
Pangasinan State University College of Fisheries Binmaley, Pangasinan	Brackishwater and marine aquaculture (grouper, siganids and tilapia)
Department of Agriculture, National Freshwater Fisheries Training and Research Center, Munoz, Nueva Ecija	Freshwater aquaculture (tilapia)
Department of Agriculture, UNDP-BFAR Training and Demonstration Center Paombong, Bulacan	Brackishwater aquaculture (milkfish and shrimp)
Palawan National Agricultural College Regional Institute of Fisheries Technology Puerto Princesa City	Brackishwater and marine aquaculture (shrimp, milkfish and seabass)
Department of Agriculture Freshwater Fisheries Research Station Los Banos, Laguna	Freshwater aquaculture (tilapia and prawns)
Department of Agriculture, National Brackishwater Aquaculture Training and Research Center, Pagbilao, Quezon	Brackishwater aquaculture (milkfish and prawns)

Annex 1. (cont.)

Panay State Polytechnic College Pontevedra, Capiz	Brackishwater aquaculture (shrimp and seabass)
Department of Agriculture Brackishwater Training Center Calape, Bohol	Brackishwater and marine aquaculture (milkfish, shrimp and oyster)
Department of Agriculture Guiuan Fishery Camp Guiuan, Eastern Samar	Marine aquaculture (lobster and abalone)
Mindanao State University, Tawi-Tawi College of Technology and Oceanography Bongao, Tawi-Tawi	Marine aquaculture (grouper and abalone)
Specialized agencies	
Southeast Asian Fisheries Development Center (SEAFDEC), Aquaculture Department Tigbauan, Iloilo	Brackishwater and marine aquaculture (milkfish, shrimp grouper, siganids, molluscs, *etc.*)
Meralco Foundation, Inc. Agro-Aquatic Development Center Jala Jala, Rizal	Freshwater aquaculture (tilapia)

Source: Guerrero (1991)

Annex 4. List of some completed research studies at the University of the Philippines in the Visayas (UPV) between 1986-1992

Chiu, Y.N., V.L. Estilo, N.S. Sumagaysay and M.A.S. Sastrillo. 1986. Philippines milkfish feeding (Trial I): Effect of supplementary feeding, increased stocking rates and aeration on milkfish yields in brackishwater ponds. Technical Report, Brackishwater Aquaculture Center, College of Fisheries, UPV, Leganes, Iloilo, Phils.

Chiu, Y.N. 1986. Effects of practical type-diets on the growth and survival of milkfish II. The effect of different levels of fish meal on the growth and survival of milkfish. A Terminal Report, Brackishwater Aquaculture Center, College of Fisheries, UPV, Iloilo, Phils.

Chiu, Y.N. 1986. Evaluation of rice bran and two pelleted diets as supplementary feed for milkfish (*Chanos chanos* Forsskal) reared in brackishwater ponds. A Terminal Report, College of Fisheries, UPV, Iloilo, Phils.

Annex 4. (cont.)

Chiu, Y.N. 1987. Increasing milkfish (*Chanos chanos* Forsskal) yields in brackishwater ponds through increased stocking rates and supplementary feeding. A Terminal Report, Brackishwater Aquaculture Center, College of Fisheries, UPV, Iloilo, Phils.

Chiu, Y.N. 1988. Development of a supplementary feeding program for milkfish II. Effects of dietary levels of energy and fiber on protein utilization of supplemental feeds of milkfish (*Chanos chanos* Forsskal) reared in brackishwater ponds. A Terminal Report, Brackishwater Aquaculture Center, College of Fisheries, UPV, Iloilo, Phils.

Corre, V.L. 1986. Use of fermentation-preserved supplemental feeds for red tilapia. Technical Report, College of Fisheries, UPV, Iloilo, Phils.

Corre, V.L. 1986. Use of fermentation-preserved supplemental feeds for Nile tilapia (year 2). A Terminal Report, Brackishwater Aquaculture Center, College of Fisheries, UPV, Iloilo, Phils.

Corre, V.L. 1988. Effects of different ratio of plant and animal protein on the growth and survival of Nile tilapia in brackishwater fishponds. A Terminal Report, Brackishwater Aquaculture Center, College of Fisheries, UPV, Iloilo, Phils.

Corre, V.L. 1988. Effects of different ratio of plant and animal protein on the growth and survival of Nile tilapia in brackishwater fishponds (year 3). A Terminal Report, Brackishwater Aquaculture Center, College of Fisheries, UPV, Iloilo, Phils.

Corre, V.L. 1991. Feeds and feeding strategies for low cost sea bass production. A Terminal Report, College of Fisheries, UPV, Iloilo, Phils.

Fortes, R.D. and V.L. Corre. 1986. Effects of fertilizers and feeds as nutrient sources on *Oreochromis niloticus* production in Philippine brackishwater ponds, p. 121-124. *In*: J.L. Maclean, L.B. Dizon and L.V. Hosillos (eds.) The First Asian Fisheries Forum. Asian Fisheries Society, Manila, Philippines.

Marasigan, E. 1989. Effects of different levels of trash fish in the formulated diet on the growth of sea bass. A Terminal Report, College of Fisheries, UPV, Iloilo, Phils.

Marasigan, E. 1992. Development of artificial feeds for sea bass culture II. Effect of different levels of protein on the growth and survival of sea bass fry. College of Fisheries, UPV, Iloilo, Phils.

Naret, E. 1991. Effect of supplementary diets containing various levels of protein on the growth and survival of *Siganus guttatus* in ponds planted with lumut. Fisheries Research Journal of the Philippines 16 (1-2): 15-21, Jan.-Dec. 1991.

Pena, M.L.D. de la. 1986. Effects of practical type-diets III. Evaluation of various leguminous seeds as protein sources for milkfish *(Chanos chanos)* juveniles. Brackishwater Aquaculture Center, College of Fisheries, UPV, Iloilo, Phils.

Pena, M.L.D. de la, Y.N. Chiu, and F. Ganchero. 1986. Effects of practical type-diets on the growth and survival of milkfish grown from fry to fingerling under controlled environmental conditions III. Evaluation of various leguminous seeds as protein sources for milkfish,*Chanos chanos* juveniles, p. 19-25. *In* Asian Fisheries Science I, Asian Fisheries Society, Manila, Phils., 1987.

Seneriches, M.L.M., Y.N. Chiu, and A.S. Camacho. 1986. Effect of practical type-diets on the growth and survival of milkfish grown from fry to fingerling under controlled environmental conditions II. The effect of fish meal on the growth and survival of milkfish (*Chanos chanos*) fry. Technical Report, Brackishwater Aquaculture Center, College of Fisheries, UPV, Leganes, Iloilo, Phils.

Seneriches, M.L. and Y.N. Chiu. 1987. Effect of fish meal on the growth, survival and feed efficiency of milkfish (*Chanos chanos*) fry. Brackishwater Aquaculture Center, College of Fisheries, UPV, Iloilo, Phils. Presented at the 7th PCARRD Annual Review of Completed and Ongoing Research Projects for Fisheries, La Granja, La Carlota City, Negros Occ., Philippines, April 22-25, 1987.

Thomforde, H. and Y.N. Chiu. 1986. Effects of chicken manure and supplementary feed on production of milkfish and red tilapia in polyculture in brackishwater earthen ponds at different stocking densities. Graduate Thesis, College of Fisheries, UPV, Iloilo, Phils.

Tubongbanua, E.S. 1986. Preliminary study on the development of artificial feeds for sea bass, *Lates calcarifer* (Bloch). College of Fisheries, UPV, Iloilo, Phils. Presented at the Workshop on "The Management of Wild and Cultured Sea bass/Barramundi." Darwin, N.T., Australia, September 24-October 1, 1986.

Annex 5. List of completed research studies of SEAFDEC/AQD (1986-1991)

Acosta, B.O. 1986. Comparative effects of organic and inorganic fertilizers on the growth of plankton and production carps in ponds.

Alava, V.R. 1989. Alternate sources of protein for fish meal in *Lates calcarifer* diets.

Alava, V.R. 1987. A reference fatty acid and amino acid profile for *Lates calcarifer* juveniles.

Avila, E.M. 1990. Food consumption of sea bass (*Lates calcarifer*) in captivity. II. Group feeding.

Baliao, D. and R.F. Agbayani. 1986. Effects of formulated feeds on intensive prawn production following the modular pond system.

Bautista, M.N. Effect of dietary calcium/phosphorus ratio on shell quality of *Penaeus monodon*.

Annex 5. (cont.)

Borlongan, I.G. and M.M. Parazo. 1991. Effect of dietary lipid sources on growth, survival and fatty acid composition of sea bass, (*Lates calcarifer*) Bloch fry. Bamidgeh 43(3): 95-102.

Borlongan, I.G. and R.M. Coloso. 1993. Requirements of juvenile milkfish (*Chanos chanos* Forsskal) for essential amino acids. J. Nutr. In press.

Catacutan, M. 1986. Degestibility of feedstuff for milkfish.

Catacutan, M.R. 1991. Apparent digestibility of diets with various carbohydrate levels and the growth of *Penaeus monodon*. Aquaculture 95:89-96.

Catacutan, M.R. 1991. Growth and fatty acid composition of *Penaeus monodon* juveniles fed various lipids. Bamidgeh 43:47-56.

Catacutan, M.R. Practical diet development for sea bass, *Lates calcarifer*: dietary protein and energy ratio.

Coloso, R.M. Nutrition and feed development for sea bass, I. Requirement for sulfur amino acids by seabass *(Lates calcarifer* Bloch) juveniles.

Coloso, R.M. and C. da Costa Reis. *In vitro* determination of protein digestibility of milkfish (*Chanos chanos* F.) (Thesis Research).

Corre, K.G. Effects of combined pelleted feed and trash fish on the growth and survival of prawn.

Eusebio, P.S. 1991. Effect of dehulling on the nutritive value of some leguminous seeds as protein sources for tiger prawn *(Penaeus monodon*) juveniles. Aquaculture 99:297-308.

Fermin, A.C. Nursery of sea bass *Lates calcarifer* in floating cages. I. Effects of varying stocking densities and supplemental feeding.

Gerochi, D.D. 1987. Growth and survival of siganids at various stocking densities fed with different natural food sources.

Gerochi, D.D. 1988. Effect of supplemental diet on siganids cultured in ponds with filamentous green algae.

Kohno, H. 1987. Effects of stocking density and food on growth and survival of groupers.

Kume, T. 1989. Development of practical diet for grouper larvae.

Millamena, O.M. 1987. Refinement of SEAFDEC-formulated diet for pond-reared *Penaeus monodon* broodstock: effect of lecithin.

Annex 5. (cont.)

Millamena, O.M. and J.H. Primavera. 1986. Lipid requirement for *P. monodon* broodstock: essential fatty acids and phospholipids.

Millamena, O.M. and P. Tackaert. 1989. The use of *Artemia* biomass as a dietary ingredients in formulated diets of *Penaeus monodon* post-larvae (Ph.D. Thesis Research).

Parazo, M.M. 1990. Effect of dietary protein and energy level on growth, protein utilization and carcass composition of rabbitfish, *Siganus guttatus*. Aquaculture 86:41-49.

Pascual, F.P. 1986. Evaluation of various SEAFDEC formulated diets for juvenile *P. monodon*.

Pascual, F.P. and M.R. Catacutan. 1986. Vitamin premix level requirement of *P. monodon* juveniles.

Pascual, F.P., N.S. Sumagaysay and I.G. Borlongan. 1991. Modular method of rearing milkfish with artificial feed, p. 159-167. *In* S.S. de Silva (ed.) Fish nutrition research in Asia. Proceeding of the Fourth Asian Fish Nutrition Workshop. Asian Fish. Soc. Spec. Publ. 5. Asian Fisheries Soceity, Manila, Philippines.

Pascual, F.P., O.M. Millamena and V.D. Penaflorida. 1986. Arginine requirement of *P. monodon* juveniles.

Penaflorida, V.D. 1989. Evaluation of plant protein as partial replacement for animal protein in diets of *Penaeus indicus* juveniles.

Ponce, A.H. Polyculture of *Gracilaria* (Gigartinales, Rhodophyceae) and *Lates calcarifer* Bloch in floating net cages.

Recodo, A., Histophysiological effects of various histidine levels in the diet on the hepatopancreas of *Penaeus monodon* Fabricius. M.S. Thesis, College of Science, Univ. of the Phils., Diliman, Quezon City. 104 p.

Santiago, C.B. 1988. Effects of dietary protein levels on gonadal development of bighead carp (*Aristichthys nobilis*) reared in floating cages: reproductive performance of broodstock.

Santiago, C.B. 1988. The effects of dietary lipids on reproductive performance, tissue composition and growth of Nile tilapia (*Oreochromis niloticus*) broodstock.

Santiago, C.B. 1990. Growth, feed utilization and body composition of young red tilapia given diets with varying protein - energy levels.

Santiago, C.B. and J.B. Pantastico. 1986. Milkfish fingerling productian in freshwater ponds using artificial and natural food.

Annex 5. (cont.)

Santiago, C.B. and J.B. Legiralde. 1989. Effect of dietary carbohydrates, lipid and energy on growth, feed efficiency, and tissue composition of bighead carp (*Aristichthys nobilis*) (Research Thesis).

Santiago, C.B. and O. Reyes. 1986. Development of artificial diets for carp and freshwater fishes.

Sumagaysay, N.S. 1991. Effects of fiber supplemental feeds on milkfish (*Chanos chanos*) production in brackishwater ponds. Asian Fisheries Science 4:189-199.

Sumagaysay, N.S. Development of supplemental diet for milkfish (*Chanos chanos*) through assessment of natural food contribution to fish growth.

Trino, A.T. 1986. Culture of seabass (*Lates calcarifer*) in ponds: Effect of feeding rates and stocking density.

Trino, A.T. 1986. Food and feeding habits of different life stages of *P. monodon* in ponds.

Trino, A.T. 1987. Integrated culture of seabass and tilapia: The use of live tilapia as food for seabass, *L. calcarifer* Bloch.

Trino, A.T. 1989. Effect of processed feed as supplemental diet in a modified extensive culture of *Penaeus monodon*.

Trino, L. and L. Benitez. 1986. Studies on the digestive enzymes of milkfish and siganids.

Tuburan, I.B. 1989. Evaluation of supplementary diets on growth, survival, and production of *Penaeus monodon*.

Tuburan, I.B. 1990. Evaluation of unprocessed feeds in semi-intensive *Penaeus monodon* (Fabricius) pond culture.

AQUAFEEDS AND FEEDING STRATEGIES IN SINGAPORE

Renee Chou

Marine Aquaculture Section, Department of Primary Production,
300 Nicholl Drive, Singapore 1749

CHOU, R. 1993. Aquafeeds and feeding strategies in Singapore, p. 354-364. *In* M.B. New, A.G.J. Tacon and I. Csavas (eds.) Farm-made aquafeeds. Proceedings of the FAO/AADCP Regional Expert Consultation on Farm-Made Aquafeeds, 14-18 December 1992, Bangkok, Thailand. FAO-RAPA/AADCP, Bangkok, Thailand, 434 p.

INTRODUCTION

Singapore has limited land and water resources. The contribution of agriculture (including fisheries and aquaculture) to GDP is less than 1%. Fish however, is an important source of protein for the population. The *per capita* fish consumption in 1991 was 40 kg. Of the 111,955 t of fish available in 1991, 2% (2,000 t) with a farm-gate value of US$ 11.2 million was derived from aquaculture. This represents 17% of local fish supply. The rest is imported. About 98.6% of aquaculture production is from culture in the 79 coastal floating fish farms licensed in Singapore. Most of the cultured species are fed readily available and relatively cheap trash fish.

In 1991, about 100 million (valued at US$ 18.2 million) ornamental fish were produced by the 70 ornamental fish farms in Singapore. Singapore exports around 300 varieties and species to about 50 countries in the world. Most of the ornamental fish cultured are fed farm-made feeds or imported aquarium fish feeds, depending on the species. There is no commercial production of ornamental fish feeds in Singapore.

Commonly cultured finfish are the Asian seabass, (*Lates calcarifer*) groupers (*Epinephelus tauvina, Plectropomus maculatus*), snapper (*Lutjanus argentimaculatus, L. johni*). The banana shrimp (*Penaeus merguiensis*) is the only cultured shrimp species in Singapore. Ornamental fish cultured are the guppy (*Poecilia reticulata*), and other poecilid species like the mollies (*P. sphenops, P. latipinna*), swordtail (*Xiphophorus helleri*) and platys (*X. maculatus, X. variatus*), corydoras (*e.g. Corydoras aeneus*), goldfish (*Carassius auratus*), tetra (*e.g. Paracheirodon innesi*) and discus (*e.g. Symphynodon aequisasciatda*).

AVAILABILITY OF FEED INGREDIENTS

Trash fish (US$ 270-300/t) is presently the most popular feed being used in marine food fish culture. Other feed ingredients available in Singapore that are suitable for aquafeed preparation have historically been those imported for the pig and poultry industries. These are generally not of the specifications or quality required for aquaculture, especially in meeting the nutritional requirements of the carnivorous species cultured here. Thus feedstuffs specifically for aquafeed manufacture have to be imported, *e.g.* shrimp meal, squid meal, and high protein fish meal.

Tables 1 and 2 show lists of aquaculture feed ingredients that are imported by local feedmills for shrimp feed production. The same are used for fish feed production, which is still in its infant stage. These ingredients are generally of higher protein and quality than for terrestrial animal feeds and, of course, more expensive. Wheat bran, oat meal, wheat flour and the milk-based products are used in farm-made feeds for ornamental fish.

Table 1. List of feed ingredients available in Singapore (1992): prices and proximate composition

Feed ingredient (primary source*)	Price (US$/t*)	Moisture (%)	EE (%)	CP (%)	CF (%)	Ash (%)
Acetes (Indonesia)	1,818	17.10	3.20	62.70	2.40	13.00
Alpha starch (Thailand)**	1,030	5.90	n.a.	n.a.	n.a.	0.05
Buttermilk (New Zealand)	1,030	11.00	5.10	32.00	n.a.	10.00
Casein (Australia, N.Z.)	3,636	10.90	0.03	81.20	0.04	n.a.
Clam meal (India)	849	8.00	10.60	51.80	0.53	7.60
Cod liver oil (Japan)	1,818	1.00	99.00	n.a.	n.a.	n.a.
Corn gluten (USA)	606	9.20	0.65	58.10	1.10	0.54
Corn gluten (Australia)	606	7.50	0.99	51.70	2.40	1.90
Corn meal (Thailand)	194	8.90	5.40	11.50	2.63	2.50
Crab meal (Malaysia)	758	9.00	0.75	22.10	11.40	46.10
Cuttlefish powder (Taiwan)	788	11.60	10.70	59.50	0.82	13.20
Cuttlefish meal (Vietnam)	667	12.40	0.22	9.30	0.46	2.80
Egg powder (Denmark)	909	5.40	33.10	47.20	0.73	3.80
Fish prot. conc. (Mauritius)	758	10.70	5.60	70.70	0.15	9.00
Fish soluble protein (USA)	2,261	5.10	14.20	64.60	0.17	4.70
Fish meal (Singapore)	546	7.60	7.60	52.20	0.56	26.10
Fish meal (Chile)	679	9.40	7.20	65.80	0.23	14.00
Fish meal (Denmark)'	849	8.70	8.70	66.80	0.11	12.50
Fish meal (Indonesia)	667	12.30	7.30	57.70	0.34	17.30
Fish meal (Thailand)	606	8.30	5.10	53.90	0.88	27.80
Fish meal, white (Japan)	970	8.60	4.20	65.70	0.20	18.90

Table 1. (cont.)

Feed ingredient (primary source*)	Price (US$/t*)	Moisture (%)	EE (%)	CP (%)	CF (%)	Ash (%)
Fish meal, brown (Japan)	909	8.60	9.00	61.50	0.40	14.60
Gelatin (Japan)	6,667	12.10	0.19	93.60	0.30	0.07
Groundnut cake (China)	242	10.10	6.20	31.10	15.70	5.00
Herring meal (Canada)	788	15.10	4.80	60.80	0.47	11.80
Krill meal (Denmark)	1,818	3.90	11.10	54.90	5.50	10.60
Meat & bone meal (N. Z.)	576	6.20	10.80	48.70	1.80	29.80
Mussel (India)	3,030	10.80	8.30	49.00	0.52	6.90
Mussel (China)	3,636	16.90	10.40	50.30	0.30	7.50
Oatmeal (Australia)	n.a.	9.90	3.70	16.00	7.40	3.90
Sago starch (Malaysia)***	303	10.90	n.a.	0.47	n.a.	0.02
Sesame cake (China)	333	8.40	6.00	22.80	30.40	5.20
Shrimp, whole (Indonesia)	3,030	18.80	2.60	56.10	2.20	17.80
Shrimp, head (Malaysia)	606	14.20	2.40	38.20	6.40	29.20
Shrimp head meal (Malaysia)	606	16.30	2.40	30.20	7.58	42.20
Shrimp head meal (Vietnam)	667	9.80	1.40	31.40	18.80	31.80
Shrimp powder (Malaysia)	667	12.00	3.70	56.70	4.92	22.60
Skimmed milk (New Zealand)	1,030	6.30	0.60	33.40	0.30	7.70
Soybean meal (China)	303	11.00	0.61	42.90	5.60	4.90
Soya oil (Singapore)	285	n.a.	100.00	n.a.	n.a.	n.a.
Squid liver meal (Taiwan)	727	6.00	15.30	45.00	n.a.	12.40
Squid meal (India)	3,455	11.10	6.40	68.80	1.21	4.90
Squid meal (Indonesia)	752	6.00	16.10	46.60	2.48	9.80
Sunflower, extr.(Indonesia)	485	9.50	0.49	28.60	24.20	7.10
Wheat bran (Indonesia)*****	212	11.10	3.70	13.30	12.40	5.60
Wheat flour (Singapore)****	279	13.10	1.10	10.60	0.47	0.35
Wheat flour (Malaysia)	303	13.10	1.60	16.60	0.60	0.52
Wheat gluten (Australia)	1,515	8.20	1.70	74.80	0.64	0.09
Whole wheat (Australia)	2,000	10.30	1.30	10.20	1.59	0.96
Yeast, bakery (Taiwan)	1,636	8.80	n.a.	40.00	1.30	7.20
Yeast, brewery (Malaysia)	667	7.90	0.60	48.20	0.10	5.70

* S$ 1.65 = US$ 1.00 (December 1992); original ingredient source may be different from country of import

Carbohydrate (%): ** = 93.7, *** = 83.6, **** = 66.0, ***** = 66.6

Table 2. List of feed ingredients available in Singapore (1992): mineral and amino acid content

Feed ingredient (primary source*)	Ca (%)	P (%)	K (%)	Lysine (%)	Methionine (%)	Cystine (%)
Acetes (Indonesia)	2.38	1.46	0.97	3.53	1.38	0.96
Buttermilk (New Zealand)	1.29	0.89	n.a.	2.40	0.70	0.40
Clam meal (India)	0.52	0.70	0.21	n.a.	n.a.	n.a.
Corn gluten (USA)	0.03	0.29	0.23	1.07	0.87	0.56
Corn gluten (Australia)	0.18	0.31	0.31	n.a.	n.a.	n.a.
Corn meal (Thailand)	0.04	0.54	n.a.	n.a.	n.a.	n.a.
Crab meal (Malaysia)	18.00	1.58	0.52	1.42	0.15	0.02
Cuttlefish powder (Taiwan)	0.98	0.91	0.56	4.48	2.28	0.46
Cuttlefish meal (Vietnam)	0.25	0.15	0.31	n.a.	n.a.	n.a.
Egg powder (Denmark)	0.21	0.63	0.51	n.a.	n.a.	n.a.
Fish prot. conc. (Mauritius)	2.38	1.93	0.90	5.30	2.10	1.66
Fish soluble protein (USA)	0.34	0.63	0.92	3.14	1.64	0.72
Fishmeal (Singapore)	7.14	2.71	0.62	2.93	1.08	1.55
Fishmeal (Chile)	3.39	2.54	0.81	5.13	1.67	0.99
Fishmeal (Denmark)	2.13	1.76	1.35	n.a.	n.a.	n.a.
Fishmeal (Indonesia)	5.37	3.16	0.80	6.45	1.81	0.64
Fishmeal (Thailand)	8.62	2.96	0.78	2.95	1.42	0.56
Fishmeal, white (Japan)	5.45	2.98	0.56	4.18	1.87	0.73
Fishmeal, brown (Japan)	3.76	2.67	0.94	n.a.	n.a.	n.a.
Gelatin (Japan)	0.02	0.01	n.a.	n,a.	n.a.	n.a.
Groundnut cake (China)	0.38	0.58	0.89	n.a.	n.a.	n.a.
Herring meal (Canada)	2.80	1.92	0.37	n.a.	n.a.	n.a.
Krill meal (Denmark)	2.19	1.51	0.50	n.a.	n.a.	n.a.
Meat & bone meal(New Zealand)	10.40	5.11	0.41	3.60	0.43	0.53
Mussel (India)	0.52	0.70	0.20	5.32	1.37	0.41
Mussel (China)	0.28	1.11	0.64	4.54	1.31	0.33
Sesame cake (China)	0.44	0.41	0.96	0.64	0.23	0.09
Shrimp, whole (Indonesia)	2.21	0.84	0.60	n.a.	n.a.	n.a.
Shrimp, head (Malaysia)	7.53	1.37	0.52	3.11	1.18	0.21
Shrimp head meal (Malaysia)	10.01	2.21	0.43	1.56	0.49	0.27
Shrimp head meal (Vietnam)	12.10	1.83	0.40	n.a.	n.a.	n.a.
Shrimp powder (Malaysia)	4.37	1.66	0.44	5.83	1.52	0.35
Skimmed milk (New Zealand)	1.31	1.00	n.a.	2.43	0.93	0.41
Soybean meal (China)	0.23	0.71	1.84	2.34	0.59	0.62
Squid liver meal (Taiwan)	0.26	0.67	n.a.	n.a.	n.a.	n.a.
Squid meal (India)	0.41	0.78	0.29	6.47	1.91	0.57
Squid meal (Indonesia)	0.15	1.31	2.13	2.25	0.81	0.85
Sunflower, extr.(Indonesia)	0.59	0.77	1.20	n.a.	n.a.	n.a.
Wheat bran (Indonesia)	0.14	1.17	0.56	0.19	0.23	n.a.
Wheat flour (Singapore)	0.03	0.13	0.11	0.32	0.11	0.14
Wheat flour (Malaysia)	0.04	0.19	0.18	0.41	0.31	0.20
Wheat gluten (Australia)	0.22	0.10	1.25	0.86	0.57	n.a.
Whole wheat (Australia)	0.05	0.22	0.30	n.a.	n.a.	n.a.
Yeast, bakery (Taiwan)	0.62	0.77	n.a.	n.a.	n.a.	n.a.
Yeast, brewery (Malaysia)	0.09	1.38	n.a.	n.a.	n.a.	n.a.

* original ingredient source may be different from country of import

AQUAFEED MANUFACTURING INDUSTRY

Complete statistics on imported aquafeeds are unavailable. However, in 1992, about US$ 66 million of animal feeds and feedstuffs were imported, of which US$ 56 million worth was re-exported. Local exports were estimated to be worth US$ 26 million. Of this, about 15% were aquafeeds, mainly for shrimp, worth US$ 3.9 million.

Marine foodfish and shrimp

The 3,500 t of dry pelleted complete feeds (hatchery and postlarval, starter, grower, and finisher feeds) for penaeid shrimp, valued at US$ 3.8 million, manufactured by the four local feedmill companies in Singapore in 1992, was about 50% of the 1991 production level. Nearly all was formulated for tiger shrimp (*Penaeus monodon*), but used to culture banana shrimp in Singapore. A list of feed mills is given in Annex 1, while the cost of their feeds is given in Table 3. Table 4 presents the proximate analyses stated by the manufacturers of these feeds. Most of these feeds are exported to other ASEAN countries.

Shrimp feeds are also imported (statistics not available) from the region for local use. Their composition and price are competitive with locally produced feeds, since no import tax is levied.

Table 3. Cost of shrimp feeds in Singapore in 1992

Name of Company (brand)	Price of shrimp feed (US$/kg)				
	PL	Starter	Grower	Finisher	Hatchery
Gold Coin Services Pte., Ltd. (Gold Coin)	-	-	-	-	Zoeal 103.0 Mysis 99.4 Early PL 94.6 Late PL 92.1
Bio-Feed Industries Pte., Ltd. (Bio-Feed)	1.15-1.46	1.09-1.27	0.97-1.09	0.97-1.09	None
Sin Heng Chan (S) Pte., Ltd. (Red Circle)	1.27	1.18	1.03	0.97	None
Virginia Industries Pte., Ltd. (Virginia)	1.15	1.06	1.03	None	None

Table 4. Proximate analysis and size range of shrimp feeds available in Singapore

Feed brand (shrimp species)	Size (mm)	Moisture (max %)	CP (min %)	EE (min %)	Ash (max %)	CF (max %)
Gold Coin						
Hatchery (*P.monodon*)*		12.0	50.0	15.0	25.0	2.0
Bio-Feed						
PL No.1 (*P.monodon*)	0.3-1.0	12.0	40.0	4.0	18.0	3.0
PL No.2 (*P.monodon*)	1.0-2.0	12.0	39.0	4.0	18.0	3.0
Starter No.3 (*P.monodon*)	2.0-2.5	12.0	38.0	4.0	18.0	3.0
Grower No.4 (*P.monodon*)	2.5x5.0	12.0	36.0	4.0	18.0	3.0
Finisher No.5 (*P.monodon*)	2.5x10.0	12.0	36.0	4.0	18.0	3.0
Red Circle						
PL No.1 (*P.monodon*)	0.3-1.0	12.0	39.0	4.0	17.0	3.5
Starter No.2 (*P.monodon*)	2.0-2.0	12.0	39.0	4.0	17.0	3.5
Grower No.3 (*P.monodon*)	2.2x2.5	12.0	38.0	4.0	17.0	3.5
Grower No.4 (*P.monodon*)	2.2x4.0	12.0	38.0	4.0	17.0	3.5
Finisher No.5 (*P.monodon*)	2.2x5.0	12.0	36.0	4.0	17.0	3.5
Finisher No.6 (*P.monodon*)	2.2x7.0	12.0	36.0	4.0	17.0	3.5
Virginia						
PL No.1 (*P.monodon*)	0.5	12.0	39.0	6.0	15.0	3.0
PL No.2 (*P.monodon*)	1.8	12.0	38.0	6.0	15.0	3.0
Starter (*P.monodon*)	2.2	12.0	37.0	6.0	15.0	3.0
Grower (*P.monodon*)	2.2	12.0	36.0	6.0	15.0	3.0
PL No.1 (*P.japonicus*)	0.5	12.0	55.0	6.0	15.0	3.0
PL No.2 (*P.japonicus*)	1.8	12.0	54.0	6.0	15.0	3.0
Starter (*P.japonicus*)	2.2	12.0	53.0	6.0	15.0	3.0
Grower (*P.japonicus*)	2.2	12.0	52.0	6.0	15.0	3.0

* Zoeal Feed : < 30 μm
Mysis Feed : 30-90 μm
Early PL Feed : 150-250 μm
Late PL Feed : 250-400 μm

Ornamental fish (mainly freshwater species)

As stated earlier, most ornamental fish are fed farm-made feeds. No commercial feeds are produced locally. About 7 t is imported from the region, with an estimated value of US$ 182,000.

ON-FARM FEED FORMULATION AND MANUFACTURING

Marine foodfish

On-farm feeds are prepared by a few farmers for marine finfish such as seabass, snappers and groupers during the monsoon months, when trash fish supplies are limited by the inability of local trawlers to set sail during rough weather. Formulation is simple, comprising fish meal (30%), commercially available vitamin and mineral premixes (0.5%) and corn gluten binder (69.5%). The resultant formulation has a crude protein level of about 57%, and is used without ground trash fish.

Alternatively a formulation including ground trash fish (30-50%), a binder meal comprising fish meal (30-50%), plant ingredients such as soybean and rice bran (30%), corn gluten or wheat flour (20-30%), and vitamin and mineral supplements (1-2%) is used. No special equipment is needed (only ordinary meat grinders). The mixture is hand-mixed with ambient or hot water to gelatinize the starch, formed into dough balls and fed by hand to the fish. One farm feeds a crude formulation of fish meal, plant material (soybean meal, bran) and starch binder without the trash fish and has reported that the feed is well accepted and beneficial to fish like seabass.

Commercial preparations of complete seabass nursery mash and seabass nursery and grow-out dry pelleted feeds are also available from feedmills in Singapore. One hatchery operator produces nursery stage seabass (1.3 cm total length, 0.5g) on a commercial (imported) dry mash which he combines with water and feeds (by hand) as wet pellets, in place of ground trash fish and live food. He reports that this saves on labour and promotes more uniform growth. The amount imported is still negligible. Fish growth and performance are not compromised. Dry pelleted seabass feed manufactured by one Singapore feedmiller is being tested in collaboration with the Primary Production Department (PPD).

Shrimp are fed dry, commercially available pelleted feeds in Singapore; no farm-made feeds are in use.

Ornamental fish

Ornamental fish like the guppy, platy, swordtail and molly are fed by hand on farm-made feeds. Farm-made diets for the guppy (adults and fry) consists of about 70-80% ground wheat bran or oats, 10-15% of either fish meal, poultry starter crumbles, eel mash or ground dried shrimp, and 5-20% skimmed milk or buttermilk powder and soybean meal. The dry composition of such diets is usually 15-34% crude protein, 48-77% nitrogen-free extract and 3-7% ether extractable fat. The diets are supplemented with tubifex (*Tubifex* spp.) worms (for adult guppy) and the cladoceran, *Moina* spp. (for fry).

Farm-made diets for the platy and swordtail are similar, consisting of a mixture of fish meal and wheat bran/flour at 60-80% and 20-40% respectively. The dry composition of these diets is 18-21% protein, 61-63% nitrogen-free extract and crude fat around 5%.

In contrast, the molly, which is essentially herbivorous, is fed bread crusts (about 12% crude protein) to supplement the zooplankton, phytoplankton and algae in the ponds. About 90% of the farmers practise this, as such feeds are cheaper than imported commercial feeds.

No special equipment is needed for preparing aquarium fish feeds on the farm, as the feeds are mixed by hand.

ON-FARM FEEDING STRATEGIES

Marine foodfish

Trash fish is fed twice daily, usually in the morning and towards evening, and at slack tide to minimize loss. The fish are fed small quantities at a time to prevent wastage. Fingerlings are fed up to 10% of their weight at first, 8% in early grow-out, and 5-3% in late grow out, until they attain the marketable weight of 500-700g. Broodstock (>3kg) are fed at 3%.

Fish are fed farm-made feeds at 5-3% of the fish body weight since they contain less moisture than trash fish. Farm-made broodstock feeds have an advantage in that the dough mix can be supplemented with fish oils known to enhance egg quality. Grow-out stocking rates are 4-5 kg/m^3, while broodstock are stocked at 13 kg/m^3.

Ornamental fish

Ornamental fish, reared in ponds, fixed netcages in ponds and tanks stocked at 2,000-4,000/m^3, are fed farm-made feeds, broadcast by hand as a wet mash. Farm-made feeds are given 2-3 times per day (morning, at noon and in the evening), and live food supplement once a day (mornings).

MAJOR INSTITUTIONS INVOLVED IN AQUAFEED RESEARCH AND DEVELOPMENT

The Primary Production Department (PPD) is involved mainly in marine foodfish (seabass and banana shrimp) feed development. The Department's Marine Aquaculture Section (MAS) represents Singapore in the 5-year ASEAN-EEC Aquaculture Development and Coordination Programme (AADCP), which began in 1990. Component 4 of the AADCP involves the PPD (ASEAN institute) and its twinned European institute, the Institut Francais de Recherche pour l'Exploitation de la Mer (IFREMER), with inputs from the Institut National de la Recherche Agronomique (INRA). Component 4's objective is to develop adequate marine aquafeeds, specifically for seabass and banana shrimp.

The Zoology Department of the National University of Singapore has worked on microencapsulated feeds for larval seabass, and on the nutritional requirements of some aquarium fish species *e.g.* guppy.

A commercial feedmill is also engaged in aquafeed research and development, particularly on value-added microfeeds and enhancer diets for penaeid shrimp and seabass feed.

CURRENT NATIONAL REGULATIONS

According to the Feeding Stuffs Act (1985), the Director of Primary Production is responsible for giving licences to manufacture, prepare, market and import animal feeds in Singapore. This includes the inspection of feedmills and of the quality of the feeds produced. Purchasers of animal feed can (upon payment of the appropriate fee) have samples taken by PPD and analysed, and receive a certificate of the results.

CURRENT PROBLEMS AND CONSTRAINTS

Even though Singapore does not have indigenous raw materials for feed manufacture, its excellent central location and good communication and transport infrastructure allows feedmillers and agents to select and import feedstuffs and feeds of good quality from most parts of the world. Aquafeed manufacturers are challenged by the highly competitive climate in the region (directed mostly at tiger shrimp feed production) although labour costs can be relatively high and there is lack of open information on the nutritional requirements of fish and shrimp species grown in the region. The latter is common to the region and limits optimisation of feed formulations (hence their performance).

TRENDS IN AQUAFEED MANUFACTURING AND USE

Some feed companies in Singapore are expanding their operations into the region, setting up production facilities overseas, while retaining their local feedmills and offices as technical resource centres. This is to take advantage of Singapore's good communication, banking and industrial services infrastructure. In Singapore, these companies are venturing into seabass feed development, the production of shrimp hatchery feeds and the development of value-added feeds.

In 1991, 6,520 t of aquafeeds (mainly shrimp nursery and grow-out feeds, valued at US$ 7.4 million), were manufactured in Singapore. In 1992, production was about half this level. One feedmill is now producing value-added hatchery shrimp feeds (*viz* for shrimp zoeal, mysis, early postlarval and late postlarval stages), and has transferred its nursery and grow-out feed production abroad where it has a feedmill within the proximity of major users. This ensures product freshness and saves on labour and transport.

Prospects and plans for changes in aquafeed manufacture for the future are therefore dependent on prevailing conditions in the region.

In the case of farm-made feeds, they will probably continue to be secondary to trash fish, unless the availability and/or price of the latter becomes prohibitive or until nutritionally adequate and tested commercial feeds become available or more economical to use.

Annex 1. List of companies producing shrimp feeds in Singapore

Name of Company (brand)	Address	Production (t/year) 1992	Type of shrimp feed
Gold Coin Services Pte., Ltd. (Gold Coin)	14 Jalan Teong Jurong, Singapore 2261 Tel: (65) 266 1500 Fax: (65) 265 5232	6	*P. monodon*
Bio-Feed Industries Pte., Ltd. (Bio-Feed)	10 Benoi Place Singapore 2262 Tel: (65) 861 5454 Fax: (65) 862 2252	2,000	*P. monodon*
Sin Heng Chan (S) Pte., Ltd. (Red Circle)	Lot A5052 Jurong Port Road Singapore 2261 Tel: (65) 264 1410 Fax: (65) 264 1418	120	*P. monodon*
Virginia Industries Pte., Ltd. (Virginia)	22 Tuas Ave. 13 Lot A 2263 Singapore 2263 Tel: (65) 862 0271 Fax: (65) 862 0455	1,400	*P. monodon* *P. japonicus*

AQUAFEEDS AND FEEDING STRATEGIES IN THAILAND

Pisamai Somsueb

National Inland Fisheries Institute
Kasetsart University Campus, Bangkok 10900, Thailand

SOMSUEB, P. 1993. Aquafeeds and feeding strategies in Thailand, p. 365-385. *In* M.B. New, A.G.J. Tacon and I. Csavas (eds.) Farm-made aquafeeds. Proceedings of the FAO/AADCP Regional Expert Consultation on Farm-Made Aquafeeds, 14-18 December 1992, Bangkok, Thailand. FAO-RAPA/AADCP, Bangkok, Thailand, 434 p.

INTRODUCTION

In the past, fish production in Thailand depended on natural resources but a national reservoir construction programme, aimed at the development of agriculture, irrigation and industry, led to a decline in natural fish stocks. The reservoirs are serving agricultural production and prevent flooding but have an adverse effect on fisheries by reducing spawning and larval rearing grounds. Moreover, industrial development and the expansion of agriculture has resulted in increased water pollution. All these factors have directly and indirectly reduced capture fisheries production. The Department of Fisheries is attempting to increase fish production by promoting aquaculture. Today, aquaculture production seems to surpass the natural catch.

Table 1. Production of captured and cultured freshwater fish in Thailand (t)

Year	Total production	Captured	Cultured
1978	141,496	102,129	39,367
1979	133,176	103,714	29,462
1980	144,995	110,490	34,505
1981	164,581	116,468	48,113
1982	133,562	87,733	45,829
1983	155,447	108,481	46,966
1984	161,819	111,409	50,410
1985	167,453	92,199	75,254
1986	187,763	98,438	89,325
1987	177,142	87,360	89,782
1988	183,607	81,479	102,128
1989	200,771	109,100	91,671
1990	230,966	127,200	103,766*
1991	n.a.	n.a.	108,954*
1992	n.a.	n.a.	114,402*

* estimate

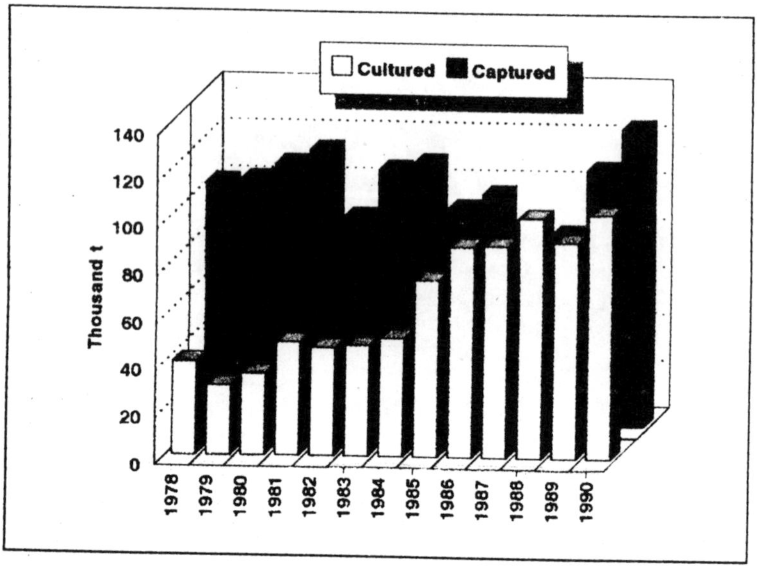

Figure 1. Captured and cultured freshwater fish production in Thailand

Between 1978 and 1990, freshwater fish production from aquaculture increased significantly and is estimated to have done so again in 1991 and 1992. In 1978 the production of freshwater capture fisheries was about 102,000 t, but was reduced to less than 82,000 t by 1988. However, there has been an increase since then. Production from inland aquaculture had increased from 39,000 t in 1978 to nearly 104,000 t by 1990 (Table 1 and Figure 1).

As shown in Table 2, between 1978 and 1990 the number of freshwater fish farms increased from 24,000 to nearly 62,000 while their total area grew from 24,000 ha to nearly 46,000 ha in 1988 but has decreased since then. The main area for freshwater fish culture is the central part of Thailand. Here the number of farms is only 25% of the total, with an average area of 2.0 ha/farm, but the volume of fish produced in this area is nearly 70% of national production (Table 3). The Northeast is the second most important area, which has the highest number of farms (45%) with an average culture area of 0.26 ha/farm. In the North there are about 15,600 farm units (25%) with an average area of 0.29 ha/farm. The least important area is the South, where the number of farms is only about 3,000 (5% of the total) with an average area of 0.05 ha/farm. Production of freshwater fish in the northern and southern parts of the country is generally low, compared with the central region, as shown in Figure 2.

Table 2. Number of freshwater fish farms and area under aquaculture in Thailand

Year	North		Northeast		Central		South		Total	
	farm (unit)	area (ha)	farm (unit)	area (ha)	farm (unit)	area (ha)	farm (unit)	area (ha)	farm (unit)	area (ha)
1978	4,415	438	9,725	3,466	7,328	20,328	2,195	60	23,663	24,292
1979	4,402	447	11,052	2,410	7,378	21,511	2,448	94	25,280	24,462
1980	5,129	554	13,217	2,805	8,512	21,769	2,626	62	29,484	25,190
1981	5,837	663	14,140	3,359	9,248	22,353	2,451	73	31,676	26,448
1982	6,156	809	15,962	3,821	10,011	23,086	2,025	78	34,154	27,794
1983	6,355	899	16,965	10,900	10,402	25,505	2,029	94	35,751	37,398
1984	6,986	1,033	17,785	4,410	11,634	26,580	1,830	100	38,235	32,123
1985	8,579	1,381	20,647	4,182	13,080	31,286	2,015	100	44,321	36,949
1986	9,052	1,529	22,332	5,838	13,731	31,103	2,154	118	47,269	38,588
1987	11,845	2,059	23,947	6,461	17,945	37,891	2,431	143	56,168	46,554
1988	13,715	2,673	25,968	7,054	17,412	35,959	2,344	146	59,439	45,832
1989	14,462	2,632	27,539	7,513	15,468	31,317	3,028	176	60,497	41,638
1990	15,598	2,895	28,080	7,278	15,275	30,827	3,027	162	61,980	41,162

Table 3. Number of freshwater fish farm units and production in different parts of Thailand

	1984	1985	1986	1987	1988	1989	1990
North							
No. of farms	6,986	8,579	9,052	11,845	13,715	14,462	15,598
Production (t)	3,408	8,319	7,595	7,886	11,504	9,464	18,832
Northeast							
No. of farms	17,785	20,647	22,332	23,947	25,968	27,539	28,080
Production (t)	3,802	5,292	5,802	7,047	9,354	7,328	9,043
Central							
No. of farms	11,634	13,080	13,731	17,945	17,412	15,468	15,275
Production (t)	42,572	60,750	74,622	73,255	79,740	71,493	72,018
South							
No. of farms	1,830	2,015	2,154	2,431	2,344	3,028	3,027
Production (t)	629	893	1,302	1,594	1,530	3,395	3,873
Totals (No.)	38,235	44,321	47,269	56,168	59,439	60,497	61,980
(t)	50,411	75,254	89,321	89,782	102,128	91,680	103,766

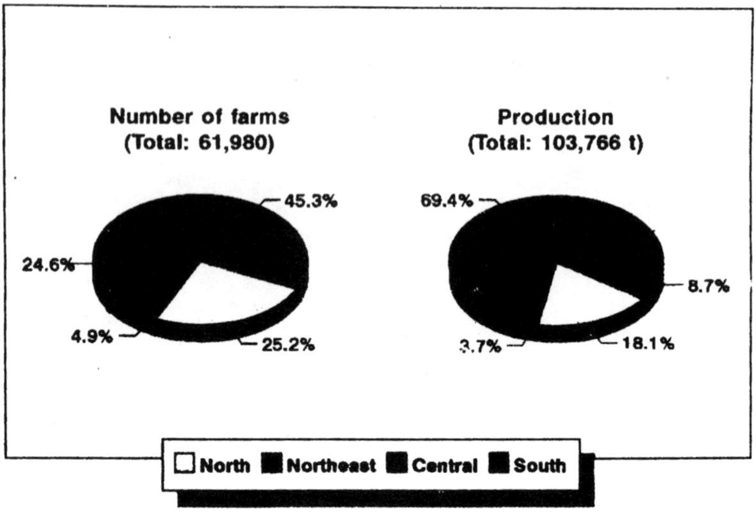

Figure 2. Freshwater fish farms and their production in different parts of Thailand in 1990

There are many economically important cultured aquatic species in Thailand, such as *Tilapia* sp., *Puntius gonionotus*, *Pangasius sutchi*, *Trichogaster pectoralis*, *Clarias* sp., *Channa striatus*, *Macrobrachium rosenbergii*, *etc.* Some marine species are also important for aquaculture, providing export commodities and high income. Farmers prefer to culture seabass (*Lates calcarifer*) and shrimp (*Penaeus monodon*) because of their high farm-gate value. The area under shrimp culture had reached nearly 76,000 ha by 1989 but had declined to 65,000 ha by 1990 (Figure 3). The national production was nearly 108,000 t in 1990 (Table 4).

Most of the seabass is cultured in cages and is limited to certain locations. About 68% of seabass farms are in five southern provinces. Shrimp farms produce the highest value among coastal farms. Because of the high farm-gate value of marine shrimp, many farmers shifted to growing shrimp. This caused environmental problems because shrimp farms invaded the coastal mangroves. The rapid growth of shrimp farms without proper regulations and control resulted in water pollution and serious disease outbreaks. The Department of Fisheries has realized the problem and is educating shrimp farmers in order to introduce proper farm management techniques. Today, many shrimp farms in the eastern and southern part of Thailand still have high production.

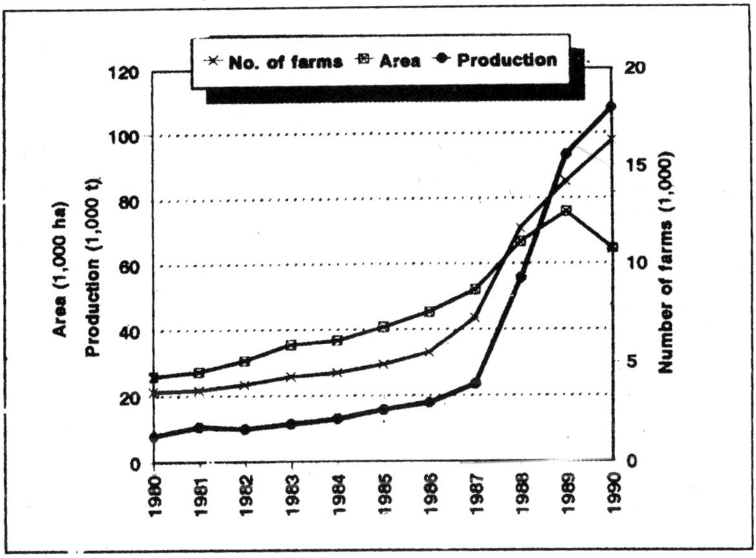

Figure 3. Number, area and production of shrimp farms in Thailand

Table 4. Number, area and production of shrimp farms in Thailand

Year	Number of farm units	Area (ha)	Production (t)
1980	3,572	26,036	8,063
1981	3,657	27,459	10,728
1982	3,943	30,793	10,091
1983	4,327	35,537	11,550
1984	4,519	36,792	13,007
1985	4,939	40,769	15,841
1986	5,534	45,368	17,886
1987	7,264	52,148	23,566
1988	11,836	66,731	55,633
1989	14,235	75,928	93,434
1990	16,299	64,606	107,969

AVAILABILITY OF FEED INGREDIENTS

Most of the raw materials used in aquafeeds can be produced in the country but domestic production of some important ingredients, such as soybean, fish meal and sesame meal, is insufficient because they are also used for manufacturing poultry feed. Table 5 shows domestic production and imports of feed ingredients.

Table 5. List of feed ingredients available in Thailand: production and imports in 1990 (1,000 t)

Feed ingredient	Domestic Production					Imports
	Northeast	North	Central	South	Total	
Oil crops						
Soybean	86	393	50	1	530	340
Peanut	46	84	23	8	161	147
Sesame seed	← 29 →				29	41
Coconut				2	2	-
Oil palm	← 33 →			1,153	1,186	-
Food crops						
Maize	1,069	1,769	860	24	3,722	723
Cassava	12,126	1,934	5,645	-	19,705	-
Sugar cane	7,797	9,650	23,214	-	40,661	-
Mung bean	18	246	34	5	303	-
Sorghum	6	128	103	-	237	-
Broken rice	← 1,079 →				1,079	-
Fish products						
Fish meal	← 437 →				437	17
Cod liver oil						793

The most important growing areas for oil crops (soybean, peanut and sesame) are the north and the northeast. Oil palm production is concentrated in the south. Although domestic production of soybean in 1990 was 190,000 t, the demand for it is even higher, therefore significant amounts have to be imported (340,000 t in 1990). Domestic food crop production almost satisfies the demand. The most important growing areas are in the central, northern and northeastern part of the country. Only maize is imported at about 723,000 t/yr.

Problems in producing feed ingredients may arise as the area under culture is not expanding while the demand increases every year. The average yield in the planted area was low because of the low quality of the seed; this results in low quality and quantity of production. The government tries to solve this problem by coordinating with the private sector in improving seed quality, maintaining cash flow and assisting farmers to complete their crop cycle. Availability of feed ingredients of good quality and in high quantity is expected in the next few years.

Another important feed ingredient, fish meal, is used in high amounts in animal feeds, so the demand is high. Fish meal, especially with a protein content over 60%, has to be imported. Low domestic fish meal production is the result of many factors, such as decreases in the fishing area and the catch.

The price and composition of the major feed ingredients available in Thailand is presented in Table 6, while compositional data for some traditional ingredients is given in Table 7.

Table 6. Price and composition of major feed ingredients in Thailand in 1992

Feed ingredients	Price (US$/t)	Moisture (%)	CP (%)	EE (%)	CF (%)	Ash (%)	Ca (%)	P (%)
Broken rice	160-240	12.57	7.02	1.12	0.43	0.75	0.01	0.16
Cassava root	120-200	10.34	2.14	0.62	2.05	2.69	0.09	0.07
Corn bran	160-280	12.50	7.58	3.62	5.63	2.21	0.04	0.26
Corn grain	200-240	12.56	8.71	4.13	1.71	1.29	0.01	0.26
Coconut oil meal	160-320	10.00	21.00	6.00	12.00	7.00	0.20	0.20
Fish meal (60% C.P.)	640-800	8.00	55.20	8.01	1.00	26.00	7.70	3.80
Kapok oil meal	160-200	10.24	32.58	2.38	21.62	7.22	0.40	1.03
Peanut oil meal	280-400	7.59	49.91	5.79	3.95	8.18	0.14	0.65
Rice bean (extract)	120-200	11.08	15.73	2.35	10.26	12.87	0.17	2.26
Sesame meal (black)	200-280	7.46	33.84	16.42	5.57	12.23	2.52	1.22
Shrimp head meal	480-600	10.26	30.59	3.77	10.55	n.a.	8.16	1.60
Soybean meal	320-480	9.50	44.30	5.30	5.70	6.00	0.83	0.66
Sunflower seed meal	320-400	7.57	21.64	7.61	29.05	6.75	0.63	0.67
Wheat flour	560-600	13.10	15.40	1.90	0.50	0.90	0.06	0.13

Table 7. Composition of some traditional feed ingredients in Thailand

Feed ingredients	Moisture (%)	CP (%)	EE (%)	CF (%)	Ash (%)	Ca (%)	P (%)
Cassava leaves	12.32	27.39	7.17	10.90	7.02	1.20	0.30
Chicken viscera meal	8.00	56.50	22.30	2.70	4.11	3.60	1.54
Ipomoea	89.73	2.86	2.77	n.a.	1.15	-	
Lemna minor	9.50	29.00	4.21	7.26	20.30	0.02	0.63
Leucaena leaves	10.05	23.54	7.70	7.68	9.71	2.52	0.17
Leucaena seed	10.30	27.80	7.06	10.25	4.04	0.39	0.05
Oil lard meal	10.04	55.89	8.69	1.78	5.87	0.76	0.51
Parboiled rice bran	9.07	13.32	27.25	9.72	9.44	0.14	2.08
Sago stem	10.14	1.23	1.80	13.32	8.88	0.84	0.02
Sesbania (leaves)	7.85	29.56	4.85	7.47	10.55	2.30	0.30
Tofu waste	90.53	2.60	0.82	1.83	0.39	0.06	0.02
Water hyacinth	92.73	1.65	0.26	1.27	1.37	0.13	0.06
Winged bean	9.31	6.79	0.23	36.95	8.89	0.44	0.09

AQUAFEED MANUFACTURING INDUSTRY

There are 13 feed mills producing aquafeeds in Thailand. Most of these originally produced livestock and poultry feeds. Nine of the mills only produce shrimp feeds; the others manufacture both shrimp and fish feeds (Annex 1).

About 88% of the 382,000 t of aquafeeds produced in the country in 1990 were shrimp feed; 11% were for carnivorous fish species and only 1% for herbivorous fish. Feed exports are only 3% of the total. Most of the feed factories produce shrimp feeds because the demand is high. Because of the high farm-gate value of shrimp, farmers have the capability to buy factory-made feeds. Although carnivorous fish (*e.g.* catfish) are popular farmed fish, their farm-gate value is low compared to shrimp. The price of catfish is only one-third to two-thirds of the price of shrimp. Low fish prices have forced farmers to use traditional, farm-made moist feeds. Since the farm-gate value of herbivorous fish is even lower (US$ 0.4-0.8/kg), farmers try to decrease production costs by using household byproducts and plants instead of dry feeds. However, if herbivorous fish culture is promoted to a commercial scale and produces good quality fish, farm-gate value could increase, rendering the use of commercial feeds more feasible.

Aquafeed produced in the country is available for five different aquatic species groups. These are marine shrimp, freshwater prawns, catfish and tilapia, for which complete diets are manufactured, and herbivorous fish, for which supplementary feed is available. Nutritional values and prices of these commercial feeds are presented in Table 8.

Table 8. Proximate composition and cost of commercial aquafeeds produced in Thailand in 1992

Feeds	Moisture (max)	CP(%) (min)	EE(%) (min)	CF(%) (max)	Ash(%) (max)	Cost (US$/kg)
Shrimp feeds (black tiger shrimp)						
Postlaval stage I (small crumble)	12	37	3	3	10	1.28
Postlaval stage II (medium crumble)	12	37	3	3	10	1.28
Juvenile stage I (large crumble)	12	36	3	3	10	1.28
Juvenile stage II (sinking pellet)	12	36	3	3	10	1.12
Grower, 12-120 g (sinking pellet)	12	35	3	3.5	10	1.08
Finisher, 120 g (sinking pellet)	12	35	3	3.5	10	1.08
Freshwater shrimp feeds (*Macrobrachium* sp.)						
Juvenile (small crumble)	12	30	4	8	8	0.72
Starter (sinking pellet)	12	30	4	8	8	0.68
Grower (sinking pellet)	12	30	4	8	8	0.64
Finisher (sinking pellet)	12	20	3	8	8	0.56
Catfish feeds						
Fry, 1-2 weeks (powder)	12	40	10	8	15	1.44
Fingerling, 2 weeks (floating pellet)	12	30	4	8	15	0.72
Grow-out, 1-3 months (floating pellet)	12	30	4	8	15	0.64
Finisher, >3 months (floating pellet)	12	25	4	8	15	0.56
Tilapia feeds						
Fingerling (floating pellet)	12	32	4	8	8	0.60
Grow out (floating pellet)	12	28	4	8	8	0.56
Finisher (floating pellet)	12	24	4	8	8	0.52
Supplemental feed for herbivores						
Grow-out/finisher (floating pellet)	12	15.5	4	8	8	0.40

ON-FARM FEED FORMULATION AND MANUFACTURING

More than 50% of cultured freshwater fish is used for household consumption. Most species are of low farm-gate value, therefore simple types of feeds are used. Only a few fish farms have their own equipment to produce farm-made feeds. Farmers are trained in feed formulation and preparation by government or private agencies. Information on fish nutrition is generated mainly by government institutions, primarily by the Feed Quality Control and Development Division (FQCD) and the National Institute of Coastal Aquaculture (NICA) of the Department of Fisheries which are the core institutions for basic and applied research (Figure 1).

Typical feed formulae which have been extended to farmers by FQCD are shown in Tables 9-14. Some traditional feeds used by the farmers are shown in Tables 15 and 16; these consist of two or more ingredients mixed together. Premixes are sometimes used. Ingredients are roughly mixed and ground in a meat mincer to form a slurry or a paste. Minced feeds are distributed to the ponds by hand, shovels or buckets for most of the species. For some species (like snakehead) the moist feed has to be placed on wooden platforms in order to ensure proper feeding.

Table 9. Feed formulae for catfish in Thailand

Ingredient	(%)	(%)
Fingerling feed (38% protein)		
Fish meal	56	
Rice bran	12	
Soybean meal or peanut meal	12	
Glutinous flour	14	
Fish oil	4	
Vitamin and mineral premix	2	
Grower feed (30% protein)		
Fish meal		15
Soybean meal		45
Broken rice or maize		24
Rice bran or trash fish or chicken viscera		10
Bone meal or dicalcium-phosphate		1
Leuceana leaf meal		3
Vitamin and mineral premix		2
Totals	100	100

Table 10. Feed formula for snakehead (40% protein) in Thailand

Ingredient	(%)
Trash fish	50.0
Fish meal	17.5
Soybean meal	7.5
Rice bran	17.0
Broken rice	7.0
Vitamin and mineral premix	1.0
Total	100.0

Table 11. Thai feed formulae for freshwater prawns from fingerling to 3 months (30% protein)

Ingredient	(%)	(%)
Formula No. 1		
Fish meal	15	
Shrimp head meal	42	
Soybean meal	10	
Coçonut oil cake meal or trash fish	15	
Wheat flour or tapioca flour or rice bran	13	
Vitamin and mineral premix	5	
Formula No. 2		
Trash fish		35
Fish meal		17
Broken rice		15
Rice bran		18
Soybean meal		12
Vitamin and mineral premix		3
Total	100	100

Table 12. Feed formula for freshwater prawns in Thailand 1 month before harvest (25% protein)

Ingredient	(%)
Fish meal	10.0
Shrimp head meal	25.0
Soybean meal	10.0
Broken rice	25.5
Rice bran	25.5
Vitamin and mineral premix	4.0
Total	100.0

Table 13. Feed formulae for herbivorous fish in Thailand

Ingredient	(%)	(%)
Fingerling (30% protein)		
Fish meal	30	
Fine rice bran	45	
Peanut meal	24	
Vitamin and mineral premix	1	
Grower (24% protein)		
Fish meal		16
Peanut meal		24
Soybean meal		14
Fine rice bran		30
Broken rice		15
Vitamin and mineral premix		1
Totals	100	100

Table 14. Recommended feed formula for shrimp in Thailand

Ingredient	(%)
Fish meal	30.0
Broken rice	12.0
Shrimp head meal	10.0
Squid meal	6.0
Yeast	2.0
Soybean meal	25.0
Rice bran	10.0
Binder	1.0
Dicalcium phosphate	1.0
Vitamin premix	0.8
Vitamin C	0.2
Mineral premix	2.0
Total	100.0

Table 15. **Traditional snakehead and catfish feed formulae used by farmers in Thailand**

Ingredient	(%)	(%)
Snakehead (40% protein)		
Trash fish	90	
Rice bran	10	
Catfish feed (30% protein)		
Trash fish		75
Rice bran		25
Totals	100	100

Table 16. Farmer's feed formula for shrimp in Thailand

Ingredient	(%)
Trash fish	30.0
Fish meal	22.0
Shrimp head meal	8.0
Soybean meal	12.0
Peanut meal	8.0
Rice bran	6.0
Squid meal	4.0
Wheat flour	4.0
Dicalcium phosphate	1.5
Zeolite	1.0
Fish oil	0.5
Vitamin and mineral premix	2.9
Vitamin C	0.1
Total	100.0

ON-FARM FEEDING STRATEGIES

When culturing carnivorous species in running water systems, stocking rates are about three times higher than in herbivorous fish culture. The stocking densities of clarias catfish in ponds are extremely high, varying from 75 to 400/m^2, representing the highest stocking densities in fish culture. The stocking density of shrimp is currently only 20-40/m^2. Farmers pay attention and take much care in rearing shrimp because they are sensitive to environmental changes and the investment in shrimp culture is high.

Farmers can make more profit by practising integrated farming or polyculture where few feed inputs are required and the natural productivity can be maximized. Integrated farming means culturing fish in ponds under chicken, duck or pig houses. Suitable fish species for integrated farming are *Tilapia* sp., *Pangasius* sp. and *Puntius* sp. The proper stocking ratios of the animals in integrated farming are 1,250 chicken or ducks or 30-60 pigs together with 12,500 fish/ha. Polyculture is most successful when fish species with different feeding behaviour are selected, such as tilapia, silver carp, striped catfish and bighead carp. The stocking rate for these four species in polyculture could be 3:3:3:1 respectively at a stocking rate of about 10/m^2.

Normally the daily feeding rate of fish is about 10% of body weight from fry to fingerling stages; it is then decreased to 5% of body weight in the juvenile and adult phases. Some shrimp farmers re-calculate feed requirement every two weeks in order to minimize feed loss and to decrease feeding costs.

The feeding frequency of fish is usually 3-4 times a day in the larval/ fingerling stage and twice a day from the fingerling stage to marketable size (Table 17). In the case of marine shrimp feeding frequency is much higher, because they eat gradually. Farmers feed shrimp with commercial sinking pellets but, in the final month before harvesting, trash fish or mussels are used once a day in the evening to induce maturation.

Extensive fish culture normally utilizes fertilizers in the nursery ponds. Herbivorous fish culture relies on natural food in the pond promoted by the addition of organic and inorganic fertilizers. Farmers prefer to apply chicken manure at a rate of 900-1,500 kg/ha/week. If cattle manure is used, relatively less is applied.

Water exchange usually starts in the grow-out phase or when the water quality deteriorates. In shrimp ponds during the first month of culture no water exchange is needed; later about 15-30% of pond volume is exchanged and paddle wheels or air jet aerators are employed to increase dissolved oxygen. About 4 air jet aerators are required for a pond with an area of 0.6 to 1.0 ha.

Table 17. Stocking density, feeding rate and feeding frequency of major cultured species in Thailand

Species age	Stocking density (pc/m²)	Feeding rate (%)	Feeding frequency (#/day)	Period of culture (months)	Inputs
Herbivorous fish					
fry	100-200	10	3-4	1	Fertilizer : 900-1,500kg/ha/week
grow-out	30-50	3-5	2	5-8	
Snakehead					
fingerling→ harvest	350-400	satiation	1-2	8-10	
Catfish					
fingerling→ harvest	30-50	3-5	2	3-4	poultry viscera animal wastes
Freshwater fish					
fingerling→ harvest	20-40	3-5	2	6-8	water exchange after 3 months of culturing
Marine shrimp					
postlarvae→ 1 month	30-50	10	2	1	water exchange 5-10%
1 month→ harvest	25-40	3-5	4	3-4	water exchange 15-30% and use air jet or air flow aerators (4 in a 0.6-1.0 ha pond)

MAJOR INSTITUTIONS INVOLVED IN AQUAFEED RESEARCH AND DEVELOPMENT

In Thailand, fish nutrition research is undertaken primarily by two institutions (Figure 4) under the control of Department of Fisheries, namely the Feed Quality Control and Development Division (FQCD) and the National Institute for Coastal Aquaculture (NICA). These two institutions coordinate their work with various fishery stations located throughout the country. FQCD is the centre for feed quality control and implements intensive research work on the nutrient requirements of freshwater fish. Its work includes the development of cost-effective feed formulae, using locally available feed ingredients. The nutritional research of NICA concentrates on marine fish species. NICA coordinates with those fishery stations which are involved in coastal aquaculture. NICA's experimental work focuses on determining the nutrient requirements of marine fish and shrimp and on feed development for them. At universities, research work is mainly carried out by students under the supervision of faculty staff. Every summer, students from many universities come to local fishery stations and institutes involved in fish nutrition research to broaden their experience and to use the facilities.

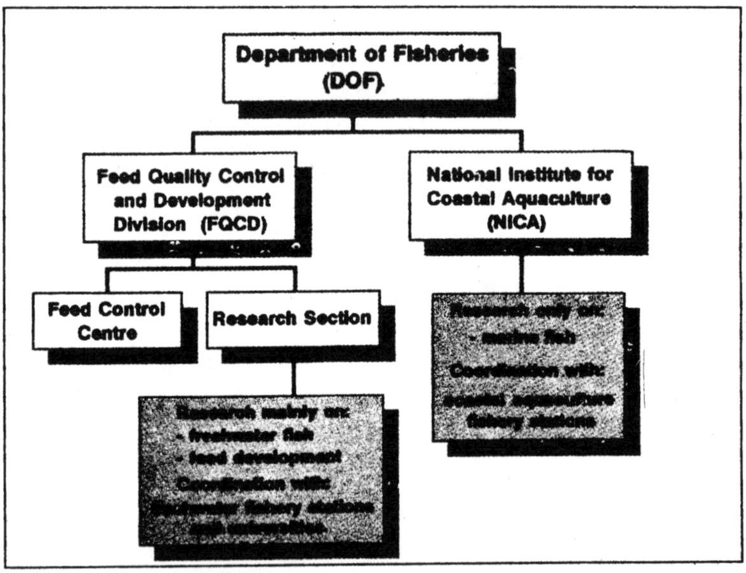

Figure 4. Organization of aquafeed research and development in Thailand

The facilities for feed and feed ingredient analysis at FQCD and NICA include equipment for proximate analysis, gas chromatography (GC) for fatty acid analysis, and high performance liquid chromatography (HPLC) for amino acid and vitamin analyses.

CURRENT NATIONAL REGULATIONS

Regulations controlling aquafeed manufacturing, marketing and feed quality in Thailand were issued by the Ministry of Agriculture and Cooperatives in 1992. These are similar to the regulations controlling poultry feed manufacturing, marketing and quality. The responsible unit in the Ministry is the Department of Fisheries (DOF) which, in turn, has assigned the Feed Quality Control and Development Division (FQCD) to implement control. The regulations are shown in Annex 2.

CURRENT PROBLEMS AND CONSTRAINTS

Total production of cultured freshwater fish in 1992 was about 114,000 t (Table 1). Assuming that this resulted from feeding dry commercial feeds with an average AFCR of 1.5:1, approximately 171,000 t of feed would have been required. However, only 46,000 t of freshwater fish feed are produced annually. The difference indicates that not all fish farmers depend on commercial feeds. The main constraint faced by farmers in the use of dry formulated feeds is that the cost may exceed the market value of some freshwater species. For instance, the market value of striped catfish, tilapia and tawes averages US$ 0.36/kg, US$ 0.44/kg and US$ 0.64/kg respectively, while the cost of feeding (assuming AFCR is 1.5:1) is US$ 0.57/kg of fish produced. This constraint leads to little or no use of formulated feeds and to the use of more fertilizers or traditional feed resources to reduce production costs. Therefore, the use of formulated feeds is limited to species with high market value.

Some farmers refuse to use dry commercial feeds as they do not believe that dry feeds could satisfy the needs of fish. The traditional feeds are moist feeds, such as trash fish mixed with rice bran. Regardless of nutritional value, farmers believe that wet feeds would satiate fish better than dry feeds.

Fattiness in the flesh of some species, such as striped and walking catfish, is favoured by fish consumers. Therefore, high energy feed sources are required rather than dry feeds. Poultry viscera, broken rice and kitchen wastes are still the feeds of choice for several farmers raising catfish.

TRENDS IN AQUAFEED MANUFACTURING AND USE

Aquafeed manufacturing and use is expected to increase in the next few years because the use of traditional or farm-made feeds is highly site- and time-specific. Although preparing farm-made feeds is practical and cost effective for certain farmers, it may not be so for others, due to the unavailability of unconventional raw materials. The quality and quantity of farm-made feeds cannot be controlled properly, resulting in unpredictable fish production. Furthermore, the ingredients traditionally used for farm-made feeds are becoming less and less available. Farm-made feeds also disintegrate rapidly in water causing problems to the aquatic environment. For these reasons, aquaculture cannot expand rapidly unless more conventional types of feeds are used.

Shrimp feed manufacturing is steadily increasing every year. There are also many other economically important aquatic species which are interesting to farmers, such as seabass, groupers and frogs. Although the nutritional requirements of these species are not totally known, many feed manufacturers have already produced and marketed feeds for them.

The number of feed mills in Thailand is not expected to increase, however, because new ones may not be able to compete with the existing large mills. Feed production is expected to increase through the expansion of existing feed mills.

Annex 1. List of aquafeed manufacturers

Name	Address	Type of feed produced
Charoen Pokphan Feedmill	313 C.P. Tower, Silom Road, Bangkok 10500 Tel.: (66-2) 231-0231/5	Shrimp Carnivorous fish Herbivorous fish
Laemthong Aquatech	1126/1 Petchburi Tadmai Road, Ratchatavee, Bangkok 10400 Tel.: (66-2) 252-3777	Shrimp.
Bangkok Feed Mill	36 Soi Yenchit, Chan Road, Yannawa, Bangkok 10120 Tel.: (66-2) 211-1080/10	Carnivorous fish Herbivorous fish
Aquafeed Business	5/73 Soi Amornphan 5, Sukhapibal 2 Road, Bangkhum Bangkok Tel.: (66-2) 214-4661	Shrimp
P. Charoenphan Feed Mill	69/6-13 Suksawadi Road, Rardburana, Bangkok 10140 Tel.: (66-2) 463-0040 463-0050	Shrimp
Centaco Aqua-Feeds	7/3 Paholyothin Road, Klonglauna, Pathumtani Tel.: (66-2) 516-8811/5	Shrimp
Premier Feeds	79 Bangna-Trad Road, Bangpakong, Chachoengsao 24130 Tel.: (66-38) 353-4219	Shrimp Catfish
Apitoon Marine Products	1-21/49-50 Thai Tower Bldg. (15th Flr.), South Sathorn Road, Yannawa, Bangkok 10120 Tel.: (66-2) 285-0051	Shrimp
Unicord Feed	544-546 Floor 13 Bamrungmaung, Pomprab, Bangkok 10700 Tel.: (66-2) 255-0110	Shrimp

Annex 1. (cont.)

Siamfeed	111/99-101 Soi Lardprao 132 , Lardprao Road, Bangkapi, Bangkok 10240 Tel.: (66-2) 375-2938/40	Shrimp
Thailux Enterprises	101/3-4 Narads Road, Bangrak, Bangkok 10500 Tel.: (66-2) 236-7347	Shrimp
Aquastar	90/12 North Sathorn Road, Silom, Bangkok 10500 Tel.: (66-2) 238-1720	Shrimp
Grobest Corporation	119/12 Vipavadee Road, Phayathai, Bangkok 10400 Tel : (66-2) 278-0563	Shrimp

Annex 2. Thai aquafeed regulations (1992)

The Minister shall have power to publish notification in the government gazette prescribing:

(i) the name, category, type and/or description of the animal feed;

(ii) the quality standard of the animal feed according to the species, genus, family or age of the animal as well as rules and procedures of production or exposition of such animal feed for sale;

(iii) the name, category, type and/or description of the animal feed to be permitted or prohibited to import for sale;

(iv) the name, category, type and/or description of the feed additive allowed to be mixed in the production of an animal feed for sale as well as the minimum and maximum ratio or quantity of such additive in the mixture;

(v) the production procedure, equipment or facilities used in the production and storage of animal feed produced or exposed for sale from deterioration;

(vi) the quality or standards and use thereof, as well as materials which are prohibited to be used, of containers for animal feed for sale.

FQCD shall appoint a committee in order to give advice or opinions (on the description of aquafeeds, consideration of appeals, suspension of licenses, *etc*) to the Director-General and has the power to issue written orders requiring any person to give a statement or submit any relevant document or evidence to supplement its consideration.

No-one may produce or import for sale any aquafeed, unless a license has been granted for production or import. Any licenses to produce or import for sale of a particular type of animal feed must first be registered with the competent office and receive a certificate of registration from the Director-General of the Department of Fisheries.

AQUAFEEDS AND FEEDING STRATEGIES IN VIETNAM

Le Thanh Luu

Research Institute for Aquaculture No. 1
Dinh Bang, Tien Son, Ha Bac, Vietnam

LUU, L.T. 1993. Aquafeeds and feeding strategies in Vietnam, p.386-396. *In* M.B. New, A.G.J. Tacon and I. Csavas (eds.) Farm-made aquafeeds. Proceedings of the FAO/AADCP Regional Expert Consultation on Farm-Made Aquafeeds, 14-18 December 1992, Bangkok, Thailand. FAO-RAPA/AADCP, Bangkok, Thailand, 434 p.

INTRODUCTION

Aquaculture in Vietnam plays an important role in meeting the protein needs of the people. In 1991, the production of aquaculture (300,000 t) significantly increased (Table 1 and Figure 1) due to the application of the semi-intensive approach; this resulted in considerable enhancement of pond productivity (Table 2). Table 1 also shows that aquaculture dominates the production of freshwater fish and prawns, constituting 81% in 1990 and 91% in 1991. Freshwater fish production comprised 35-45% Chinese carps, 25-30% Indian major carps, 20-35% indigenous species (*e.g.* clariad and pangasid catfishes, mud carp, snakehead, *etc.*), 1-2% tilapia and 12-13% prawns.

Table 1. Fish production in Vietnam (1,000 t)

	1986	1987	1988	1989	1990	1991	1992*
Total production	707.6	754.4	811.5	881.5	890.6	1,068.0	1,080.0
- of which							
- marine fish	563.4	591.6	629.1	683.2	691.0	739.0	740.0
- freshwater fish	144.2	162.8	182.4	198.3	199.6	329.0	340.0
- aquaculture	127.5	141.8	154.3	164.9	162.0	300.4	n.a.
Aquaculture as % of total production	18.0	18.8	19.0	18.7	18.2	28.1	n.a.
Aquaculture as % of freshwater fish production	88.4	87.1	84.6	83.2	81.2	91.3	n.a.

* 1992 data are planned

Table 2. **Average yield of aquaculture in Vietnam**

	1987	1988	1989	1990	1991
Total area used for aquaculture (ha)	249,000	311,534	296,682	295,751	309,760
Average yield (kg/ha/yr)	569	495	556	548	970

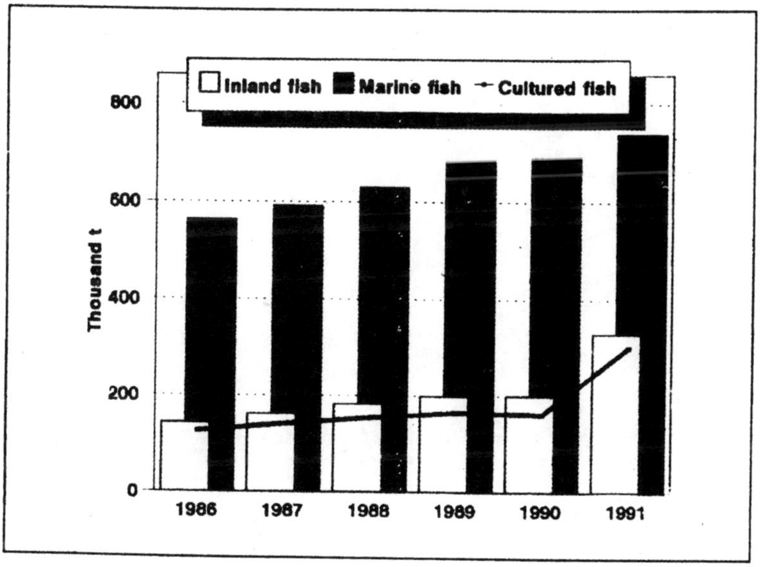

Figure 1. Growth of fish production in Vietnam

Unfortunately, there are no statistical data on the contribution of aquafeeds to aquaculture production. Aquaculture in Vietnam is still based on the traditional polyculture approach within which the application of semi-intensive systems with some supplementary feeding began quite recently.

AVAILABILITY OF FEED INGREDIENTS

Like many tropical countries, Vietnam has an abundance of feed ingredients for animals and fish. In the list of feed ingredients available in Vietnam there are more than 50 entries of which nine are considered major raw materials. The data given in Table 3 indicate significant production of rice, other cereals, tubers and other feed ingredients. These crops, *e.g.* maize, soybean, cassava, sweet potato, *etc*. are harvested in different seasons depending on the geographical location of each region. The cost of these products is lowest in the harvesting period and increases afterwards. After harvest the price of cereals often increases by 10-15%. The approximate cost of major animal feed ingredients is given in Table 4. However, prices may be 10-20% less if the products are purchased locally and in large amounts.

Table 3. Annual production of major feed ingredients in Vietnam (1,000 t)

Ingredients	1976	1980	1985	1990	1991
Rice	11,827.0	11,647.4	15,874.8	19,225.1	19,426.8
Maize	386.8	428.8	587.1	671.0	651.6
Sweet potato	1,484.6	2,417.6	1,777.7	1,929.0	2,104.5
Cassava	1,815.7	3,323.0	2,939.8	2,275.8	2,389.9
Potato	260.2	872.2	188.6	365.3	281.3
Soybean	20.7	32.1	79.1	86.6	98.7
Peanut	100.1	95.2	202.4	213.1	211.7
Rice bran	2,365.4	2,329.4	3,174.8	3,845.0	3,885.2
Fish meal				4.9	8.1

Table 4. Prices of major feed ingredients in Hanoi/Habac Province, Vietnam, in 1992

Ingredients	US$/kg
Rice	0.14 - 0.19
Rice bran	0.11 - 0.13
Maize	0.10 - 0.13
Dried cassava	0.06 - 0.09
Soybean	0.30 - 0.37
Fish meal	0.37 - 0.60

Fish meal is considered the main animal protein resource. However, domestic production of fish meal is limited (it was 8,500 t in 1992) due to the high cost of raw materials and transport. About 80-85% of the fish meal is produced in the south and only 15-20% in the north of the country.

Only a few feed ingredients (such as vitamins A, B, C, D and lysine) are imported; they are used to produce premixes. The amount of imported vitamins of all kinds is estimated be about 0.1-0.15 t/yr, while the amount of imported lysine ranges from 3.0 to 3.5 t/yr.

AQUAFEED MANUFACTURING INDUSTRY

Unfortunately, industrial-scale manufacturing of aquafeeds in Vietnam is limited, although there are many medium-sized factories producing animal feeds. In the north of Vietnam during the past two decades 20 feed mills were constructed with an installed capacity totalling 94,000 t. In the south the number of feed mills is 10, with a total installed capacity of 83,200 t per year. The actual production of all these feed mills reached 22,600 t, which is about 13% of the installed capacity (Annex 1). As much as 65% of the produced animal feeds are used for chicken, 30% for pig raising and the rest for dairy cows. None of these factories has produced aquafeeds.

In the fisheries sector 14 local feed mills with an installed capacity of about 2,300 t/yr have been constructed to produce feeds primarily to meet the demand, which is still low, from shrimp culture (Annex 2). At present these plants are producing only about 800 t/yr. These feeds are used by farmers in pelleted form as supplementary feeds.

Averages of the most important biochemical parameters of feed samples collected from various factories and shops are shown in Table 5. Experiments with locally produced grow-out feeds in *Penaeus monodon* ponds have shown that yields up to 0.5-2.0 t/ha/crop may be achieved with fairly high food conversion rates (3.0-4.0:1).

ON FARM-FEED FORMULATION AND MANUFACTURING

In Vietnam there are few laboratories involved in research on fish and shrimp nutrition. Research projects are scattered and very limited due to shortage of funds, without which nothing can be done. The experimental feeds formulated by the Research Institute for Aquaculture No. 2 have rather high protein content and cost about US$ 780-960/t. Meanwhile, a group of scientists from the Research Institute for Aquaculture No. 1, using a simple computer programme "FORM", have formulated various types of feeds for carp broodstock, for rohu and mrigal, for *Penaeus monodon, P. merguiensis* and *Macrobrachium rosenbergii*. Some of these formulae are given in Tables 6-8.

Table 5. Average composition of shrimp feed in Vietnam in 1992

Composition	Content (%)
Protein	25.0 - 35.0
Lipid	4.0 - 7.0
Carbohydrate	30.0 - 55.0
Fibre	8.0 - 12.0
Moisture	8.0 - 10.0
Vitamin premix	0.1 - 0.2

Table 6. Shrimp rearing feed formulated by RIA No. 1.

Ingredients	(%)
Fish meal (35-38% protein)	74.58
Soybean meal	5.00
Rice meal	13.42
Buffalo fat	5.00
Vitamin premix	2.00
Total	100.00

Biochemical parameters:	Protein:	35.0 %
	ME:	2,561 kcal/kg
	Lipid:	7.9 %
	Carbohydrate:	11.6 %
	Crude fibre:	0.3 %
	Calcium:	5.5 %
	Total P:	2.2 %
	Moisture:	11.8 %
	Ash:	35.8 %
Approximate cost:		US$ 0.48/kg

Table 7. Carp fingerling rearing feed formulated by RIA No. 1

Ingredients	(%)
Fish meal (40-42% protein)	15.00
Meat and bone meal	20.00
Soybean	13.00
White corn	5.00
Rice bran	30.52
Azolla	15.00
Vitamin premix	1.48
Total	100.00

Biochemical parameters:	Protein:	25.7 %
	ME:	2,600 kcal/kg
	Lipid:	11.8 %
	Carbohydrate:	26.7 %
	Crude fibre:	5.2 %
	Calcium:	2.5 %
	Total P:	1.3 %
	Moisture:	11.2 %
	Ash:	17.2 %
Approximate cost:		US$ 0.26/kg

Table 8. Carp broodstock diet formulated by RIA No. 1

Ingredients	(%)
Fish meal	10.00
Extracted peanut meal	18.00
White corn	10.00
Rice bran	61.80
Vitamin premix	0.20
Total	100.00

Biochemical parameters:	Protein:	21.1	%
	ME:	2,773	kcal/kg
	Lipid:	11.2	%
	Carbohydrate:	32.6	%
	Crude fibre:	6.7	%
	Calcium:	0.8	%
	Total P:	1.1	%
	Moisture:	12.8	%
	Ash:	15.1	%
Approximate cost:		US$ 0.15/kg	

On their own farms farmers are using raw materials or home-made feeds to feed shrimp, crabs and cage cultured fish. In the north, farmers use soybean milk, and cooked chicken and pig feeds to feed larvae during the rearing period. Grass carp in cages are fed cassava, cassava leaf or sweet potatoes. Raw trash fish and krill, as well as small, soft-shelled molluscs are used for feeding shrimp and crabs in ponds. In the south farmers also use trash fish for feeding cultured *Pangasius* and other fish species cultured in cages and *Penaeus monodon* in ponds. Some farmers have been encouraged to use commercial feeds for shrimp in their grow-out ponds and have achieved success. However, farmers still prefer to use their farm-made feeds as they seem to be cheaper than commercial feeds.

For preparing farm-made feeds farmers are using very simple grinders and pelleting machines. In rural areas, where there is no electric energy supply, farmers have to use generators to operate their machinery. The operational scheme of farm-made aquafeed processing in Vietnam is as follows: ingredients are sun-dried, ground, stored separately, mixed and then pelleted (Figure 2).

Figure 2. Large mincer for producing aquafeeds in Vietnam

ON-FARM FEEDING STRATEGIES

As commercial feeds are only used by a few farmers, official guidelines or instructions on feeding rates have not been issued for any of the cultured species. However, from our experiments, a daily rate of 5-7% of body weight for shrimp has been recommended for ponds in warm weather and 3-5% in cold weather. Shrimp are fed twice per day; 20% of the daily ration is fed in the morning and 80% in the evening.

Farmers tend to feed their cages or ponds according to the availability of feeds. They also learn through experience to reduce or to increase the feeding rate by checking the food remaining in feeding trays.

MAJOR INSTITUTES INVOLVED IN AQUAFEED RESEARCH AND DEVELOPMENT

The three institutions in Vietnam involved in aquafeed research and development are the following:

Research Institute for Aquaculture No 1, (Vien Nghien Cuu Nuoi Trong Thuy San 1) Ministry of Fisheries, Dinh Bang, Tien Son, Ha Bac (Telephone: 271368);

- *Research Institute for Aquaculture No 2,* (Vien Nghien Cuu Nuoi Trong Thuy San 2) Ministry of Fisheries, 117 Nguyen Dinh Chieu, Ho Chi Minh City (Telephone 299592); and

- *Research Center,* State Committee for Science and Technology (Trung Tam Nghien Cuu Thuc Nghien), 1 Mac Dinh Chi, Ho Chi Minh City.

CURRENT NATIONAL REGULATIONS

No regulations, technical recommendations or instructions concerning different aspects of aquafeed manufacturing or use have been issued in Vietnam so far .

CURRENT PROBLEMS AND CONSTRAINTS

As was pointed out above, commercial aquafeeds are not yet widely used in Vietnam. The reasons are many, which include:

- the price of aquafeeds is very high (Table 9), requiring rather high inputs from the farmers practicing aquaculture, while the price of their products in domestic markets seems to be rather low (Table 10);

Table 9. Price of aquafeeds in Haiphong, Vietnam, in 1992

Protein content (%)	US$/kg
16 - 20	0.24 - 0.27
21 - 25	0.28 - 0.31
26 - 30	0.32 - 0.38
31 - 35	0.39 - 0.46
36 - 40	0.47 - 0.62

Table 10. Price of selected fish and crustacean species in Vietnam, in 1992

Species (size)	US$/kg
Grass carp (2 kg/pc)	0.57 - 0.62
Silver carp (0.5-1.0 kg/pc)	0.38 - 0.43
Rohu, mrigal (0.5-1.0 kg/pc)	0.62 - 0.71
Common carp (0.5-1.0 kg/pc)	0.67 - 0.81
White shrimp (30-50 g/pc)	3.81 - 4.76
Mud crab (150-250 g/pc)	4.76 - 5.71

- government support is not yet strong. As a result research and development funds are very limited;

- there is a lack of economic data on the benefits of feed utilization, therefore no specific guidance has been given to the farmers.

TRENDS IN AQUAFEED MANUFACTURING AND USE

Shrimp and crab culture, as well as fish culture in cages, commenced only a few years ago. No doubt recent Government policy, which allows farmers to have control over water areas for aquaculture with low taxation rates, will encourage them to apply semi-intensive methods. Up to two years ago only cooperatives were allowed to use water for aquaculture. The annual demand for aquafeeds will certainly increase. According to our estimate, aquafeed demand will double every year and by the end of the century the area used for semi-intensive aquaculture systems will reach about 50-55% of the total area under fish and shrimp culture. Cage culture of carp as well as suitable marine fish species (*e.g.* seabass and groupers) will develop on a larger scale.

LIST OF REFERENCES

Anon. 1992. Statistical Data of Vietnam Agriculture, Forestry and Fishery (1976-1991): the Pivotal areas of Commodity Production. Statistical Publishing House, Hanoi. 349 p.

Ministry of Fisheries. 1992a. p. 21. *In* Annual Report for 1991. Hanoi, Vietnam.

Ministry of Fisheries. 1992b. National fisheries policies and investment priorities during the period 1991-2000. Paper presented at the ICD Workshop: Fishery Industry Development, May, 1992.

Annex 1. List of feed mills producing animal feeds in Vietnam

Factory name*	Planned capacity (t)	Actual production (t)
Huong Canh	20,000	200
An Khanh	10,000	3,000
Phap Van	2,000	2,000
Chua Boc	2,000	1,000
Thuy Phuong	2,000	1,000
Cau Dien	4,000	-
Tu Liem	2,000	-
Thanh Tri	2,000	-
Dong Anh	2,000	600
Son Tay	6,000	-
Ninh Giang	4,000	-
Quan Goi	6,000	-
Cho Chua	6,000	-
Tan Dinh	4,000	500
Vo Cuong	4,000	500
Kien An	6,000	-
Phu Luong	4,000	500
Hoa An	4,000	500
Lang Son	2,000	300
Ham Rong	2,000	-
Da Nang	2,400	500
Quy Nhon	2,400	500
Nha Trang	2,400	500
My Tho	10,000	1,000
An Giang	10,000	1,000
Dong Nai	10,000	1,000
Song Be Vifoco	10,000	1,000
Thu Duc Durafee	9,000	1,000
Ho Chi Minh City Vifaco	18,000	4,000
An Phu	9,000	2,000
Total	177,200	22,600

* according to name of location

Source: Department of Science and Technology, Ministry of Agriculture

Annex 2. List of feed mills producing aquafeeds in Vietnam

Factory name*	Planned capacity (t)	Actual production (t)
Hai Phong (1988)	150	5 - 10
Nghe An (1991)	150	30
Thua Thien-Hue	100	50
Da Nang (1990)	500	50 - 60
Quang Nam (1991)	20	20
Binh Dinh (1990)	200	150
Khanh Hoa # 1 (1990)	50	40
Khanh Hoa # 2 (1991)	100	80
Ho Chi Minh City # 1 (1989)	150	80 - 100
Ho Chi Minh City # 2 (1990)	500	40 - 50
Ba Ria (1991)	100	30
Ben Tre (1991)	100	50
Cuu Long (1991)	50	40
Minh Hai (1991)	100	100
Total	2,270	810 (max)

* according to name of location

Source: Department of Science and Technology, Ministry of Fisheries

A SUMMARY OF INFORMATION ON AQUAFEED PRODUCTION IN ELEVEN ASIAN COUNTRIES

Michael B. New[1] and Imre Csavas[2]

ASEAN-EEC Aquaculture Development and
Coordination Programme (AADCP)
P.O. Box 1006, Kasetsart Post Office
Bangkok 10903
Thailand

[2] FAO Regional Office for Asia and the Pacific
Maliwan Mansion
Phra Atit Road
Bangkok 10200
Thailand

NEW, M.B. and I. CSAVAS. 1993. A summary of information on aquafeed production in eleven Asian countries, p. 397-419. *In* M.B. New, A.G.J. Tacon and I. Csavas (eds.) Farm-made aquafeeds. Proceedings of the FAO/AADCP Regional Expert Consultation on Farm-Made Aquafeeds, 14-18 December 1992, Bangkok, Thailand. FAO-RAPA/AADCP, Bangkok, Thailand, 434 p.

INTRODUCTION

The information summarized in this paper originates mainly in the country papers presented at the FAO/AADCP Expert Consultation on Farm-Made Aquafeeds (New *et al.* 1993). Information from some of the working papers presented was also utilized. The countries included in the consultation were Bangladesh, Cambodia, China, India, Indonesia, Malaysia, Nepal, the Philippines, Singapore, Thailand and Vietnam.

Some background information about each of the countries was provided in the country papers, concerning demographic, topographic, agricultural, fisheries and aquacultural matters, by each of the authors. Being of a general nature, this material is not reviewed in our paper, which concentrates on the information presented which related directly to aquafeed production and use.

INGREDIENTS

Those who wish to produce either commercial or farm-made aquafeeds have access to all the normal ingredients used in the manufacture of feeds for poultry, pigs and cattle. However the growth of more intensive forms of aquaculture, requiring the use of supplementary or "complete" feeds, places increased competition on limited resources. This is exacerbated in the case of some ingredients because much of the more intensive types of aquaculture involves carnivorous species, thus making substantial demands on animal proteins, particularly marine proteins. The cost of aquaculture feeds will reflect the severity of competition and the availability of conventional feedstuffs in each country. Some countries, such as Thailand, have most ingredients locally available, importing only high quality fish meals for shrimp feeds, and vitamins, minerals and other minor ingredients. Others, like the Philippines and Singapore, import virtually all their major feedstuff ingredients though, in the case of Singapore, demand is low.

The local abundance of certain raw materials is naturally reflected in their use for aquafeed production. This is particularly noticeable in Bangladesh and India, especially because of the importance of carps amongst those species which are fed, in the use of a wide range of oil cakes. While vitamin premixes designed especially for aquafeed are readily available in Thailand they are not yet so in China or in most of the other developing countries in Asia. There is a tendency to utilize vitamin and mineral premixes designed for other species, particularly in farm-made aquafeeds. Rice byproducts, not surprisingly, are a dominant feature of the cereal content of aquafeeds in the countries surveyed. Detailed information on the composition and current costs of conventional feedstuff ingredients are presented in the relevant country papers. Some countries prohibit, as in the case of Bangladesh, or control the import of major feed ingredients. Import duties are variously applied but exemptions, as in the case of India. where an incentive for the development of foreign exchange-earning aquaculture is provided, exist.

Besides the conventional ingredients used by commercial feedstuff manufacturers, a wide variety of less conventional raw materials are locally available. These valuable ingredients are unsuitable for use in commercial feedstuff manufacture but are often ideal for farm-made aquafeeds. Farm-made aquafeed production can adapt more easily to ingredients which are only available seasonally, or in small quantities, or are moist. Their use for farm-made aquafeeds prevents their wastage and may remove an environmental hazard. Some of these less conventional ingredients are listed in Table 1, while others appear in the list of ingredients actually used in farm-made aquafeeds shown in

Tables 2 and 3. Some of those described variously as slaughterhouse wastes, viscera or offal and regarded as waste products in one country may be used for human food, in another (*e.g.* blood), while others (*e.g.* gut contents) are often totally discarded. Others, such as earthworms, may be developed as human food resources in the future. Thus, products which some think of as "waste" products, may not be regarded as such in other places. Ultimately supply <u>and</u> demand will, as always, govern the use of such less conventional ingredients in aquafeeds. A waste is no longer a waste when it has a use, and aquaculture is just one more factor in the demand situation. Some of the more interesting less conventional ingredients which can be, or are, used in farm-made aquafeeds include silkworm pupae, frog wastes, wastes from tofu and noodle manufacture, culled eggs, bread crusts, terrestrial and aquatic vegetation, wet fish and crustacean processing wastes, and "trash" fish.

Table 1. Availability of less conventional (1) ingredients for potential (2) aquafeed use in Asia in 1992

COUNTRY	BAN	CAM	CHI	INDI	INDO	MAL	NEP	PHI	SIN	THA	VIE
MIXED											
Kitchen wastes	+										
ANIMAL											
Slaughterhouse wastes, including viscera/offal (3)	+	+		+						+	
Earthworm meal				+							
Trash fish (3)						+		+	+	+	+
Fish silage	+			+							
Crust. wastes/meals	+			+							
Silkworm pupae	+			+							
Frog wastes	+										
Squid wastes				+	+						
Molluscan wastes				+							
PLANT											
Soybean curd (tofu waste)					+					+	
Waste noodles										+	
Sago palm										+	
Sweet potatoes											+
Mung bean										+	
Pulse brans	+										
Leaf meals					+			+		+	
Seaweed meals					+						

(1) not normally used by commercial feedstuff manufacturers

(2) as mentioned by authors of country papers (New *et al.*, 1993); similar ingredients also exist in other countries but were not listed

(3) several country papers note competition for these ingredients as food for humans

Table 2. Major ingredients used in farm-made fish feed production in Asia in 1992

COUNTRY	BAN	CAM	CHI	INDI	INDO	MAL	NEP	PHI	SIN	THA	VIE
BROODSTOCK FEEDS (1)											
Fish meal	+			+							+(2)
Groundnut extractions											+(2)
Pulses				+							
Broken rice				+							
Rice bran											+(2)
White corn											+(2)
Vit./min.premixes											+(2)
NURSERY FEEDS (1)											
Cooked animal feeds											+
Fish meal	+						+				
Soybean milk											+
Soybean meal							+				
Silkworm pupae	+										
Molasses	+										
Vit./min. premixes	+										
GROW-OUT FEEDS											
Fish meal			+					+	+	+	+(2)
Trash fish or tilapia		+				+		+	+	+	+
Fish processing wastes		+									
Crustacea									+		
Fish oil										+	
Animal viscera						+				+	
Culled eggs						+				+	
Meat and bone meal											+(2)
Sunflower cake				+							
Soybean cake/meal					+				+	+	+(2)
Tofu waste					+					+	
Mustard oil cake	+			+							
Rapeseed cake			+								
Sesame oil cake	+										
Unspecified oil cakes			+				+				
Groundnut cake				+						+	
Cottonseed cake			+	+							
Sweet potatoes											+
Leaf meals/leaves					+					+	+
Grasses						+	+				
Aquatic vegetation		+						+			+(2)

Table 2. (cont.)

COUNTRY	BAN	CAM	CHI	INDI	INDO	MAL	NEP	PHI	SIN	THA	VIE
Cassava										+	+
Bread crusts/wastes						+			+		
Noodle wastes						+				+	
Rice bran	+	+	+	+	+	+	+	+	+	+	+(2)
Broken rice		+					+			+	
Oats									+		
Wheat flour							+	+	+	+	
Wheat bran	+		+				+		+		
Maize (corn)						+				+	+(2)
Corn gluten									+		
Starch binder									+		
Vit./min. premixes				+				+	+	+	+(2)

(1) ingredients listed are additional to those used in grow-out feeds in the respective country

(2) experimental diets

Table 3. Major ingredients used in farm-made crustacean feed production in Asia in 1992

COUNTRY	BAN	CAM	CHI	INDI	INDO	MAL	NEP	PHI	SIN	THA	VIE
MARINE SHRIMP											
Trash fish			+							+	+
Fish meal	+			+						+	+(1)
Crust. meals/wastes				+						+	
Krill											+
Squid meat/meal				+						+	
Snail meal				+							
Other molluscs			+	+							+
Animal viscera	+										
Blood meal				+							
Meat and bone meal				+							
Liver meal				+							
Chicken eggs										+	
Fish oil				+						+	
Animal fat											+(1)
Vegetable oil				+							
Yeast										+	
Maida				+							
Mustard oil cake				+							
Soybean cake/meal				+						+	+(1)
Groundnut cake/meal				+						+	
Extracted groundnut				+							
Sunflower cake				+							
Sesame cake				+							

Table 3. (cont.)

COUNTRY	BAN	CAM	CHI	INDI	INDO	MAL	NEP	PHI	SIN	THA	VIE
Tapioca				+							
Broken rice										+	
Rice meal				+							+(1)
Rice bran	+			+						+	
Wheat flour	+			+						+	
Wheat bran				+							
Wheat gluten										+	
Millet				+							
Maize (corn) meal				+							
Brown sugar										+	
Zeolite										+	
Binders										+	
Vit./min. premixes				+						+	+(1)
CRABS											
Trash fish						+		+			+
Krill											+
Molluscs											+
FRESHWATER PRAWNS											
Fish meal										+	
Trash fish										+	
Crustacean wastes										+	
Clam meat				+							
Mussel meat				+							
Snail meat				+							
Animal viscera						+					
Compound animal feeds										+	
Groundnut cake				+							
Soybean meal										+	
Coconut oil cake										+	
Rice bran				+						+	
Broken rice										+	
Maize (corn) meal										+	
Wheat flour										+	
Tapioca flour										+	
Vit./min. premixes										+	

(1) experimental diets

Scientists are aware of the need to replace animal proteins, particularly fish meal, with plant proteins in the formulation of aquafeeds for carnivorous species. Encouraged by a trade association, work in this field is on-going throughout the region, including China and Thailand, on the use of soybean products. Fresh fish, while an "unconventional" ingredient for commercial feedstuff manufacturers, whose equipment cannot handle it, is nevertheless "conventional" to farmers growing marine fish in Malaysia, Singapore and Thailand, for example. This is also true in Japan, where fresh fish is either added to "complete" commercial aquafeeds to improve the latters' performance or is added to commercial ingredient "concentrates" designed for farmers making their own feed. Even as alternative compounded feeds for marine finfish are developed and become commercially available to a limited extent, farmers prefer to use trash fish, to which their stock so obviously respond in terms of palatability, growth rate, survival and product quality. Trash fish is also a major ingredient for farm-made feeds for freshwater fish, notably catfish and snakehead in Thailand. While trash fish remains locally available, at a price which makes profit achievable, it will continue to be used. However, its use may be restricted to those farmers who are able to "corner" the market because they have sufficient financial resources, or have been farming for a long time. Other farmers may not be able to obtain supplies of trash fish regularly enough to utilize this ingredient. The same applies to some other ingredients, such as poultry viscera. Supply, demand and influence play significant roles in the utilization of less conventional feedstuff ingredients.

AQUAFEED MANUFACTURING INDUSTRY

Thailand undoubtedly has the most sophisticated commercial aquafeed manufacturing industry amongst the eleven countries reviewed, having developed this on the foundations of an existing large animal feedstuff manufacturing sector, abundant local ingredient sources, a dynamically expansionist aquaculture sector and extrepreneurial spirit. In terms of quantity however, Thailand appears to be topped, unsurprisingly, by China (Table 4) by a factor of almost two. Subsidiaries of Thai aquafeed companies exist in many countries of the region, including Indonesia and China. Indonesia and the Philippines also have significant commercial aquafeed manufacturing sectors; in common with Thailand, the driving force has been the demands from a rapid expansion of intensive shrimp farming. In China, production of commercial aquafeeds for freshwater finfish exceeds that for marine shrimp. This is also true of India, reflecting the importance of freshwater fish in the aquaculture of both countries. In the Indian sub-continent this may change, as marine shrimp culture increases in importance. Only Pantha (1993) and Nandeesha (1993) attempted an estimate of the production level of farm-made aquafeeds; these were for Nepal (4,000 t/yr) and India (>165,000 t/yr) for carps alone.

Table 4. Characteristics of commercial grow-out aquafeed production in Asia in 1991/2

COUNTRY	BAN	CAM	CHI	INDI	INDO	MAL	NEP	PHI	SIN	THA	VIE
Number of aquafeed mills											
shrimp feed	1	nil	200	16	43	7	nil	12	4	9	14
fish feed	1	nil	1,400	6	+	3	1	> 5	nil	13	nil
total	1	nil	1,600	>20	?	9	1	>17	4	13	14
Installed capacity of aquafeed mills (t/yr)											
shrimp					363,400						2,300
total	6,000	nil	1,870,000	n.a.	?	n.a.	n.a.	>59,000 (1)	n.a.	n.a.	2,300
Actual production (t/yr)											
freshwater fish feeds	720 (2)	nil	400,000	35,000	n.a.	n.a.	109	12,000	nil	45,840	nil
shrimp feeds	1,440 (2)	nil	300,000	20,000	60,350	n.a.	nil	46,000	3,500	336,160	810
total	2,160 (2)	nil	700,000	55,000	?	n.a.	109	58,000	3,500	382,000	810
Farm-made feed production (t/yr)											
carps				>165,000			4,000				

(1) based on 28.5 t/hr x 8 x 260 days/yr

(2) based on 1992 10-month production x 1.2

While only thirteen aquafeed mills were reported in Thailand, a similar number was recorded for Vietnam (with a fraction of the output), and more in Indonesia and the Philippines. Clearly the capacity of the mills is very different. This is particularly so in the case of China, where 1,600 commercial mills producing aquafeeds were reported. Some authors had difficulties in deciding what constituted commercial aquafeed production and what should be regarded as farm-made aquafeeds when their papers were prepared. This is not surprising since a working definition of farm-made aquafeeds was not developed until the meeting convened. This definition, recorded in paragraph 9 of the official report of the expert consultation, limits farm-made feeds to those "made for the exclusive use of a particular farming activity, not for commercial sale or profit". Thus aquafeeds made by a cooperative of farmers would not be regarded as commercial, but as farm-made feeds. Drawing an exact line between the two is difficult but the following concept may help. If aquafeed production is farmer-driven it is "farm-made"; if stimulated by the existing feedstuff industry, it is "commercial". Thus feeds made by individuals or groups of individuals who are primarily farmers are farm-made. Those manufactured by companies, whose primary business is animal feedstuff production and which have diversified into aquafeeds, or which have set up specific aquafeed mills, are commercial.

Reviewing the information summarized in Table 4, and making the assumptions of an 8-hour working day and a 260-day working year, the production rate of Chinese aquafeed mills averages just over 210 kg/hr, whereas those in Thailand average more than 14 t/hr. In the case of Vietnam the limited production of shrimp feeds averages less than 30 kg/hr. In India, freshwater fish feed production averages 2.8 t/hr. In Bangladesh only one aquafeed mill was reported, which is clearly commercial. Its production to date, on the same assumptions, averages about 1 t/hr. In Indonesia, 43 shrimp feed mills were reported, with a production of over 60,000 t/yr; average production was thus about 0.7 t/hr. It is clear that many of the apparently vast number of commercial mills making aquafeeds in China do so as a very ancillary activity or, under the definition developed during the meeting, are actually "farm-made" feed production units. Perhaps in refining the definition, scale should also enter the equation. "Small-scale aquafeed production" might be more appropriate terminology, with perhaps 2,000 t/yr (about 1 t/hr) set as the line which delineates it from "commercial" production. There are many examples of aquafeed production which is very obviously "farm-made" which exceeds 1 t/yr at peak production periods, but production is intermittent and these farms would certainly not exceed 2,000 t/yr.

In another paper (New and Csavas 1993) we attempted to derive figures for commercial aquafeed production in Asia-Pacific in 1990, with predictions for 2000. We used the definition indicated above for farm-made feeds, which were excluded from our calculations. It is interesting to compare our estimates for four countries (China, Indonesia, the Philippines and Thailand), and those of Akiyama (1992a, 1992b), with those of the authors of other papers of the consultation (Table 5). At first glance it would seem that we (New and Csavas 1993) were very conservative in our estimates, which we derived from a series of "first principle" assumptions, whereas Akiyama (Akiyama 1992a, 1992b) was much closer to the mark for shrimp feed production but wildly over-optimistic for freshwater fish feed production. The Akiyama (1992b) estimate for finfish feeds in China is 2.25 times higher than reported production; ours is about 3% of actual ! As noted earlier, if "farm-made" or "small-scale" aquafeed production had been excluded, true "commercial" production of finfish feeds in China would have been very significantly less than 400,000 t, though perhaps not as low as our estimate (13,000 t). 900,000 t (Akiyama 1992b) seems an exaggeration. Total reported finfish feed production in China, Indonesia, the Philippines and Thailand in 1991/2 was around 500,000 t, whereas our estimates (for 1990) were 151,000 t and Akiyama's (also for 1990) more than 1,000,000 t.

Table 5. Comparison of reported commercial aquafeed production with estimates (t)

COUNTRY	CHI	INDO	PHI	THA	TOTAL
Freshwater fish					
1990 estimate (1)	12,568	72,112	57,104	9,588	151,372
1991/2 reported (2)	400,000	n.a.	12,000	45,840	>457,840 (5)
1990 estimate (4)	900,000	35,000	7,000	80,000	1,022,000
Marine shrimp					
1990 estimate (1)	67,025	136,742	77,744	128,593	410,104
1990 estimate (4)	250,000	50,000	35,000	250,000	585,000
1991/2 reported (2)	300,000	60,350	46,000	336,160	742,510
1991 estimate (3)	250,000	85,000	45,000	350,000	730,000

(1) New and Csavas (1993)

(2) Wang (1993); Djunaidah (1993); Pascual (1993); Jantrarotai and Jantrarotai (1993)

(3) Akiyama (1992a)

(4) Akiyama (1992b)

(5) excludes Indonesia, for which no data available

In the case of marine shrimp feed production, the 1990 estimates of Akiyama (1992a; 1992b) ranged from 74-83% of reported production in China, Indonesia, the Philippines and Thailand for 1991/2. Our estimates for Indonesia (227%) and the Philippines (169%) were higher than reported production but lower in the case of Thailand (38%). As before, the biggest discrepancy is in the figures for the People's Republic of China, caused, we believe, by the inclusion of production data from small-scale, "farm-made" production units. Akiyama himself calls his figure of 250,000 t a "guesstimate" (Akiyama, 1992a) and both he and Wang (1993) acknowledge the difficulty in collecting data or making estimates about such a large, complex, and diverse feedstuff sector as exists in China.

Overall, reported shrimp feed production for China, Indonesia, the Philippines and Thailand in 1991/2 was nearly 750,000 t, while estimates varied from 410,000 t for 1990 (New and Csavas, 1993) to 730,000 t for 1991 (Akiyama 1992a). We believe that, had the other data and estimates concerning China excluded small-scale production, all three figures would have coincided quite closely. Taiwan (Province of China), a significant aquafeed producer, did not participate in the meeting; thus, no comparisons between estimates and reported production there are made in this review.

The difficulties of accurately estimating aquafeed production have thus been neatly demonstrated. There are only scarce, unreliable and out-of-date statistics, and estimates based either on commercial feed manufacturers unwilling to disclose proprietary information or on theoretical considerations. There has also been a lack of clear definitions - a common fault: even the meaning of the word aquaculture is contentious (New and Crispoldi-Hotta 1992). Thus the report of the meeting (New *et al.* 1993) recognized that it had only made the first, faltering steps towards gathering information on this topic and recommended that further data should be collected and collated. Despite all these difficulties, we believe that the trends we predicted (New and Csavas 1993), namely that commercial aquafeed production in Asia-Pacific would increase by nearly 70% in the decade 1990-2000, were valid. Time will tell ! We agree with Akiyama (1992b) that the largest aquafeed producing countries are expected to be China (including Taiwan, Province of China) and Indonesia. Thailand and the Philippines will remain important producers, and the Indian sub-continent will become so. Amongst the developed countries of the region, Japan will remain paramountly important as a commercial aquafeed producer, but not for marine shrimp.

ON FARM FEED FORMULATION AND MANUFACTURING

Farm-made feed manufacture is particularly active in Malaysia, the Philippines, Singapore and Thailand for marine fish; in Bangladesh, India, Indonesia and Nepal for carps; in Thailand for catfishes and snakeheads; in Bangladesh, India and Vietnam for marine shrimp; and in Thailand for freshwater prawns (Table 6). In Bangladesh, India and Nepal, farm-made feeds predominate over commercial manufacture. Though the volume of production is unclear, for reasons discussed earlier, and Wang (1993) does not report which species are supplied by farm-made feeds, we believe that feeds made on-farm or in other very small production units also predominate in China, both for marine shrimp and for finfish. We estimate (New and Csavas 1993) that, in Asia-Pacific generally, over 70% of the production of carnivorous finfish, selected non-

Table 6. Use of farm-made grow-out aquafeeds in Asia in 1992 by species groups (1)

COUNTRY	BAN	CAM	CHI	INDI	INDO	MAL	NEP	PHI	SIN	THA	VIE
FRESHWATER FISH											
carps	+			+(3)	+	+(2)	+(3)				+
tilapias	+	+			+	+(2)	+(2)	+		+	
silver barbs	+	+					+(2)				
snakeheads		+								+	+
catfishes	+			+		+(2)				+	+
ornamental									+		
others or unspecified		+	+		+	+(2)	+(2)				+
MARINE FISH											
seabass						+		+	+	+	
groupers						+		+	+	+	
snappers									+		
CRUSTACEANS											
marine shrimp	+			+						+	+
freshwater prawns				+		+(2)	+(2)			+	+
crabs						+		+			+
unspecified			+								

(1) farm-made aquafeeds may also be used for species other than those recorded in the country papers summarized

(2) experimental use, generally on government/research farms

(3) only India and Nepal attempted to quantify the amount of farm-made fish feeds produced (for carps), which were >165,000 t and 4,000 t, respectively

carnivorous finfish (common carp, tilapias and milkfish) and crustaceans is <u>not</u> produced by the use of commercial feeds. A significant, but unknown, proportion of that 70% is produced through the use of farm-made aquafeeds. Added to that is the production of other finfish, such as Indian and Chinese carps in India and Nepal, as reported by Nandeesha (1993) and Pantha (1993), as well as in China, partially through the application of farm-made feeds. Farm-made feeds are clearly very important in Asia but leave much room for improvement. As noted in the official report of the meeting (New *et al.* 1993) most of the semi-intensive farming systems rely on the use of farm-made feeds, as do some intensive systems (*e.g.* for snakehead, catfish, groupers and seabass). Farm-made feeds are common for ornamental fish in Singapore (Chou 1993).

The manufacture of farm-made feeds facilitate the use of less conventional ingredients which cannot be utilized by commercial feed manufacturers or are not in sufficient or reliable supply. Farm-made feeds, in utilizing such ingredients, thus have an added environmental value. Farm-made feeds fill a seperate niche than commercial feeds, which are particularly appropriate for intensive aquaculture. Therefore farm-made feeds are not a source of competition for feedstuff manufacturers. Indeed, farmers who become economically successful through the use of farm-made feeds often graduate to commercial feeds and are thus potential customers of feedstuff manufacturers.

As one would expect, a very wide range of ingredients are used in farm-made feeds in the countries surveyed (Tables 2 and 3). The one ingredient which is ubiquitous is rice bran. In fish feeds, fish meal, trash fish and fish processing wastes are the commonest marine proteins used. Only in Singapore were crustacean ingredients (ground dried shrimp) mentioned and only for ornamental fish feeds. Conversely, several marine proteins are used in crustacean feeds - krill, shrimp meal, shrimp shell meal, squid meal, clam meat and mussel meat, for example. Animal viscera of various types are used both in fish and crustacean feeds to a limited extent. Their most obvious use is in snakehead feeds in Thailand.

Many plant proteins are used in fish and crustacean feeds, especially in India, but also in Bangladesh, China and Thailand. Soybean cake or meal is a common ingredient. The use of cereals and cereal byproducts is normal in all formulations; the type depends on local availability. Apart from natural binders, *e.g.* fish flesh, wheat starch, and corn gluten, only one paper (Boonyaratpalin and New 1993) mentions the use of binders, in marine shrimp feeds in Thailand. Many countries use vitamin and mineral premixes in farm-made aquafeeds but these are rarely designed for the species to which they are fed.

The farm-made feeds given as examples in the country papers range from single ingredients (mainly trash fish), and simple mixtures (usually of rice bran and an oil cake) to quite complex formulated feeds, especially for crustacea and in some countries, notably Thailand. Farm-made feeds are also fed in a wide range of forms - powders, slurries, doughballs, moist pellets, and sun-dried or machine-dried pellets. Feeding strategies, with and without fertilization, are also presented by some authors but they are too diverse to review satisfactorily here.

The general impression is that there is a great deal of scope for improvement in the selection of ingredients, in formulation, in processing techniques, in storage (of ingredients and finished feeds) and, not least, in the feeding of farm-made feeds which could be encouraged. This is reflected in the recommendations of the meeting (New *et al.* 1993). While machinery for farm-made feed processing is specially designed and available in some countries, such as China, Japan, Singapore, and Taiwan (Province of China), or has been adapted from equipment designed for other use, such as butchers' mincers and second-hand car engines (Thailand and Vietnam) it is reported unaffordable or unavailable in other countries such as Bangladesh. Sometimes hand-powered mincers are used for extrusion; some examples of cooking ingredients or complete feeds to improve digestibility and water stability were also described (*e.g.* Cambodia). However, most farm-made aquafeeds appear to be fed in an uncooked form.

Only for India (more than 165,000 t for carps) and Nepal (4,000 t, also for carps) did the authors (Nandeesha 1993; Pantha 1993) attempt to quantify the production of farm-made feeds. However, much of the 700,000 t of commercial aquafeed production in China reported by Wang (1993) would fall into our definition of farm-made aquafeeds. In our other paper (New and Csavas 1993) we estimated that almost 2 million t of Asian finfish and crustacean aquaculture production in 1990 was achieved without the use of commercially available aquafeeds, a proportion of it by farm-made aquafeeds. This estimate excluded non-carnivorous fish, except common carp, milkfish and tilapias. It is clear that farm-made feeds are also being used for other non-carnivorous species, such as Indian and Chinese carps. Looking at these various, but limited, estimates it may not be too wild to speculate that more than 1 million t of farm-made aquafeeds - some simple, some complex - are annually in use in Asia. Carrying this speculation just one (dangerous) step further, about one third of Asian finfish and crustacean production may be partially achieved through the application of all types of farm-made feeds, from single ingredients through simple mixtures to complex formulated diets. Farm-made feed manufacture and use is therefore not a topic to be ignored by those involved in aquaculture development and research. Donors please note!

RESEARCH AND DEVELOPMENT

All 11 countries, except Cambodia, are reported to have on-going research and development activities in fish and shrimp nutrition, both in the university and government sectors (Table 7). 37 governmental research stations and 35 universities involved in finfish or crustacean nutritional research, and the addresses and topics researched in many of them, are recorded in the country papers. However, little work is done on farm-made feeds, most research and development being of primary benefit to commercial feedstuff manufacturers. Thailand is one exception, where a pioneering UNDP/FAO project targetted farmers making their own feeds. Several of the farm-made aquafeed formulae mentioned by Jantrarotai and Jantrarotai (1993) and Boonyaratpalin and New (1993) have been modified as a result of the work of that project.

Table 7. Characteristics of aquafeed research, development and control in Asia in 1992

COUNTRY	BAN	CAM	CHI	INDI	INDO	MAL	NEP	PHI	SIN	THA	VIE
MAJOR R & D CENTRES											
government	1	nil	7	5	3	4	3	8	1	2	3
university	1	nil	1	17	+	>1	nil	14	1	+	+
FEEDSTUFF REGULATIONS											
aquafeed specific		-	?	-	+	-	-	+	-	+	-
general feedstuffs	+		?	+	+	?	+	+	+	+	?
IMPORT PROHIBITION											
ingredients	Yes(1)	?	No	No(2)	No	No(3)	?	No	No	No(4)	?(5)
finished feeds	Yes	?	No	No	No	No	?	No	No	No	?

(1) except chemical-like substances

(2) duty free import if farm is export-oriented; if not, 25% duty applied

(3) 13% tax applied

(4) fish meal and soybean quotas

(5) only vitamins and amino acids currently imported

AQUAFEED REGULATONS

Specific regulations concerning aquafeed manufacture and sale appear to exist only in Indonesia, the Philippines and Thailand, although this topic was not covered in the case of China (Wang 1993). Aquafeed production is, however, controlled by general feedstuff regulations in most countries (Table 7). Only Cambodia is known not to have feedstuff regulations, since there is no reported feedstuff production (Nuov and Nandeesha 1993). It was not reported (Luu 1993) whether Vietnam has general feedstuff regulations; the total animal feedstuff industry there is still small (less than 25,000 t/yr). Where good general feedstuff regulations exist those should normally suffice for commercial aquafeeds also, since they should cover not only analytical characteristics but control the use of drugs (in particular withdrawal time before marketing the farmed product). Special feedstuff regulations for aquafeeds, such as those for the Philippines and Thailand, while sensitive to the issues raised by controls exerted by importing countries on the quality of farmed shrimp, may be over-zealous.

TRENDS IN AQUAFEED MANUFACTURING AND USE

The value of utilizing aquafeeds, already accepted in ASEAN countries and in China, is becoming better known in India where a very significant growth in production is expected, particularly for shrimp culture. Trash fish availability is expected to become a major constraint in Cambodia (Nuov and Nandeesha 1993), where carnivorous and herbivorous species are expected to increase in aquaculture importance. Cambodia regards research and development on on-farm aquafeed manufacture and use to be a priority area, while Vietnam expects aquafeed demand to double every year in this decade. Demand for aquafeeds in Bangladesh is very low and not yet conducive to commercial production; farm-made feeds are thus likely to retain and increase their importance.

Hopefully, the results of research (in Asia and elsewhere) into the nutritional requirements of aquacultured species in semi-intensive culture, in the presence of natural feeds, will lead to improvements in feeding management, with consequential financial benefits. While many grow-out feeds are available, locally-made feeds for the younger developmental stages do not exist in some countries (*e.g.* China, Indonesia).

Fish meal is imported by most countries in the region, including China, Indonesia, and the Philippines, or the higher quality batches made locally are exported (India) leaving only poor quality batches available for local use. Though a fish meal producer, Thailand also imports high quality fish meal for use in shrimp feeds. China finds the need for foreign exchange to purchase fish meal excessive. Indonesia has not optimally exploited its fisheries bycatch and also has to import fish meal. Competition for fish meal for other animal feedstuff manufacture, and global demand for this commodity, may become a problem for intensive aquaculture, though efforts are being intensified to limit the essentiality of fish meal in aquaculture diets.

Zaher and Mazid (1993), Nandeesha (1993) and Pantha (1993) expected farm-made feed manufacture for carps in Bangladesh, India and Nepal to remain critically important. This will probably also be true in China. Feeds for intensive marine shrimp culture will remain the domain of the commercial feed manufacturers, though scope exists for farm-made feeds for semi-intensive culture, particularly in Bangladesh and India. Djunaidah (1993) sees Indonesian fish and shrimp farmers becoming increasingly dependent on commercial aquafeeds. Many freshwater prawn farmers in Thailand and Vietnam are expected to continue to use farm-made feeds and these may become accepted in other countries where freshwater prawn farming is expected to increase, such as India, Indonesia and the Philippines. Pascual (1993) predicts a shift in importance from commercial shrimp feed to finfish feed production in the Philippines and expects more farms involved in semi-intensive aquaculture to use farm-made feeds than do so now. Utama (1993) sees a bright future for the local feed milling industry in Malaysia as (intensive) aquaculture expands, although almost all marine shrimp feeds are currently imported. Pantha (1993) is confident that existing local animal feed manufacturers will be able to absorb any future demand for commercial aquafeeds in Nepal. Pascual (1993) expects imported aquafeeds to lose ground to local manufacture in the Philippines and feed costs to be reduced through the use of more locally available ingredients, including tuna fish meal. Shrimp feeds in the Philippines are currently considerably more expensive than in the other countries reviewed (Table 8).

Table 8. Average unit cost of aquafeeds in Asia in 1992 (US$/kg)

COUNTRY	BAN	CAM	CHI	INDI	INDO	MAL	NEP	PHI	SIN	THA	VIE
COMMERCIAL, IMPORTED											
Marine shrimp											
grower				1.20				1.14			
COMMERCIAL, LOCAL											
Marine shrimp											
larval									97.3		
nursery	1.14							1.47	1.24	1.28	
starter #1	1.02							1.37	1.14	1.28	
starter #2	1.02			0.22-						1.12	
grower	1.02			0.96	1.34			1.27	1.03	1.08	
grower (supplementary)	-							1.17	-	-	
finisher	1.02							1.22	1.00	1.08	
Freshwater prawns											
starter #1										0.72	
starter #2										0.68	
grower										0.64	
finisher										0.56	
Carps											
nursery	0.64						0.23				
grower	0.38			0.13	0.26		0.17			0.40	
Catfishes											
fry										1.44	
nursery	0.64									0.72	
grower	0.38									0.64	
finisher										0.56	
Tilapias											
starter								0.46		0.60	
grower					0.26			0.43		0.56	
finisher								0.41		0.52	

Table 8. (cont.)

COUNTRY	BAN	CAM	CHI	INDI	INDO	MAL	NEP	PHI	SIN	THA	VIE
Milkfish											
starter								0.51			
grower								0.47			
finisher								0.46			
Unspecified											
16-20% crude protein											0.26
21-25% crude protein											0.30
26-30% crude protein											0.35
31-35% crude protein											0.43
36-40% crude protein											0.55
FARM-MADE											
Catfishes											
grower (moist)										0.21	
grower (dry)										0.30	
Snakehead											
grower (moist)										0.25	
Carps											
nursery	0.13									0.48	
grower	0.08										
Tilapias											
grower	0.08										
Unspecified f/w fish											
grower			0.29								
Freshwater prawns											
grower										0.48	

Chou (1993) predicts that Singaporean feed companies, while retaining their offices and mills in Singapore as technical resource centres, will expand operations into other countries in the region. Seabass grow-out feeds and shrimp larval diets are expected to be major features. Farm-made feeds, either trash fish or compounded, will remain important for marine fish culture in the region until reliable and economical commercial feeds become available. Somsueb (1993) sees current farm-made feeds as environmentally unsound (thus highlighting the need for research to be directed towards improvements in the formulation and manufacture of such feeds). She also predicts that increasing commercial aquafeed production in Thailand will result in the enlargement of production capacity in existing mills rather than new entrants into the industry. This may also be true in the commercial aquafeed sector of other countries in the region.

The future of farm-made aquafeeds would seem to have a future as varied as the eleven countries on which papers were presented and the views of the authors concerned. There certainly will be national differences but our own belief is that farm-made feeds will continue to have major impact regionally and may became more important as the trend towards more environmentally acceptable, semi-intensive aquaculture develops.

CONCLUSIONS

The FAO/AADCP Regional Expert Consultation on Farm-Made Aquafeeds highlighted, for the first time, the importance of farm-made feeds in Asia-Pacific. Such feeds are also being utilized and have further potential in other regions, for example in Africa. It is hoped that the proceedings of the meeting (New *et al.* 1993) will help to persuade donors and research workers that more attention should be paid to this topic. Public sector funding should assist small-scale farmers rather than feed manufacturers. More work needs to be done on supplemental, rather than "complete" feeds, and on other means of increasing production through less intensive aquaculture. Semi-intensive aquaculture is becoming increasingly more environmentally acceptable than intensive aquaculture; its feed needs therefore must be addressed. There are signs that at least some feedstuff manufacturers are aware of this trend, as reflected in the marketing of feeds designed for less intensive systems of production (Pascual 1993) and which result in less nutrients released into the environment.

While there are many potential benefits from improvements in the formulation, manufacture and use of farm-made aquafeeds these cannot be forced on farmers. Development must be based on working with farmers to improve existing practices and there is a need for simple instructional materials including local-language booklets, videos, *etc*. The use of farm-made feeds is accepted by farmers in some countries but almost unknown in others; scope for TCDC therefore exists.

Detailed recommendations were drawn up during the meeting and are recorded in the proceedings (New *et al.* 1993). Amongst the more important were recommendations that the proceedings should be used to alert governments, international aid agencies and potential donors to the importance of farm-made aquafeeds to small-scale aquaculture and the need for support for feed development to place more emphasis on this area. Simple and cheap methods of increasing the nutritional value of locally available ingredients need to be developed, as well as the availability or adaptation of cheap processing machinery encouraged. Improved methods of processing and storing farm-made aquafeeds are needed. Feed advisers, when formulating farm-made aquafeeds, should not mimic commercial feed specifications but address the specific needs of the farms concerned, and the role of natural food organisms. Locally available ingredients should be used and the use of vitamin premixes, binders and other expensive ingredients minimized, where possible. The participants in the consultation also recognized that the potential for improved profitability in small-scale aquaculture was greater through improvements in feeding strategy than in attempting to perfect dietary composition and recommended further work on methods of feed presentation, biomass assessment, reduction of feed wastage and ingredient quality assessment.

LIST OF REFERENCES

Akiyama, D.M. 1992a. "Guesstimated" world production of shrimp feed. World Shrimp Farming: 17.11.37.

Akiyama, D.M. 1992b. Future considerations for the aquaculture feed industry, p. 5-9. *In* D.M. Akiyama and R.K.H. Tan (eds.) Proceedings of the Aquaculture Feeds Processing and Nutrition Workshop, Thailand and Indonesia ,19-25 September 1991. American Soybean Association, Singapore.

Boonyaratpalin, M. and M.B. New. 1993. On-farm feed preparation and feeding strategies for marine shrimp and freshwater prawns, p. 120-134. *In* M.B. New, A.G.J. Tacon and I. Csavas (eds.) Farm-made aquafeeds. Proceedings of the FAO/AADCP Regional Expert Consultation on Farm-Made Aquafeeds, 14-18 December 1992, Bangkok, Thailand.

Chou, R. 1993. Aquafeeds and feeding strategies in Singapore, p. 354-364. *In* M.B. New, A.G.J. Tacon and I. Csavas (eds.) Farm-made aquafeeds. Proceedings of the FAO/AADCP Regional Expert Consultation on Farm-Made Aquafeeds, 14-18 December 1992, Bangkok, Thailand.

Djunaidah, I.S. 1993. Aquafeeds and feeding strategies in Indonesia, p. 255-281. *In* M.B. New, A.G.J. Tacon and I. Csavas (eds.) Farm-made aquafeeds. Proceedings of the FAO/AADCP Regional Expert Consultation on Farm-Made Aquafeeds, 14-18 December 1992, Bangkok, Thailand.

Jantrarotai, W. and P. Jantrarotai. 1993. On-farm preparation and feed strategies for catfish and snakehead, p.101-119. *In* M.B. New, A.G.J. Tacon and I. Csavas (eds.) Farm-made aquafeeds. Proceedings of the FAO/AADCP Regional Expert Consultation on Farm-Made Aquafeeds, 14-18 December 1992, Bangkok, Thailand.

Luu, L.T. 1993. Aquafeeds and feeding strategies in Vietnam, p. 386-396. *In* M.B. New, A.G.J. Tacon and I. Csavas (eds.) Farm-made aquafeeds. Proceedings of the FAO/AADCP Regional Expert Consultation on Farm-Made Aquafeeds, 14-18 December 1992, Bangkok, Thailand.

Nandeesha, M.C. 1993. Aquafeeds and feeding strategies in India, p.213-254. *In* M.B. New, A.G.J. Tacon and I. Csavas (eds.) Farm-made aquafeeds. Proceedings of the FAO/AADCP Regional Expert Consultation on Farm-Made Aquafeeds, 14-18 December 1992, Bangkok, Thailand.

New, M.B. and A. Crispoldi-Hotta, 1992. Problem in the application of the FAO definition of aquaculture. FAO Aquaculture Newsletter (FAN) 1:5-8.

New, M.B. and I. Csavas. 1993. Aquafeeds in Asia - a regional overview, p.1-23. *In* M.B. New, A.G.J. Tacon and I. Csavas (eds.) Farm-made aquafeeds. Proceedings of the FAO/AADCP Regional Expert Consultation on Farm-Made Aquafeeds, 14-18 December 1992, Bangkok, Thailand.

New, M.B., A.G.J. Tacon and I. Csavas (eds.) 1993. Farm-made aquafeeds. Proceedings of the FAO/AADCP Regional Expert Consultation on Farm-Made Aquafeeds, 14-18 December 1992, Bangkok, Thailand, 434 p.

Nuov, S. and M.C. Nandeesha. 1993. Aquafeeds and feeding strategies in Cambodia, p.181-200. *In* M.B. New, A.G.J. Tacon and I. Csavas (eds.) Farm-made aquafeeds. Proceedings of the FAO/AADCP Regional Expert Consultation on Farm-Made Aquafeeds, 14-18 December 1992, Bangkok, Thailand.

Pantha, M.B. 1993. Aquafeeds and feeding strategies in Nepal, p.297-316. *In* M.B. New, A.G.J. Tacon and I. Csavas (eds.) Farm-made aquafeeds. Proceedings of the FAO/AADCP Regional Expert Consultation on Farm-Made Aquafeeds, 14-18 December 1992, Bangkok, Thailand.

Pascual, F.P. 1993. Aquafeeds and feeding strategiens in the Philippines, p.317-353. *In* M.B. New, A.G.J. Tacon and I. Csavas (eds.) Farm-made aquafeeds. Proceedings of the FAO/AADCP Regional Expert Consultation on Farm-Made Aquafeeds, 14-18 December 1992, Bangkok, Thailand.

Somsueb, P. 1993. Aquafeeds and feeding strategies in Thailand, p.365-385. *In* M.B. New, A.G.J. Tacon and I. Csavas (eds.) Farm-made aquafeeds. Proceedings of the FAO/AADCP Regional Expert Consultation on Farm-Made Aquafeeds, 14-18 December 1992, Bangkok, Thailand.

Utama, C.M. 1993. Aquafeeds and feeding strategies in Malaysia, p.282-296. *In* M.B. New, A.G.J. Tacon and I. Csavas (eds.) Farm-made aquafeeds. Proceedings of the FAO/AADCP Regional Expert Consultation on Farm-Made Aquafeeds, 14-18 December 1992, Bangkok, Thailand.

Wang, P. 1993. Aquafeeds and feeding strategiens in China, p.201-212. *In* M.B. New, A.G.J. Tacon and I. Csavas (eds.) Farm-made aquafeeds. Proceedings of the FAO/AADCP Regional Expert Consultation on Farm-Made Aquafeeds, 14-18 December 1992, Bangkok, Thailand.

Zaher, M. and M.A. Mazid. 1993. Aquafeeds and feeding strategies in Bangladesh, p.161-180. *In* M.B. New, A.G.J. Tacon and I. Csavas (eds.) Farm-made aquafeeds. Proceedings of the FAO/AADCP Regional Expert Consultation on Farm-Made Aquafeeds, 14-18 December 1992, Bangkok, Thailand.

LIST OF PARTICIPANTS

BANGLADESH

Mr. Muhammad Zaher
Senior Scientific Officer
Fisheries Research Institute
Freshwater Station
Mymensingh-2201
BANGLADESH

Tel.: (880-91) 4874

CAMBODIA

Mr. Sam Nuov
Senior Fisheries Officer
Department of Fisheries
Ministry of Agriculture
Phnom Penh
CAMBODIA

Current address : Asian Institute of Technology (AIT)
AFE Division
G.P.O. Box 2754
Bangkok 10501
THAILAND

Tel.: (66-2) 516-0110/29, 516-0130/44
Tlx.: 84-276 TH
Fax.: (66-2) 516-2126

CHINA

Ms. Wang Ping
Section Chief
Division of Science and Technology
Bureau of Aquatic Products
Ministry of Agriculture
11 Nongzhanguan Nanli
Beijing 10026
P. R. CHINA

Tel.: (86-1) 500-2280
Tlx.: 22233 MAGR CN
Fax.: (86-1) 500-1886, 500-2448

INDIA

Dr. M. C. Nandeesha
Department of Aquaculture
College of Fisheries
University of Agricultural Sciences
Mangalore - 575 002
Karnataka
INDIA

Tel.: (91) 824-24636; 824-27834
Cable: FISHCOL
Fax.: (91) 824-22395; 824-35068; 824-35079

Current address : Fisheries Advisor
Partnership for Development
in Kampuchea
PADEK Phnom Penh
G.P.O. Box 2420
Bangkok 10501
THAILAND

Fax/Tel.: (855) 23-26224

INDONESIA

Ms. Iin Siti Djunaidah
Brackishwater Aquaculture Development Centre
P.O.Box 1
Jepara
Central Java
INDONESIA

Tel.: (62) 291-91125
Fax.: (62) 291-91724

MALAYSIA

Mr. Che Musa Che Utama
Fisheries Research Officer
National Prawn Fry Production and Research Centre
Kampung Pulau Sayak
08500 Kota Kuala Muda
Kedah
MALAYSIA

Tel.: (60-4) 474-021/23
Fax.: (60-4) 474-470

NEPAL

Mr. Madhab Bahadur Pantha
Chief
Fisheries Development Division
Department of Agriculture Development
Harihar Bhawan, Pulchowk
Kathmandu
NEPAL

Tel.: (977-1) 522-121, 521-597, 527-844
Fax.: (977-1) 527-090

PHILIPPINES

Dr. Felicitas P. Pascual
Nutritionist
30, Fajardo Extension Jaro
Iloilo City 5000
PHILIPPINES

Tel.: (63-33) 77-548
Fax.: (63-33) 270-275

Ms. Dawn R. Jamandre
G.P.O. Box 409
Iloilo City
PHILIPPINES

Tel.: (63-33) 76-270
Fax.: (63-33) 201-569

THAILAND

Dr. Mali Boonyaratpalin
Director
Feed Quality Control and Development Division
c/o National Inland Fisheries Institute
Kasetsart University Campus
Bangkok 10900
THAILAND

Tel.: (66-2) 579-2151, 579-9525
Fax.: (66-2) 561-1727

Dr. Wimol Jantrarotai
Fisheries Biologist
National Inland Fisheries Institute
Kasetsart University Campus
Bangkok 10900
THAILAND

Tel.: (66-2) 579-9525, 579-8145
Fax.: (66-2) 561-1727

Mr. Prasert Sitasit
Feed Ingredient and Quality Control Officer
Feed Quality Control Centre
c/o National Inland Fisheries Institute
Kasetsart University Campus
Bangkok 10900
THAILAND

Tel.: (66-2) 579-9525, 579-9528
Fax.: (66-2) 561-1727

Ms. Pisamai Somsueb
Fisheries Biologist
National Inland Fisheries Institute
Kasetsart University Campus
Bangkok 10900
THAILAND

Tel.: (66-2) 579-9525, 579-8145
Fax.: (66-2) 561-1727

Dr. Supit Thongrod
Fish Nutritionist
National Inland Fisheries Institute
Kasetsart University Campus
Bangkok 10900
THAILAND

Tel.: (66-2) 579-9525, 579-6742
Fax.: (66-2) 561-1727

VIETNAM

Dr. Le Thanh Luu
Vice-National Project Director VIE/86/011
Research Institute for Aquaculture No. 1
Dinh Bang, Tien Son
Ha Bac
S. R. VIETNAM

Tel.: (84-14) 271-368
Fax.: (84-42) 59-257 (c/o FAO Hanoi)

AMERICAN SOYBEAN ASSOCIATION (ASA)

Dr. Dean M. Akiyama
Technical Director - Aquaculture
541 Orchard Road
11-03 Liat Towers
SINGAPORE 0923

Tel.: (65) 737-6233
Tlx.: (023) 740-7820 BEANUC
Fax.: (65) 737-5849

ASEAN-EEC AQUACULTURE DEVELOPMENT AND COORDINATION PROGRAMME (AADCP)

Mr. Sompong Hiranwat
Director
AADCP
P.O. Box 1006, Kasetsart Post Office
Bangkok 10903
THAILAND

Fax. : (66-2) 561-3023

Current address : Director
Policy and Planning Division
Department of Fisheries
Ministry of Agriculture and Cooperatives
Kasetsart University Campus
Bangkok 10900
THAILAND

Tel.: (66-2) 561-1974
Fax.: (66-2) 579-8560

Mr. Hassanai Kongkeo
Technical Officer
AADCP
P.O.Box 1006, Kasetsart Post Office
Bangkok 10903
THAILAND

Tel.: (66-2) 561-3022
Fax.: (66-2) 561-3023

Mr. Michael B. New
Former Coordinator
AADCP
P.O.Box 1006, Kasetsart Post Office
Bangkok 10903
THAILAND

Tel.: (66-2) 561-3022
Fax.: (66-2) 561-3023

ASIAN INSTITUTE OF TECHNOLOGY (AIT)

Dr. Amararatne Yakupitiyage
Assistant Professor
Fish Nutrition and Feed Technology
Division of Agricultural and Food Engineering
Asian Institute of Technology
G.P.O.Box 2754
Bangkok 10501
THAILAND

Tel.: (66-2) 516-0110/29, 516-0130/44
Tlx.: 84-276 AIT TH
Fax.: (66-2) 516-2126

BAY OF BENGAL PROGRAMME (BOBP)

Mr. Charles L. Angell
Senior Aquaculturist
Small-Scale Fisherfolk Communities
in the Bay of Bengal
c/o FAO, P. O. Box 5039 (New Market)
Dhaka 1205
BANGLADESH

Tel.: (880-2) 241-383, 241-437
Tlx.: 64-2979 FAO BJ
Fax.: (880-2) 813-446

Mr. T. W. Bostock
Post-Harvest Adviser
Bay of Bengal Programme
91 St. Mary's Road, Abhiramapuram
Madras 600 018
INDIA

Tel.: (91-44) 836-294, 836-296, 836-188
Tlx.: 41-8311 BOBP
Fax.: (91-44) 836-102

BP NUTRITION

Dr. T. V. Gayathri Rao
Information Specialist
Aquaculture Research Centre
P.O. Box 5
Songkhla 90000
THAILAND

Tel.: (66-74) 321-054
Fax.: (66-74) 321-055

FOOD AND AGRICULTURE ORGANIZATION OF THE UNITED NATIONS (FAO)

Dr. Kee-Chai Chong
(present under FAO/RAPA Special Service Agreement)
Chief of Party/Research Management Advisor
INDECO/WINROCK Agricultural Research Management Project
Jalan Salak 22
Bogor 16151
INDONESIA

Tel.: (62-251) 324-865
Tlx.: 48-369 AARD IA
Fax.: (62-251) 328-489

Mr. Imre Csavas
Regional Aquaculture Officer
FAO Regional Office
Phra Atit Road
Bangkok 10200
THAILAND

Tel.: (66-2) 281-7844
Tlx.: 82-815 FOODAG TH
Fax.: (66-2) 280-0445

Mr. Laurent Debas
Associate Professional Officer
FAO Regional Office
Phra Atit Road
Bangkok 10200
THAILAND

Tel.: (66-2) 281-7844
Tlx.: 82-815 FOODAG TH
Fax.: (66-2) 280-0445

Dr. Veravat Hongskul
Regional Fishery Officer
FAO Regional Office
Phra Atit Road
Bangkok 10200
THAILAND

Tel.: (66-2) 281-8744
Tlx.: 82-815 FOODAG TH
Fax.: (66-2) 280-0445

Dr. Sena S. De Silva
(present under FAO/RAPA Special Service Agreement)
Associate Professor
Faculty of Aquatic Sciences
Deakin University
P.O.Box 423
Warrnambool
Victoria 3280
AUSTRALIA

Tel.: (61-55) 633-527
Fax.: (61-55) 633-531

Dr. Albert G. J. Tacon
Fishery Resources Officer (Feed Specialist)
Inland Water Resources and Aquaculture Service
FIRI
FAO Headquarters,
Viale delle Terme di Caracalla
00100 Rome
ITALY

Tel.: (39-6) 5797-6470
Tlx.: 610-181 FAO I
Fax.: (39-6) 5797-3020

Mr. John Wood
(present under FAO/RAPA Special Service Agreement)
Food and Feed Technologist
Grain and Food Processing Section
Grain Technology Department
Natural Resources Institute
Central Avenue
Chatham Maritime
Kent ME4 4TB
UNITED KINGDOM

Tel.: (44-634) 880-088
Tlx.: 263-907/8 LDN G
Fax.: (44-634) 880-066/77

SOUTHEAST ASIAN FISHERIES DEVELOPMENT CENTER (SEAFDEC)

Dr. Relicardo M. Coloso
Head
Research Division
SEAFDEC AQD
P. O. Box 256
5000 Iloilo City
PHILIPPINES

Tel.: (63-33) 81-261
Fax.: (63-33) 81-340

VIETNAM NATIONAL SEA PRODUCTS EXPORT CORPORATION SEAPRODEX)

Mr. Nguyen Quoc Khanh
Director
SEAPRODEX Shrimp Feed Mill
263 Phan Chu Trinh
Danang
S. R. VIETNAM

Tel.: (84-151) 21-948, 23-247
Tlx.: 515-713 SPDEX VT
Fax.: (84-151) 23-769

Mr. Nguyen Dung Nam
Engineer
SEAPRODEX
263 Phan Chu Trinh
Danang
S. R. VIETNAM

Tel.: (84-151) 21-948, 21-936
Tlx.: 515-713 SPDEX VT
Fax.: (84-151) 23-769

INDEX